Transport of *Escherichia coli* in saturated porous media

Jan Willem Anton Foppen

Thesis Committee: prof.dr. M. Hassanizadeh (UU)
prof.dr. G. Amy (UNESCO-IHE)
prof.dr. N. Tufenkji (McGill University)
dr. W.F.M. Röling (VU)

In Dutch: Transport van *Escherichia coli* in verzadigde poreuze media

Published by A.A. Balkema Publishers, a member of Taylor & Francis Group plc.
www.balkema.nl and www.tandf.co.uk

ISBN13: 978-0-415-44477-4

Front cover photo: Washing qat with microbiologically contaminated groundwater, abstracted from the Quaternary Alluvium and Cretaceous Sandstone aquifers in the Sana'a Basin in Yemen.

VRIJE UNIVERSITEIT
UNESCO-IHE Institute for Water Education

Transport of *Escherichia coli* in saturated porous media

ACADEMISCH PROEFSCHRIFT

ter verkrijging van de graad Doctor aan
de Vrije Universiteit Amsterdam en
het UNESCO-IHE Institute for Water Education te Delft,
op gezag van de rector magnificus
prof.dr. L.M. Bouter
en de rector prof.dr. R. Meganck,
in het openbaar te verdedigen
ten overstaan van de promotiecommissie
van de faculteit der Aard- en Levenswetenschappen
en het UNESCO-IHE Institute for Water Education
op vrijdag 8 juni 2007 om 15.00 uur
in het Auditorium van UNESCO-IHE

door

Jan Willem Anton Foppen

geboren te Amstelveen

promotoren:	prof.dr. S. Uhlenbrook
	prof.dr. P.J. Stuyfzand
copromotor:	dr. J.F. Schijven

Aan Bien

DANKWOORD

Kan je een land bedanken? Eigenlijk niet, maar ik doe het toch: Jemen bedankt. De ruige bergen, mooie mensen en de suq van Sana'a zijn onvergetelijk. Verbreding van m'n horizon, in alle opzichten. De bezoeken en bezoekjes van familie en vrienden verhogen de Jemen-nageniet-index enorm. Samen djambija dansen bij Bir Ali, qat kauwen in Wadi al-Furs of naar de bronnen van de Wadi Surdud. Die momenten koester ik nog steeds, dank daarvoor.

Niet alleen Jemen als land was prachtig, maar ook de open, opmerkelijke, verrassende, lachwekkende, hilarische en serieuze momenten met Achmed Al-Qaisy, Abdu Baqi, Abdu Laleem, Lutfia Siddiq, Mohammed Mayas en de chauffeurs Achmed, Abdu and Yahia, oftewel de jemenitische SAWAS project team members. Ze zitten, niemand uitgezonderd, in een gebied van positieve gevoelens ergens in m'n hersenen. De droge en humoristische commentaren van Klaas (de Groot) bij al deze momentjes waren treffend.

Als team leader van het SAWAS project, stimuleerde Gideon (Kruseman) me om de resultaten van de chemische en microbiologische sampling rondes in en om de stad en de grondwater modellering van het Sana'a Basin tot een proefschrift om te smelten. Abdallah (Babaqi), mijn directe counterpart op de Sana'a Universiteit tijdens het SUS Project, na SAWAS, heeft altijd al gevonden, dat ik een proefschrift moest schrijven. Gideon en Abdallah hebben een gevoelige snaar geraakt. Dank voor de aanmoedigingen. Alf sjukran.

Mike (Hall) gaf de promotie daadwerkelijk vorm: "Schrijf een aantal artikelen over transport van microorganismen, gebruik je Jemen data als basis en zorg ervoor dat het hout snijdt. Ik ben je promotor." Dat klonk tenminste overzichtelijk. Helaas overleed Mike en werden de taken overgenomen door Stefan (Uhlenbrook) en Pieter (Stuyfzand). Naar mijn volle tevredenheid, kan ik wel stellen. Jullie zijn kritisch, constructief, efficient, open-minded, maar ook praktisch. Dank. Da's precies wat ik allemaal nodig had.

Nog kritischer en contructiever is Jack (Schijven). Veel van de hoofdstukken in dit boekje zijn pas de review in gegaan na de "interne review" door Jack. Menig hoofdstuk heb ik twee, drie keer volledig opnieuw geschreven, omdat het niet klopte volgens Jack. Uiteraard had-ie gelijk. Jack's filter was absoluut noodzakelijk om onze manuscripten geaccepteerd te krijgen. Anders gezegd, zonder Jack's betrokkenheid was dit proefschrift er waarschijnlijk niet geweest. Mijn dank is enorm.

Datzelfde geldt voor Joop (de Schutter). Het afdelingshoofd, dat doet wat-ie zegt. Vanaf het eerste ogenblik heeft Joop zich betrokken getoond en de perfecte randvoorwaarden gecreeerd: een Ph.D. projekt nummer. Dank, Joop, voor je altijd positieve, realistische en enthousiaste opstelling. Ik hoop, dat het IHE nog lang niet van je af is.

Natuurlijk heb ik de data niet allemaal zelf verzameld. Robert (Hoogeveen) en Henk-Jan (Wiesenekker) hebben enorm goed werk verricht in Sana'a en omgeving. In het lab op het IHE zijn Anthony (Mporokoso), Ervin (Kolici), Abongwa (Pride), Mazen (Naaman), Simon (Okletey), Mulusew (Kebtie) en Yunus (Liem) zeer actief geweest. Met kolommen, petri-schaaltjes, zeta-meters en microscopen om data te verzamelen, processen te bestuderen, te modelleren, en tenslotte te beschrijven in een mooie full-color UNESCO-IHE M.Sc. thesis. Sommige van deze studies waren zo goed, dat ze met wat aanpassingen gepubliceerd konden worden. Speciale dank voor Manon (van Herwerden), die 9 maanden lang onafgebroken met kolommen in de weer in geweest. Op het eind van haar stage op het IHE was ze de absolute koningin van de kolom experimenten.

Ik heb het lab al even genoemd, maar de mensen nog niet. Een verzameling karakteristieke figuren, waar je altijd kan aankloppen als je experimenten door wat voor oorzaak dan ook niet lukken. Of je nu een dekseltje op de spectrofotometer zoekt, een celletje voor de zeta-meter of een hele nieuwe licht microscoop met camera en software, ze denken enthousiast met je mee, staan altijd voor je klaar, en weten vaak een oplossing te bedenken. Fred (Kruis), Peter (Heerings), Lyzette (Robbemont), Don (van Galen) en Frank (Wiegman): dank. Zonder jullie ondersteuning was het niks geworden.

Niet minder karakteristiek en net zo behulpzaam zijn Paula (Franken) en Gina (Koornaar) van de bieb. Voor de komst van ScienceDirect en de online DTU-verbinding vroeg ik menig artikeltje aan, dat probleemloos geleverd kon worden. Bij deze los ik m'n belofte in: Paula en Gina, bedankt.

Natuurlijk dank ik m'n ouders. Ze hebben me gevormd en gesteund, door dik en dun. Zelfs toen ik onaangekondigd van studie veranderde of toen Bien en ik zomaar, min of meer onaangekondigd, naar Jemen vertokken. Tot twee keer toe. Merijn en Louse, jullie konden als geen ander me weer terug op aarde zetten, na alweer een 'enerverende' dag in het lab. Sorry dat ik te veel achter de computer heb gezeten en niet meedeed met stiekballen, knutselen, voetballen of konijntjes aaien. Ik zal het goed maken. Beloofd.

Bien, bedankt en dikke kus. Je hebt me enorm gesteund door dik en dun en dat is niet altijd even makkelijk geweest. I know bubba. Miljoen excuus. Mag ik dit boekje aan jou opdragen? En dan moesten we maar eens op vakantie gaan en vooral genieten van elkaar. Ook daarna, vind je niet?

CONTENTS

CHAPTER 1

INTRODUCTION

1.1 Groundwater contamination originating from wastewater infiltration

According to the **Framework for Action on Water and Sanitation** (WEHAB Working Group, August 2002 at the Johannesburg 2002 World Summit on Sustainable Development), poor water quality continues to pose a major threat to human health and faecal contamination in water is still the pollutant that most seriously affects the health of children. Diarrhoea, cholera, typhoid and schistosomiasis are the leading water-borne diseases. Some 200 million people world-wide have schistosomiasis, of whom 20 million suffer severe consequences. Diarrhoeal diseases, a result of lack of adequate water and sanitation services, in the past 10 years have killed more children than all people lost to armed conflict since World War II.

In general, aquifer passage reduces pathogenic microorganism concentrations, and numerous successes have been reported in cases, like artificial recharge schemes or riverbank filtration projects, where microorganisms were completely removed. However, studies in the USA have shown that up to half of all US drinking water wells tested had evidence of fecal contamination and an estimated 750,000 to 5.9 million illnesses and 1400-9400 deaths per year may result from contaminated groundwaters (Macler and Merkle, 2000). Figures on illnesses and deaths as a result from contaminated groundwaters for the developing world are not available. However, in Africa around 80% of the population in the largest cities (in Asia: around 55%) have on-site sanitation, such as septic tanks, pour-flush, VIP latrines or simple pits (World Health Organization, 2000-2003), and, according to Foster (2000), in developing countries in Asia, South America and Africa for an estimated 1,300 million persons living in urban areas, the main source of drinking water is groundwater.

Groundwater may be contaminated, when wastewater infiltrates into the soil and recharges groundwater via leaking sewerage systems, leakage from manure, wastewater or sewage sludge spread by farmers on fields, waste from animal feedlots, waste from healthcare facilities, leakage from waste disposal sites and landfills, or artificial recharge of treated waste water. If the distance from source of pollution to point of abstraction is small, there is a real chance of abstracting pathogens. To predict the presence of pathogens in water, usually a separate group of microorganisms is used. The common descriptive term for this group of organisms is fecal indicator organisms (Medema et al., 2003), from which *Escherichia coli* (or *E. coli*) and thermotolerant coliform bacteria are two important members. *E. coli* is widely preferred and used as an index of fecal contamination (World Health Organization, 2003), because its detection is relatively simple, fast and reliable, and the organism is routinely measured in water samples throughout the world. The same applies to thermotolerant ('fecal') coliforms. These coliforms are a less reliable index of fecal contamination than *E. coli*, although under most circumstances their concentrations are directly related to *E. coli* concentrations (World Health Organization, 2003). Viruses may be considered more critical to groundwater quality than *E. coli*. Because of their smaller size, stability, and negative charge, they may be transported even further through the ground, and because of their infectiousness they represent a major threat to public health.

However, the detection and enumeration of viruses, including bacteriophages requires more technical skills and laboratory infrastructure than for *E. coli.*

1.2 Colloid Filtration Theory (CFT)

CFT is a commonly used approach for describing attachment of *E. coli* in saturated porous media (Matthess et al., 1988; Powelson and Mills, 2001; Pang et al., 2003; chapter 5 of this study), and since a part of this research is using the CFT as a starting point, the theory will be briefly explained here. The one-dimensional macroscopic mass balance equation for mobile bacteria suspended in the aqueous phase without the interference of biological factors such as growth and decay can be expressed as (Cameron and Klute, 1977; Corapcioglu and Haridas, 1984, 1985):

$$\frac{\partial C}{\partial t} = D\frac{\partial^2 C}{\partial x^2} - v\frac{\partial C}{\partial x} - \frac{\rho_{bulk}}{\theta}\frac{\partial S}{\partial t} \qquad (1)$$

where C is the mass concentration of suspended bacteria in the aqueous phase (kg/m^3), D is the hydrodynamic dispersion coefficient (m^2/s) and includes the effects of random motility and chemotaxis, v is the pore water flow velocity (m/s), ρ_{bulk} is the bulk density of the porous medium (g/mL), θ is the effective porosity, S is the total retained bacteria concentration (kg/kg) and x is the distance traveled (m). Eq. (1) is also known as the advection-dispersion-sorption (ADS) equation (De Marsily, 1986) to describe transport of mass in porous media. The retained bacteria fraction can be expressed by a first-order kinetic term:

$$\frac{\partial S}{\partial t} = \frac{\theta}{\rho_{bulk}} k_a C \qquad (2)$$

where k_a (s^{-1}) is the attachment rate coefficient. If there is steady state and if the influence of hydrodynamic dispersion is negligible, for a continuous particle injection at concentration C_0 (at $x = 0$) and time period t_0 , the solution to eqs. 1 and 2 for a column initially free of particles can be written as:

$$C(x) = C_0\, exp\left[-\frac{k_a}{v}x\right] \qquad (3)$$

and

$$S(x) = \frac{t_0\theta}{\rho_{bulk}} k_a C(x) = \frac{t_0\theta C_0}{\rho_{bulk}} k\, exp\left[-\frac{k_a}{v}x\right] \qquad (4)$$

Eq. 3, developed by Iwasaki (1937), is commonly referred to as the classical colloid filtration model and has been extensively used in modeling the transport of colloids and micro-organisms in saturated porous media. Equations (1), (3), and (4) are depicted in Fig. 1a, and parameter values can be obtained from column experiments.

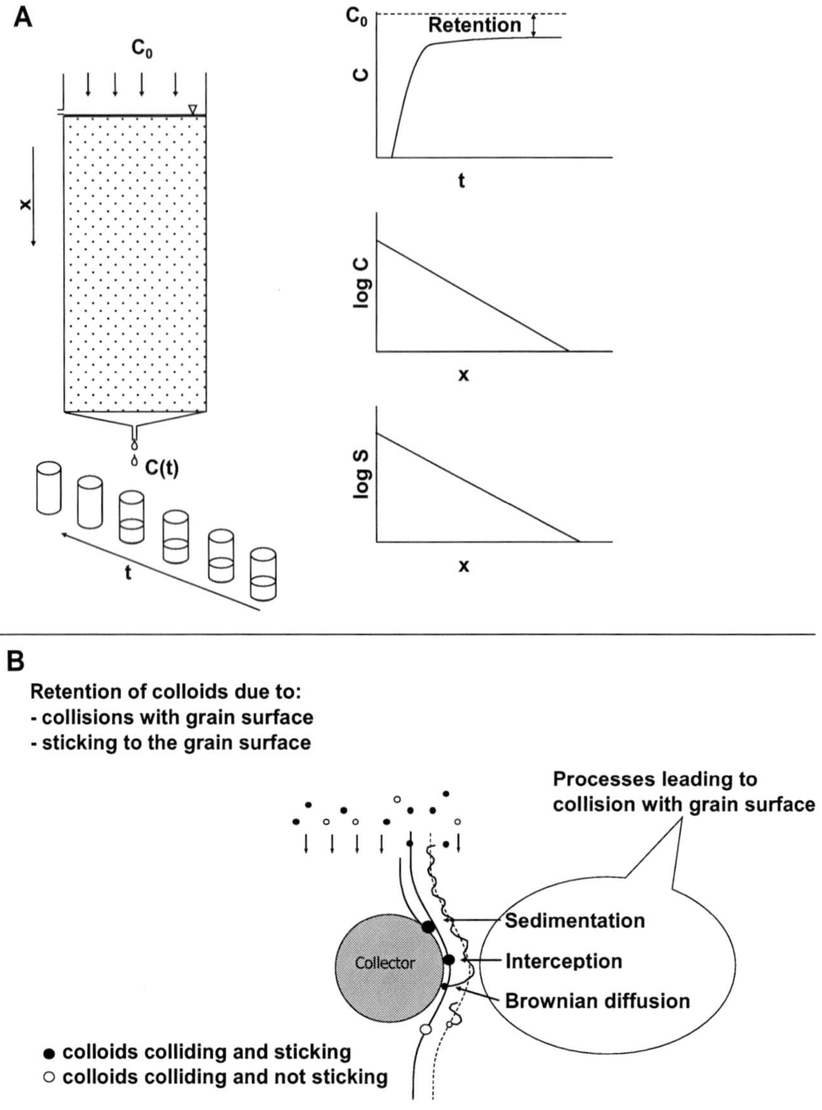

Figure 1: The Classical Colloid Filtration Theory (CFT):(A) As a result of a continuous injection of colloids at concentration C_0 (at $x = 0$) in a column initially colloid-free, colloids may be retained. (upper graph). In case hydrodynamic dispersion can be neglected, the concentration of colloids in the column in the water phase (middle graph) and attached to the solid phase (lower graph) decrease linearly on a logarithmic scale. (B) Retention of colloids is the combined result of collisions of colloids with the grain surface and sticking to that surface. Collision is determined by sedimentation, interception, and Brownian diffusion. Sticking is determined by electro-chemical forces between colloid and grain surface.

The attachment rate coefficient, k_a, is defined as (Harvey and Garabedian, 1991; Ryan and Elimelech, 1996):

$$k_a = \frac{3(1-\theta)}{2a_c} v\eta_0\alpha \tag{5}$$

where a_c (m) is the collector diameter, η_0 is the dimensionless single-collector contact efficiency, and α is the dimensionless sticking efficiency of the colloid. The single-collector contact efficiency is a parameter representing the ratio of the rate of particles approaching the collector to the rate of particles striking a collector, and is determined by gravitational sedimentation, interception, and Brownian diffusion (Fig. 1b). The sticking efficiency is defined as the ratio of the rate of particles striking a collector to the rate of particles sticking to a collector, and is mainly determined by electro-chemical forces between the colloid and the surface of the collector. These processes are explained in chapter 2, which consists of an in-depth review of literature.

1.3 Scope and objectives

Although the presence of *E. coli* and thermotolerant coliforms as representatives of the group of fecal indicator organisms in groundwater systems is indeed found many times, a comprehensive study aimed at predicting their transport was missing. Therefore the purpose of this research was to analyze and evaluate processes that are important in the retention of *Escherichia coli* and thermotolerant coliforms during transport in aquifers under saturated conditions in order to be better able to predict and characterize their fate and transport. To broaden our view, we also carried out a study on the retention of a virus (Chapter 7).

The classical CFT, discussed in the previous section, does not include a number of characteristic phenomena, that can affect the retention of *E. coli* during transport in aquifers. More specifically, these phenomena are:

- Straining;
- Geochemical heterogeneity including site competition;
- Variable deposition rate coefficients as a result of bacterial intra-population heterogeneity;
- The effect of preferential flow;

A schematic representation of these processes is given in Fig. 2.

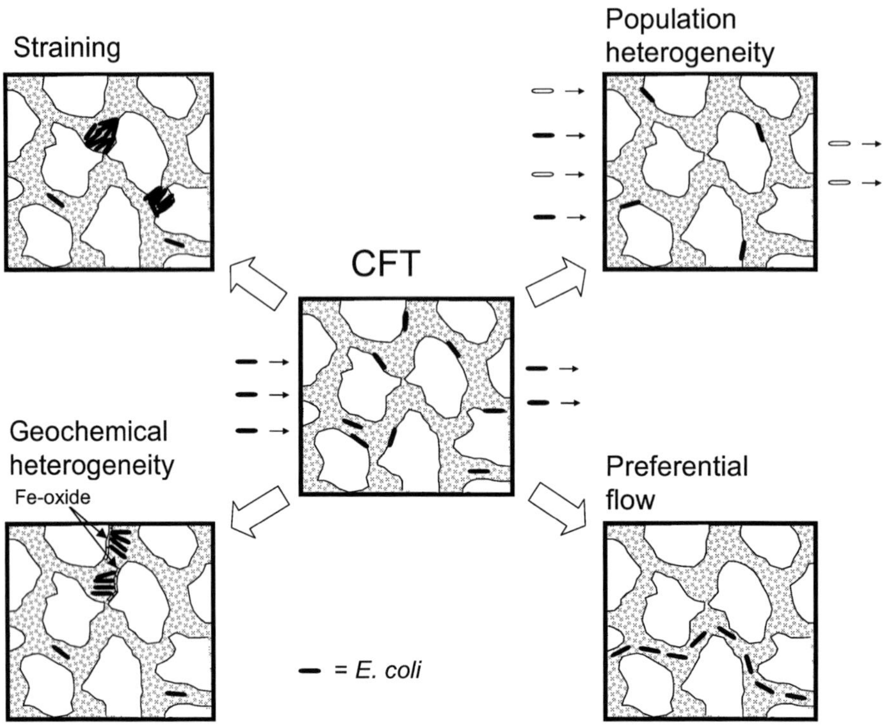

Figure 2: The relation between Colloid Filtration Theory (CFT) on the one hand and straining, population heterogeneity, geochemical heterogeneity, and preferential flow systems on the other hand

Straining is defined as the trapping of bacteria in pore throats that are too small to allow passage, and it results from pore geometry. Although various workers have defined straining ratios (ratio of colloid diameter to grain diameter), the characteristics of the straining mechanism and the relative contribution of straining to the total retention of microorganisms in saturated porous media have remained unclear. In a series of column experiments of ultra-pure quartz sand, the relation between straining and attachment, and the magnitude of straining over a wide range of grain sizes, column lengths, input concentrations, and pore water flow velocities, was studied (Chapter 4). Also, the dynamics of straining was investigated theoretically and experimentally, and the possibility that straining will cease after the smaller pores that are responsible for straining are filled (Chapter 5). For this, column experiments were carried out up to 200 pore volumes.

Various workers have assessed the role of grain charge heterogeneity, resulting from differences in mineralogical composition of the sediment. This work resulted in the development of a theory for predicting transport of colloids in heterogeneously charged porous media, based on sticking efficiency variations (e.g. Song et al., 1994; Johnson et al., 1996; Ryan and Elimelech, 1996; Elimelech et al., 2000; Bhattacharjee et al., 2002). These studies used artificial colloids at high concentrations (10^9 to 10^{14} cells/mL). We studied the effect of geochemical heterogeneity on the transport of *Escherichia coli* at lower concentrations (10^3 to 10^4 cells/mL), which are more appropriate for waste water infiltration conditions, in a series of column experiments containing various sediment mixtures of quartz sand with goethite-coated sand, grains of calcite, and grains of activated carbon (Chapter 5). Because of the importance of goethite in the natural environment, and because Dissolved Organic Carbon (DOC) sorbs to goethite almost irreversibly (Thurman, 1985), thereby blocking sites for bacteria attachment, we also studied the effect of humic acid on the attachment of *E. coli* in columns of goethite-coated sand for various chemical compositions of the bacteria suspension (Chapter 6). A study on the transport of bacteriophages in the goethite – DOC systems was included, to raise our knowledge on the effect of both goethite and DOC on the retention of viruses (Chapter 7). Bacteriophages are viruses that may infect bacteria. We used the bacteriophage PRD1, which is considered to be a model virus for human pathogenic viruses.

Because of heterogeneity among members of the bacteria population, deposition of bacteria may change upon the distance bacteria are transported in an aquifer. These deposition variations can have a profound effect on the distances *E. coli* can travel in the subsurface (Simoni et al., 1998; Tong and Johnson, 2007; Bradford et al., 2003), and therefore deposition variations of *E. coli* were examined in columns of soda-lime glass beads of various sizes. In addition, column experiments were carried out to confirm the presence of subpopulations of bacteria causing deposition variations (Chapter 8).

To evaluate the importance of bacteria removal processes and preferential flow in a real case, the waste water infiltration process below Sana'a, the capital of Yemen, was chosen. Below Sana'a, the unconsolidated shallow alluvial aquifer is recharged with untreated waste water via pit latrines, present throughout the town. In order to understand the mechanisms of microbiological transport and contamination of urban groundwater, the study of Sana'a was subdivided into three components:

a) Transient groundwater flow patterns on a regional and local scale were studied, both quantitatively and qualitatively. A transient groundwater model including a conservative contaminant transport model was constructed (with MODFLOW and MT3DMS) of a 6400 km^2 area, covering the Sana'a basin (Chapter 9);

b) Hydrochemical processes and patterns in the aquifers below the Sana'a urban area were studied, and an exploratory quantitative hydrochemical model was constructed. A major hydrochemical data collection programme was carried out in Sana'a (around 250 wells) and analysed for all major cations and anions. The exploratory hydrochemical transport model included dispersion, advection, cation-exchange, redox reactions and equilibrium reactions (Chapter 10);

c) Total retention rates in the upper aquifers below Sana'a were determined in order to explain microbiological contamination patterns in the Sana'a area. A major microbiological data collection programme was carried out in Sana'a, mainly aimed at determining *Escherichia coli* and thermotolerant coliforms concentrations in wells. In addition, data on aquifer geometry were collected. Explanation and interpretation of microbiological contamination patterns were supported by laboratory experiments (columns and microcosms), and indicative kinetic transport modelling calculations, using existing modeling codes (Chapter 11);

1.3 Outline of the thesis

Part 1 of the thesis discusses the factors determining retention of microorganisms, and is mainly focused on retention of *Escherichia coli*, and to a lesser extent on the removal of viruses. Part 1 consists of chapter 2 to 8. Chapter 2 gives a literature review on the removal of *Escherichia coli* and thermotolerant coliforms in saturated porous media. In chapter 3 the dynamics of straining upon prolonged infiltration are discussed, while in chapter 4 straining and kinetic attachment are analyzed and quantified. In chapter 5 the effects of geochemical heterogeneity on the attachment of *E. coli* is determined, and in chapter 6 and 7 the effect of the simultaneous presence of goethite and DOC in a column on attachment of *E. coli* (chapter 6) and attachment of a virus (the bacteriophage PRD1; in chapter 7) are quantified. In chapter 8, due attention will be given to the aspect of population heterogeneity and deposition rate coefficient variations.

As was stated in the previous section, to evaluate the importance of bacteria removal processes and preferential flow in a real case, the waste water infiltration process below Sana'a, the capital of Yemen, was studied. To do so, the Sana'a study was subdivided into 3 chapters (chapters 9, 10, and 11), and together they form part 2 of this thesis. In chapter 9, the transient water balance of the entire basin in which Sana'a is located, is discussed. Also, flow patterns and groundwater flow velocities are taken into consideration. In chapter 10, the focus is on hydrochemical processes upon wastewater infiltration below the urban area, and the transport of contaminants in vertical downward direction. In chapter 11, the results of the water balance and the mass balances of dissolved solutes are combined with the mass balance of *Escherichia coli* in the aquifers in order to conceptualise vertical transport mechanisms in the urban aquifer.

Finally, in chapter 12 the main results of this research are summarized. Based on these results, a new model approach for the transport of *Escherichia coli* in saturated porous media was developed and applied to two example cases. At the end of this chapter, a number of important future research directions are discussed.

PART 1

FACTORS DETERMINING THE RETENTION OF *ESCHERICHIA COLI*

CHAPTER 2

TRANSPORT AND SURVIVAL OF *ESCHERICHIA COLI* AND THERMOTOLERANT COLIFORMS IN AQUIFERS UNDER SATURATED CONDITIONS – A REVIEW

Foppen, J.W.A and J.F. Schijven, 2006. Evaluation of data from the literature on the transport and survival of *Escherichia coli* and thermotolerant coliforms in aquifers under saturated conditions. Wat. Res. Vol. 40, p. 401-426.

Abstract

Escherichia coli and thermotolerant coliforms are of major importance as indicators of fecal contamination of water. Due to its negative surface charge and relatively low die-off or inactivation rate coefficient, *E. coli* is able to travel long distances underground and is therefore also a useful indicator of fecal contamination of groundwater. In this review, the major processes known to determine the underground transport of *E. coli* (attachment, straining and inactivation) are evaluated. The single collector contact efficiency (SCCE), η_0, one of two parameters commonly used to assess the importance of attachment, can be quantified for *E. coli* using classical colloid filtration theory. The sticking efficiency, α, the second parameter frequently used in determining attachment, varies widely (from 0.003 to almost 1) and mainly depends on charge differences between the surface of the collector and *E. coli*. Straining can be quantified from geometrical considerations; it is proposed to employ a so-called straining correction parameter, α_{str}. Sticking efficiencies determined from field experiments were lower than those determined under laboratory conditions. We hypothesize that this is due to preferential flow mechanisms, *E. coli* population heterogeneity, and/or the presence of organic and inorganic compounds in wastewater possibly affecting bacterial attachment characteristics. Of equal importance is the inactivation or die-off of *E. coli* that is affected by factors like type of bacterial strain, temperature, predation, antagonism, light, soil type, pH, toxic substances, and dissolved oxygen. Modeling transport of *E. coli* can be separated into three steps: (1) attachment rate coefficients and straining rate coefficients can be calculated from Darcy flow velocity fields or pore water flow velocity fields, calculated SCCE fields, realistic sticking efficiency values and straining correction parameters, (2) together with the inactivation rate coefficient, total rate coefficient fields can be generated, and (3) used as input for modeling the transport of *E. coli* in existing contaminant transport codes. Areas of future research are manifold and include the effects of typical wastewater characteristics, including high concentrations of organic compounds, on the transport of *Escherichia coli* and thermotolerant coliforms, and the upscaling of experiments to represent typical field conditions, possibly including preferential flow mechanisms and the aspect of population heterogeneity of *E. coli*.

2.1 Introduction

In developing countries in Asia, South America and Africa for an estimated 1,300 million urban dwellers the main source of drinking water is groundwater (Foster, 2000). This groundwater may be contaminated by infiltrated wastewater, because very often a sewer system is not present and households dispose of their solid and liquid waste on-site. For instance, in Africa around 80% of the population in the largest cities (in Asia: around 55%) have on-site sanitation, such as septic tanks, pour-flush, VIP latrines or simple pits (World Health Organization, 2000-2003). In some cases, the distance from pit latrine to abstraction well is small (< 500 m; Chapter 10) and there is then a real risk of abstracting pathogens. If this groundwater is not disinfected, those who drink it are at increased risk of infection and disease.

Studies in the USA have shown that up to half of all US drinking water wells tested had evidence of fecal contamination and that an estimated 750,000 to 5.9 million cases of illness and 1400-9400 deaths per year may result from contaminated groundwaters (Macler and Merkle, 2000). In addition to leakage from on-site sanitation or from sewerage, the sources of fecal contamination could be manure, wastewater or sewage sludge spread by farmers on fields, waste from animal feedlots, waste from healthcare facilities, and leakage from waste disposal sites and landfills (Pedley et al., 2005).

2.1.1 Indicator organisms

To predict the presence of pathogens in water, a separate group of microorganisms is usually used, generally known as fecal indicator organisms (Pedley et al., 2005). Many microorganisms have been suggested as microbial indicators of fecal pollution (like enterococci, coliphages and sulphite reducing clostridial spores; Medema et al., 2003), but two of the most important indicators used worldwide are *Escherichia coli* and thermotolerant coliform bacteria (for microbiological definitions of these indicators, see section 2.2.1). Both are widely used, because their detection is relatively simple, fast, and reliable. *E. coli* is the preferred indicator of fecal contamination, as it is the only member of the thermotolerant coliform group that is invariably found in feces of warm-blooded animals and it outnumbers the other thermotolerant coliforms in both human and animal excreta (Medema et al., 2003). Thermotolerant coliforms are a less reliable index of fecal contamination than *E. coli*, although under most circumstances their concentrations are directly related to *E. coli* concentrations (Payment et al., 2003). Viruses may be considered as the most critical or limiting microorganism. Because of their small size, their mostly negative surface charge, and their high persistence in the environment, they may travel long distances in the subsurface. In addition, they can be highly infectious (Schijven, 2001). In the study by Karim et al. (2004a), water from 20 groundwater wells from 11 US states was monitored monthly for one year for the presence of culturable viruses, nucleic acid of enteric viruses (enterovirus, hepatitis A, norwalk virus, rotavirus, and adenovirus) by RT-PCR, bacteriophages, and indicator bacteria (total coliforms, *E. coli*, enterococci, and spores of *Clostridium perfringens*). Virus occurrence was common: 17 of the 20 wells tested positive by cell culture or RT-PCR for enteric viruses. If indicators were to be used to monitor the probability of the fecal contamination of groundwater resources, the authors recommended using coliforms and F-specific bacteriophages in combination, because all the virus-positive wells were detected using both these indicators. If a single indicator were to be used, they recommended coliforms, because these were found to occur most frequently (80% of the sites) and were present at 82% of the virus-positive sites.

2.1.2 Aim of this review

Although *E. coli* and thermotolerant coliforms as representatives of the group of fecal indicator organisms have often been found in groundwater systems, to date there has been no comprehensive report evaluating and discussing their transport characteristics. Various authors have reviewed the transport and survival of pathogenic and/or non-pathogenic micro-organisms originating from wastewater.

Some of the reviews concentrate on the movement of bacteria and viruses in aquifers in a qualitative way, without attempting to predict their migration (e.g. Romero, 1970; Lewis et al., 1980; Hagedorn et al., 1981; Crane and Moore, 1984; Bitton and Harvey, 1992; Stevik et al., 2004). Others mainly focus on first-order die-off rates, thereby neglecting the transport component including attachment and detachment processes (e.g. Reddy et al., 1981; Barcina et al., 1997). Murphy and Ginn (2000) mainly summarize the mathematical descriptions of the various physico-chemical and biological processes involved in the transport of bacteria and viruses, without indicating the relative importance of these processes and their occurrence in the natural environment. Merkli (1975) and Althaus et al. (1982) have presented a comprehensive bacteria transport model based on the colloid filtration theory (Herzig et al., 1970; Yao et al., 1971), including the effects of dispersion, diffusion, sedimentation, and filtration. The effects of retardation due to equilibrium adsorption were also included, as well as the physical, chemical, and biological factors determining the die-off of pathogens and indicator organisms. Their findings, however, were based on relatively few data, derived from batch experiments and equilibrium sorption.

The aim of this review is to analyze and summarize existing models, parameters, and their values used to describe and quantify the attachment and inactivation of *Escherichia coli* and thermotolerant coliforms during transport in aquifers under saturated conditions, with the aim of elucidating the relative importance of the various factors influencing transport. In order to illustrate the effect of solution chemistry and ionic strength on the variability of surface charge, we have included a set of *E. coli* zeta potential data from our own experiments. In addition, this review attempts to present a framework to help predict fecal indicator organism concentrations in aquifers. We conclude by using some example cases to demonstrate the importance of key parameters.

2.2 Morphology and surface characteristics of *E. coli*

2.2.1 Taxonomy

Escherichia coli is a gram-negative, facultatively anaerobic, straight, rod-shaped bacterium of 2.0-6.0 μm x 1.1-1.5 μm occurring singly or in pairs (Bergey et al., 1984) and is a taxonomically well-defined member of the family *Enterobacteriaceae*. Thermotolerant coliforms are defined as the group of coliforms that are able to ferment lactose at 44-45 °C. They comprise the genus *Escherichia* and, to a lesser extent, species of *Klebsiella*, *Enterobacter*, and *Citrobacter*. (Bergey et al., 1984). Of these organisms, only *E. coli* is considered to be specifically of fecal origin (Medema et al., 2003). Thermotolerant coliforms are frequently reported as fecal coliforms. However, this term is not correct (Payment et al., 2003), because not all thermotolerant coliforms may be of fecal origin: thermotolerant coliforms other than *E. coli* may originate from organically enriched water such as industrial effluents or from decaying plant materials and soils.

Most members of the *E. coli* species are considered to be harmless organisms, while some strains are responsible for illness. Three general clinical syndromes can result from infection with pathotypes: enteric/diarrhoeal disease, urinary tract infections, and sepsis/meningitis (Kaper et al., 2004). Pathogenic strains have been categorized into six groups, based on serological and virulence characteristics (Kaper et al., 2004; AWWA, 1999). Perhaps the most well-known pathogenic *E. coli* is the serotype O157:H7 of the enterohemorrhagic *E. coli* (EHEC) group.

2.2.2 The surface of *E. coli*

Bacterial attachment to surfaces is influenced by cell surface charge, hydrophobicity, size, and the presence of particular surface structures such as flagella, fimbriae, and extracellular lipopolysaccharides (Gilbert et al., 1991). Because of their physical location on the outside of *E. coli*, lipopolysaccharides (LPSs) are believed to be a key factor in the attachment of microbes to mineral surfaces, the uptake of metal ions, and microbially induced precipitation/dissolution reactions. The LPS of *E. coli* (Amor et al., 2000) consists of (i) a hydrophobic lipid A component anchored in the outer membrane, (ii) a phosphorylated, nonrepetitive hetero-oligosaccharide known as the core oligosaccharide (OS), and (iii) a polysaccharide that extends from the cell surface and forms the O antigen detected in serotyping. Smooth LPS molecules found in most clinical isolates of *E. coli* are composed of this three-part structure, whereas rough LPS lacks the O antigen and can have a truncated core OS (Kastowsky et al., 1991; Walker et al., 2004). The extent and structural diversity of LPS molecules in *E. coli* range from the highly conserved lipid A to the extreme variations reflected in more than 170 known O antigens (Hull, 1997). The core OS is conceptually divided into inner and outer core regions. The inner core is composed primarily of L-*glycero*-D-*manno*-heptose (heptose) and 3-deoxy-D-*manno*-oct-2-ulosonic (or 2-keto-3-deoxyoctulasonic) acid (KDO) residues; this part of the core oligosaccharide is phosphorylated in *E. coli*. Most of the charge on the LPS is concentrated in these phosphate groups of the inner core, and to a lesser extent on carboxylic acid groups of KDO (Kastowksy et al., 1991). The molecular interactions of these bacterial surface functional groups control attachment to surfaces. In section 2.9 of this paper (factors affecting the collision efficiency), more attention will be paid to the influence of LPS structure on the transport of *E. coli* in aquifers.

2.2.3 *E. coli* in water

Usually, when *E. coli* bacteria are introduced into water environments, they gradually die and this process is accompanied by changes in their characteristics. Daubner (1975) carried out survival experiments using *E. coli* strains freshly isolated from the excreta of healthy persons in sterile demineralized water, Danube water, and highly mineralized water. He monitored the morphological and physiological properties of *E. coli* during die-off. Five to twenty hours after *E. coli* had been transferred to the sterile water, shrinking of the cell and the reduction of cytoplasmatic content (distilled and Danube water) or damage to the integrity of the cell (mineralized medium) was observed. Kerr et al. (1999) also observed widespread damage to the majority of cells of *E. coli* O157:H7, with large spaces between cell wall and cell membrane. In addition, Daubner (1975) observed changes in the biochemical activity of the cells, the most important ones being an immediate decrease in respiration and dehydrogenase activity of the cells when introduced into the water environments. Another important biochemical activity change was the utilization of citrate as a sole carbon source. During survival of *E. coli* STCC 416 under adverse conditions, Arana et al. (2004) observed the release of proteins, dissolved free amino acids, and dissolved monomeric carbohydrates into the surrounding medium and the transition from the culturable state to the viable but non-culturable (VBNC) state. In the VBNC state *E. coli* cells maintain their integrity and some metabolic activity, but they lose the ability to grow in culture media (Barcina et al., 1997). The obvious question is what effect these morphological and physiological changes will have on bacterial adhesion. For a number of strains (none of them *E. coli*), Kjelleberg and Hermansson (1984) studied the effects of starvation on bacterial surface characteristics; they also studied the changes in adhesion characteristics. They reported a considerable decrease in cell volume in the course of the starvation process (which lasted 22 hours in their study), which corroborates the findings of Daubner (1975) and Kerr et al. (1999) for *E. coli*. Furthermore, a number of strains exhibited increases in both hydrophobicity and irreversible binding to glass surfaces, initiated after different starvation times.

2.2.4 Hydrophilic bacterial cell wall

The role of bacterial cell wall hydrophobicity in attachment has also been recognized in a number of other studies (Van Loosdrecht et al., 1987a, b; Gilbert et al., 1991; Noda and Kanemasa, 1986). Van Loosdrecht et al. (1987b) identified a positive relation between the bacterial hydrophobicity of 23 strains and their adhesion to negatively charged polystyrene disks, irrespective of their surface charge. They found that the contact angles of a 0.1 M NaCl suspension with *E. coli* NCTC 9002 and *E. coli* K-12 were 16 ° and 24 ° respectively, which is relatively small in both cases. Because of the small contact angle the hydrophobicity was low, as was adhesion to the polystyrene disks. It was therefore concluded that *E. coli* is poorly adhesive. Noda and Kanemasa (1986) also found that *E. coli* was poorly hydrophobic; as a measure of hydrophobicity they determined the attachment of a nonionic surfactant to 8 bacterial strains. They found that the *E. coli* K-12 adsorbed much less surfactant (37-69 μg surfactant/15 mg dry weight bacteria) compared to the other strains used in their experiments.

From these studies (Gilbert et al., 1991; Van Loosdrecht et al., 1987a and b; Noda and Kanemasa, 1986), it can be concluded that *E. coli* is more of a hydrophilic than a hydrophobic organism. Gilbert et al. (1991) measured the hydrophobicity at regular intervals during the growth phase of *E. coli* ATCC 8739 and observed a significantly lower hydrophobicity during the mid-exponential growth phase than during the early growth and the stationary end phases. Gilbert et al. (1991) also measured the ability of *E. coli* to adhere to nitric-acid cleansed glass surfaces during the bacterial growth phase. A clear relationship was identified between decreased zeta potential (more surface negativity) and decreased attachment ($r^2 = 0.91$). However, hydrophobicity did not correlate with adhesion ($r^2 = -0.004$). Based on these findings, Gilbert et al. (1991) refined the general hypotheses by van Loosdrecht et al. (1987a) and arrived at the important conclusion that for relatively hydrophilic organisms, such as *E. coli*, the hydrophobicity did not determine adhesion. Instead, the primary modulator of adhesion appeared to be the zeta potential. Scholl et al. (1990) and Scholl and Harvey (1992) also concluded that for relatively hydrophilic organisms, the major factor controlling the initial adhesion of bacteria is the surface charge of the minerals in the aquifer.

2.2.5 The surface charge of *E. coli*

Cell surface charge, e.g., as a result of charged functional groups on the LPS of *E. coli*, is not measured directly, but is usually determined via the so-called electrophoretic mobility, which is a measure of the zeta potential (defined as the potential drop across the mobile part of the double layer that is responsible for electrokinetic phenomena; Stumm and Morgan, 1996). To illustrate the variability of charge on the outer membrane of *E. coli*, we determined in our laboratory the zeta potential of *E. coli* ATCC 25922 in various solutions. To do so, we prepared bacterial suspensions consisting of various electrolyte concentrations and stored these for 24 hours in the dark at room temperature (20° C) prior to measuring electromobility with a zeta-meter similar to the one made by Neihof (1969). Movement of bacteria was visible on a video screen attached to a camera mounted on top of a light microscope (Olympus EHT) in phase contrast mode. Velocity measurements on each particle were made over 100 μm in both directions by reversing the current (by applying a voltagc potential varying between 20 to 90 V across the cell). Particle mobility values were obtained from measurements on at least 20 particles. Velocity measurements were used to calculate the zeta potential with the Smoluchowski-Helmholtz equation. The results (Fig. 1) indicated that in solutions of monovalent ($NaHCO_3$ and NaCl) and divalent ($CaCl_2$ and $MgSO_4$) electrolytes with ionic strength values ranging from 10^{-5} mM to 400 mM, within a pH-range of 7.19-8.81, the zeta potential ranged from –20 to –170 mV. In the

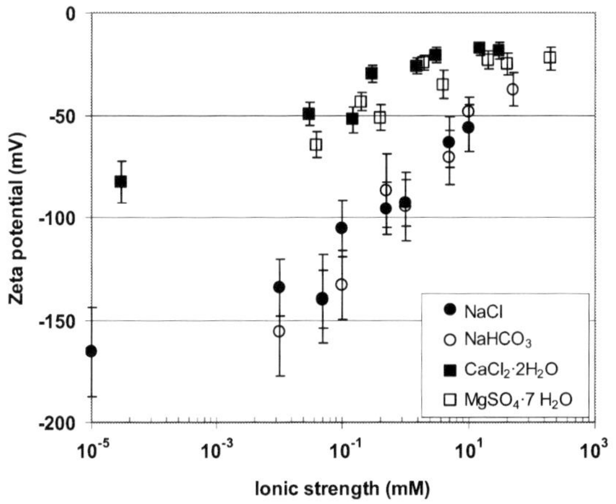

Fig. 1: *Zeta potential as a function of ionic strength of various solutions for E. coli ATCC 25922 (our measurements). Error bars indicate one standard deviation. Solution pH-values: NaCl and $CaCl_2{\cdot}H_2O$: pH = 7; $NaHCO_3$: pH = 8.35; $MgSO_4{\cdot}7H_2O$: pH varies from 7.13 (at 0.01 mmol/L) to 8.81 (at 50 mmol/L)*

divalent electrolytes, the zeta potential was considerably less negative than in the monovalent electrolytes, while pH variations did not seem to be very important. This can be explained in qualitative terms by the Schultze-Hardy rule, which states that the coagulation of a colloid is affected by that ion of an added electrolyte which has a charge opposite in sign to that of the colloidal particle, and the effect of such ion increases markedly with the number of charges it carries (Sawyer et al., 1994).

2.2.6 Non-uniform surface charge distribution

The zeta potential has two major shortcomings in characterizing the interaction between *E. coli* and collector surfaces (De Kerchove and Elimelech, 2005). Firstly, the zeta potential typically applies to hard spherical particles, whereas biological cells form a soft ion-permeable layer consisting of charged and non-charged LPS structures (section 2.2.2) around the cell. The presence of such a layer controls the spatial distribution of the electric surface potential and the resulting interaction forces between a bacterial cell and solid surfaces. De Kerchove and Elimelech (2005) tested the applicability of an electrokinetic theory for soft particles to characterize the electrophoretic mobility and adhesion kinetics of various *E. coli* K12 mutants, and concluded that the theory failed to predict attachment efficiencies. The lack of success was attributed to chemical and physical heterogeneities of the polyelectrolyte layer at the cell surface. Secondly, the zeta potential is a macroscopic parameter that reflects the net or average electrokinetic properties of the cell and is not sensitive to small scale variations in the cell surface. Therefore, lumping this heterogeneous distribution of charge into one zeta potential value might result in erroneous predictions of *E. coli* - collector surface interactions. Walker et al. (2005) demonstrated that zeta potential values for *E. coli* D21g harvested in the mid-exponential and in the stationary growth phase in solutions of various KCl concentrations were similar, while the bacterial deposition rate in columns packed with clean quartz sand (with a negative charge) at intermediate ionic strength values (0.003 – 0.03 M KCl) was at most 10 times higher for the stationary growth phase *E. coli* than for the mid-exponential *E. coli*. This was attributed to a combination of factors: 1) Mid-exponential phase cells are covered with relatively simple proteins, while stationary-phase cells are covered with more complex LPS molecules, 2) Apparently, upon approach toward the collector surface, the mid-exponential phase cells appear to be more uniformly (negatively) charged, while the stationary-phase cells will have a more uneven distribution of (negative) charge on the LPS molecules (phosphate and carboxyl groups), which resulted in less repulsion and more attachment of stationary-phase cells.

To summarize, *E. coli* bacteria have often been found in fecally contaminated groundwater, are easy to count, and are hydrophilic and strongly negatively charged. These properties make this bacterium a useful indicator of fecal contamination of groundwater, especially in developing countries lacking sufficient laboratory resources. In addition, some *E. coli* strains are enteropathogenic. Viruses may be considered more critical to groundwater quality than *E. coli*. Because of their smaller size, stability, and negative charge, they may be transported even further through the ground, and because of their infectiousness they represent a major threat to public health. However, the detection and enumeration of viruses, including bacteriophages requires more technical skills than needed for *E. coli*.

2.3 General bacteria transport model

The transport of mass in porous media may be generally described by the advection-dispersion-sorption (ADS) equation (De Marsily, 1986). Various expressions of the ADS equation have been used for the transport of colloids in general (Herzig et al., 1970; Yao et al., 1971; Corapcioglu and Haridas, 1984, 1985; Murphy and Ginn, 2000; Schijven, 2001) and, more particularly, for the transport of *E. coli* and thermotolerant coliforms (Pang et al., 2003; Powelson and Mills, 2001; Matthess and Pekdeger, 1981 and 1985; Matthess et al., 1985, 1988; chapter 5 of this study). To appropriately describe the dynamic effects of colloid deposition and possible blocking effects, the colloid transport equation is expressed in terms of particle number concentration rather than mass concentration (Sun et al., 2001; Johnson et al., 1996) along with terms for attachment, detachment, straining, and inactivation or die-off (Bhattacharjee et al., 2002)

$$\frac{\partial C}{\partial t} = \nabla \cdot (D \nabla C) - \nabla \cdot (vC) - \frac{f_s}{\pi a_p^2} \frac{\partial S}{\partial t} - k_i C \qquad (1)$$

where C is the number concentration of suspended bacteria in the aqueous phase (m^{-3}), D is the hydrodynamic dispersion coefficient (m^2d^{-1}), v is the pore water flow velocity (md^{-1}), f_s is the specific surface area of the porous medium (m^{-1}), a_p is the diameter of the bacteria (m), S is the dimensionless fractional surface coverage, which is defined as the total cross-section area of deposited bacteria per interstitial surface area of the porous medium solid matrix (Sun et al., 2001), k_i is the inactivation or die-off rate coefficient of bacteria in the fluid (d^{-1}), and t is time (day). The dispersion coefficient is defined as (Corapcioglu and Haridas, 1984 and 1985; Matthess and Pekdeger, 1981 and 1985)

$$D = D' + D_B \qquad (2)$$

where D' is the coefficient of mechanical dispersion due to the tortuosity of the pore channels and D_B is the coefficient of diffusion due to Brownian motion. D' depends on groundwater flow velocity and the dispersivity coefficient, which is a function of the inhomogeneity of the aquifer (e.g. De Marsily, 1986; Appelo and Postma, 1993). D_B can be estimated by the Stokes-Einstein equation (e.g. Corapcioglu and Haridas, 1984)

$$D_B = \frac{kT}{3\pi\mu a_p} \qquad (3)$$

where k is the Boltzmann constant (1.2806×10^{-23} JK^{-1}), T is the temperature (K) and μ is the dynamic water viscosity (Pa s; $\mu = 9.85x10^{15}T^{-7.7}$ when $273 < T(emperature) < 303$ in K; Matthess, 1982). For micron-sized particles (1-5 μm) at temperatures between 273 and 303 K, D_B is in the range of 10^{-8} m^2s^{-1}.

For bacteria retained by the solid matrix, several authors (Harvey and Garabedian, 1991; Lindqvist and Bengtsson, 1991; Lindqvist et al., 1994; Pang et al., 2003; Powelson and Mills, 2001; Matthess and Pekdeger, 1981 and 1985; Matthess et al., 1985, 1988) assume a fraction available for equilibrium sorption (S_{eq}) and a fraction available for kinetic attachment (S_{att}), some (Tan et al., 1994; McCaulou et al., 1995; Hendry et al., 1997; Hendry et al., 1999) assume only kinetic sorption (S_{att}), while others (Bradford et al., 2003; Bradford and Bettahar, 2005; chapter 5 of this study) assume kinetic sorption and a kinetic fraction available for straining (S_{str}). Taking all these processes into account, the mass balance equation for retained bacteria can be expressed as

$$\frac{\partial S}{\partial t} = \frac{\partial S_{eq}}{\partial t} + \frac{\partial S_{att}}{\partial t} + \frac{\partial S_{str}}{\partial t} - k_{is} S \quad (4)$$

$$S_{eq} = \pi a_p^2 K_D C \quad (5)$$

$$\frac{\partial S_{att}}{\partial t} = \pi a_p^2 k_a C B(S_{att}) - k_r S_{att} \quad (6)$$

$$\frac{\partial S_{str}}{\partial t} = \pi a_p^2 k_{str} C \quad (7)$$

where k_{is} is the inactivation or die-off rate coefficient of bacteria on the solid matrix (d^{-1}), K_D is an empirical distribution coefficient (m), k_a is the attachment rate coefficient (md^{-1}), k_r is the detachment rate coefficient (d^{-1}), k_{str} is the straining rate coefficient (d^{-1}), and $B(S_{att})$ is the dimensionless dynamic blocking function. Blocking is the occlusion of collector surface resulting in a declining bacteria deposition rate as bacteria accumulate on that collector surface (Ryan and Elimelech, 1996). The dynamic blocking function can be expressed in terms of the random sequential adsorption model (Ko et al., 2000)

$$B(S_{att}) = 1 - a_1 S_{att} + a_2 S_{att}^2 + a_3 S_{att}^3 + ... \quad (8)$$

with appropriate values of the coefficients a_i. It should be noted that the more common Langmuirian adsorption model may also be employed to account for the blocking behavior, in which case eq. (8) is truncated after the first-order term (Bhattacharjee et al., 2002; Ko et al., 2000). Inserting eqs. (5), (6) and (7) in eq. (4) yields the rate expression for the fractional surface coverage, S, of attached bacteria

$$\frac{\partial S}{\partial t} = \pi a_p^2 \{K_D \frac{\partial C}{\partial t} + (k_a B(S_{att}) + k_{str})C\} - k_r S_{att} - k_{is} S \quad (9)$$

Eqs. (1) and (9) are relatively simple general mass balance equations for bacteria in the fluid phase (eq. (1)) and bacteria attached on the solid matrix in aquifers (eq. (9)). More complex general colloid transport models are available. Sun et al. (2001)

and Bhattacharjee et al. (2002) consider partial solid matrix mass balances for bacteria attached to favorable and unfavorable sites. Corapcioglu and Haridas (1984, 1985) consider temporal and spatial changes in porosity due to bacterial attachment to aquifer grains. Others (e.g. Kim and Corapcioglu, 1996) consider the mass balances of an aqueous phase, a carrier phase, and a stationary solid matrix phase, with or without a contaminant present in one or all of the phases. Although it might be more appropriate from a theoretical point of view to use these more complex models, we used eqs. (1) and (9), because, as will become clearer in sections 2.4-13, the limited data available did not justify the use of more complex model descriptions.

2.3.1 Simplifying the rate expression for fractional surface coverage (eq. 9) for *E. coli* transport

Equilibrium sorption

The so-called retardation factor due to equilibrium sorption, R_F, is defined as the ratio of the mean groundwater flow velocity, v_w, to the mean transport velocity of microorganisms, v_m (Althaus et al., 1982; Harvey, 1997; Matthess et al., 1988; Sinton et al., 1997 and 2000; Pang et al., 2003; see eq. (9)). It can be approximated if the distribution coefficient, K_D, of the bacteria is known:

$$R_F = \frac{v_w}{v_m} = 1 + \pi a_p^2 K_D \tag{10}$$

In a number of studies (see Table 1), the retardation factor in the case of *E. coli* transport, either measured or fitted (from breakthrough curves), ranged between 0.1 and 2.81 (although mathematically $R_F \geq 1$); the average of all values mentioned in the Table is 1.15. Matthess et al. (1988) attributed retardation factors less than 1 to preferential transport of the microbes in the macropore systems of the aquifers. The explanation most commonly offered for this phenomenon is pore size exclusion (Harvey, 1997; Sinton et al., 1997, 2000; Sinton, 2001): larger particles such as bacteria cannot pass through smaller pores available to dissolved chemicals, but can only travel through the larger pores, where groundwater velocity is higher. However, in an injection experiment of a mixture of rhodamine WT dye, F-RNA phages, and thermotolerant coliforms into an aquifer, Sinton et al. (1997) observed that F-RNA phages (with a diameter of 26 nm) reached peak concentrations not only before the dye, but also before the thermotolerant coliform peak. Sinton et al. (1997) were unable to come up with a completely satisfactory explanation for this. One hypothesis was that the electrostatic repulsion forces might have been acting more strongly on phages than bacteria. Another more plausible explanation was that phages might have been adsorbed to particles of a size that are transported faster than bacteria (by pore size exclusion). On the basis of the data in Table 1 plus our own observations of *E. coli* in laboratory columns and in field tests, we believe that retarded breakthrough by equilibrium adsorption is of little significance.

Table 1: Retardation of Escherichia coli determined from field and laboratory experiments

Type of experiment	Retardation factor	Type of microbe	Method	Reference
Column experiment (continuous source)	0.56-2.06 0.95-1.94	*E. coli* Fecal Coliform	Fitted with CXTFIT[a)]	Pang et al., 2003;
Column experiment (pulse source)	1.02-2.81 0.97-2.76	*E. coli* Fecal Coliform		
Column experiment (continuous source)	1.03-1.22	*E. coli* ATCC25922	Fitted with CXTFIT	Powelson and Mills, 2001;
Field test (pulse source, natural gradient))	0.95 0.77	*E. coli* ATCC 15224 Fecal Coliform	Relative to tracer velocity (rhodamine WT)	Sinton et al., 1997;
Field test (pulse source, natural gradient)	0.50 0.74	*E. coli* ATCC 15224 *E. coli* K-12	Relative to tracer velocity (rhodamine WT)	Sinton et al., 2000;
Field test (pulse source)	0.75-0.93	*E. coli*	Relative to tracer velocity (^{82}Br)	Alexander and Seiler, 1983; Havemeister and Riemer, 1985, both in: Matthess et al., 1985;
Field test in fractured granite (forced gradient)	0.1	*E. coli*	Relative to tracer velocity (bromide)	Champ and Schroeter, 1988;
Column experiment with quartz	1	*E. coli*	Relative to tracer velocity	Merkli, 1975;
Column experiment with K-feldspar	1	*E. coli*		
Column experiment with marble	1	*E. coli*		
Column experiment with serpentine	1	*E. coli*		

[a)]: CXTFIT curve-fitting program (Toride et al., 1999)

Blocking

Although blocking has been an important mechanism in some studies (Lindqvist et al., 1994; Johnson and Elimelech, 1995; Liu et al., 1995), it has not been used in the literature to explain the breakthrough of *E. coli* and thermotolerant coliforms, either in laboratory column experiments or in field tests. Even in long-term experiments of 3 months and longer (Butler et al., 1954; Bouwer et al., 1974b) with treated sewage effluent in rapid infiltration basins, there was no sign of blocking. The most plausible explanation for this is that grain surface coverage is apparently low due to relatively "low" bacteria concentrations and/or is kept low due to the die-off of attached bacteria (which free up the surface area again). The net result is that a decline in bacteria deposition rate has not yet been observed.

Because of the apparent insignificant equilibrium adsorption and the apparent lack of blocking, we assumed $K_D = 0$ and $B(S_{att}) = 1$. Therefore, eq. (9) reduces to

$$\frac{\partial S}{\partial t} = \pi a_p^2 (k_a + k_{str})C - k_r S_{att} - k_{is} S \qquad (11)$$

which indicates that the transport of *E. coli* in aquifers is mainly controlled by five rate coefficients (k_a , k_r , k_{str} , k_{is} in eq. (11) and k_i in eq. (1)).

2.4 The attachment rate coefficient

2.4.1 The single collector removal efficiency

The attachment rate coefficient, k_a , can be determined by (Ryan and Elimelech, 1996; Johnson et al., 1996):

$$k_a = \frac{\eta v \theta}{4} \qquad (12)$$

where η is the dimensionless single collector removal efficiency (SCRE) and θ is the effective porosity . The SCRE is a parameter representing the ratio of the rate of particles striking a collector to the rate of particles approaching the collector multiplied with an empirical attachment (sticking) efficiency. When the geochemical composition of the sediment is known, grain surface charge heterogeneity can be included in the formulation of the SCRE by (Johnson et al., 1996; Ryan et al., 1999; Elimelech et al., 2000)

$$\eta = \eta_0 [\lambda \alpha_f + (1-\lambda)\alpha_u] = \eta_0 \alpha_{total} \qquad (13)$$

where η_0 is the single collector contact efficiency (SCCE), determined from physical considerations (e.g. Tufenkji and Elimelech, 2004a; see section 2.4.2), α_f and α_u are the sticking efficiencies to favorable and unfavorable attachment sites respectively, and λ is a dimensionless heterogeneity parameter describing the fraction of aquifer grains composed of minerals favorable for the attachment or of grains having patches favorable for attachment (Elimelech et al., 2000; Bhattacharjee et al., 2002). For *E. coli*, which at typical groundwater pH values (6-8) is mostly negatively charged, favorable attachment sites are positively charged. Examples of the minerals commonly found in sediments that are positively charged at typical groundwater pH values are goethite, (α-FeOOH, pH_{PZC} of pure goethite, not influenced by carbonate adsorption = 9.0 -9.1; Gaboriaud and Ehrhardt, 2003) and carbonates like calcite. Somasundaran and Agar (1967) determined the zero point of charge of freshly ground Iceland spar calcite, as measured by the streaming potential technique, to be within a pH range of 9.5-10.8.

Eq. (13) is important, since it directly determines the attachment rate coefficient (eq. (12)). Also, since in most of the cases in the literature the sediment is considered to be a homogeneous medium, eq. (13) is useful for determining an overall collision efficiency, α_{total} , and an overall SCRE, η_{total} , for collector or sediment in which the favorable and unfavorable fractions are unknown (see sections 2.11 and 2.14).

2.4.2 The single collector contact efficiency, η_0

The SCCE is defined as (e.g. Tufenkji and Elimelech, 2004):

$$\eta_0 = \eta_D + \eta_I + \eta_G \tag{14}$$

where η_D, η_I and η_G, represent theoretical values for the single-collector contact efficiency when the sole transport mechanisms are diffusion, interception, and sedimentation, respectively (Yao et al., 1971; Tufenkji and Elimelech, 2004). There have been many studies to determine the SCCE, η_0. When comparing a number of filtration models with the experimental data from Yao et al. (1971), Logan et al. (1995) recommended the use of the model developed by Rajagopalan and Tien (1976) to calculate aquasol removals in packed beds. This model has been used very frequently (e.g. Tobiason and O'Melia, 1988; Martin et al., 1992; Rijnaarts et al., 1996; Ryan and Elimelech, 1996). However, the major shortcoming of the Rajagopalan and Tien model is the omission of the influence of hydrodynamic and Van der Waals interactions on the deposition of particles that are dominated by Brownian diffusion. Tufenkji and Elimelech (2004) developed an equation that includes these hydrodynamic and Van der Waals interactions and defined the transport mechanisms diffusion, interception, and sedimentation as:

$$\eta_D = 2.44 A_S^{1/3} N_R^{-0.081} N_{Pe}^{-0.715} N_{vdW}^{-0.052} \tag{15a}$$

$$\eta_I = 0.55 A_S N_R^{1.675} N_A^{0.125} \tag{15b}$$

$$\eta_G = 0.22 N_R^{-0.24} N_G^{1.11} N_{vdW}^{-0.053} \tag{15c}$$

where $A_S = \dfrac{2(1-p^5)}{2-3p+3p^5-2p^6}$ and $p = (1-\theta)^{1/3}$, $N_R = \dfrac{a_p}{a_c}$ for interception, a_c is the median of the grain size number distribution (m). The Peclet number $N_{Pe} = \dfrac{v\theta a_c}{D_B}$ for the sum of convection and diffusion. The van der Waals number characterizing the ratio of van der Waals interaction energy to the particle's thermal energy $N_{vdW} = \dfrac{H}{kT}$, where H is the Hamaker constant, assumed here to be constant at $6.5\text{x}10^{-21}$ J (Walker et al., 2004). The attraction number representing the combined influence of the van der Waals attraction forces and fluid velocity on particle deposition rate due to interception $N_A = \dfrac{H}{12\pi\varpi a_p^2 v\theta}$. The gravity number $N_G = \dfrac{2ga_p^2(\rho_p - \rho_{fl})}{9\varpi v\theta}$ for sedimentation, where g is the gravitational acceleration constant (9.81 ms^{-2}), ρ_p is the particle density (kgm^{-3}) and ρ_{fl} is the fluid density (kgm^{-3}).

2.5 The straining rate coefficient

Straining is defined as the trapping of bacteria in pore throats that are too small to allow passage, and it results from pore geometry. Matthess and Pekdeger (1988) defined a so-called geometrical suffusion security δ, or straining, as

$$\delta = \frac{a_p}{0.12 a_{c,10}} \tag{16}$$

where $a_{c,10}$ is the 10th percentile of the log-normal grain size number distribution (m). These authors hypothesized that straining occurs only if $\delta \geq 1.5$, i.e. when the bacteria are larger than 18% of d_{10}. However, Bradford et al. (2002, 2003) reported that straining was more pronounced than predictions based upon eq. (16). A less empirical approach was followed by Herzig et al. (1970), who calculated the volume of spherical colloids that could be retained (σ) based on purely geometrical considerations (e.g. Corapcioglu and Haridas, 1984 and 1985; chapter 5 of this study):

$$\sigma = \frac{1-\theta}{2} \pi N \left(\frac{a_p}{a_c}\right)^2 \sqrt{\left(1+\frac{a_p}{a_c}\right)^2 - 1} \tag{17}$$

where N is the coordination number or the number of contact points between grains. Foppen et al. (2005; Chapter 3) determined the pore volume available for straining from modeling high concentration *E. coli* breakthrough curves, from geometrical considerations based on Herzig et al. (1970) and from a pore size density function, and concluded that pore volumes determined with those methods were in reasonable agreement. Neumann (1983) developed a geometrically derived contact efficiency due to straining, based on Hall (1957). Assuming the collector consists of spherical grains surrounded by six spherical collectors or grains of similar size, Neumann (1983) determined a straining contact efficiency (η_S) as

$$\eta_S = \frac{12}{\pi\sqrt{2}} \left(\frac{a_p}{a_c}\right)^{\frac{3}{2}} \approx 2.7 \left(\frac{a_p}{a_c}\right)^{\frac{3}{2}} \tag{18}$$

in which η_S can be interpreted as the probability of entering a pore in which straining can occur. The straining rate coefficient developed by Bradford et al. (2003) was proportional to $\left(\frac{a_p}{a_c}\right)^{1.42}$, which is surprisingly similar to the geometrically derived relation by Neumann. In order to arrive at a straining rate coefficient, both Neumann (1983) and Hall (1957) multiplied the straining contact efficiency by a correction parameter K to account for fluid velocity variations at pore level, grain geometry variations, and filter bed porosity. The need for a correction factor also emerged from Bradford's work (Bradford et al., 2003; Bradford and Bettahar, 2005). Bradford and his colleagues proposed a depth-dependent dimensionless power law straining function with a value between 0 and 1. Comparing eq. (18) with the straining rate coefficient determined by Foppen et al. (2005; Chapter 3) from modeling breakthrough curves of *E. coli* in the case of straining also revealed the need for a correction factor. In the experimental work of Matthess (Matthess, 1991a and b; Bedbur, 1989; Peters, 1989) eq. (18) was used to account for straining, but Matthess used the sticking efficiency, α , to correct for the variations mentioned above. However, we believe that the latter is not a good concept, because colloid filtration theory describes the probability of colliding with the surface of a collector and the probability of remaining attached, whereas in straining it is the probability of being strained in a pore that is of concern. Therefore, in order to calculate the straining rate coefficient, we propose to combine eq. (18) and eq. (12) into:

$$k_{str} = \frac{\eta_S v\theta\alpha_{str}}{4} = 0.68\left(\frac{a_p}{a_c}\right)^{\frac{3}{2}} v\theta\alpha_{str} \qquad (19)$$

where α_{str} is a dimensionless straining correction factor to account for fluid velocity variations at pore level, grain geometry variations, and filter bed porosity. It is tempting to interpret α_{str} as the probability of a colloidal particle being retained due to purely geometrical considerations, once a pore in which straining can occur has been entered. If so, then the value of α_{str} would be between 0 and 1. Experimental evidence (Bradford et al., 2003; Bradford and Bettahar, 2005, Matthess et al., 1991a and b) does indeed suggest this range of values, but this explanation needs to be confirmed by future research on straining.

2.6 Relative importance of the bacteria transport mechanisms

At Darcy groundwater velocities between 0.1 and 10 md^{-1} and a grain size of 1 mm, the diffusion and sedimentation components are the dominant bacterial transport mechanisms (see Fig. 2, upper graph). When Darcy flow velocity increases, the diffusion and sedimentation components decrease, and interception and straining remain unaltered. When the grain size decreases to 0.02 mm (the approximate boundary grain size value between silt and clay; Fig. 2, lower graph), then diffusion, interception, and straining are the dominant transport mechanisms.

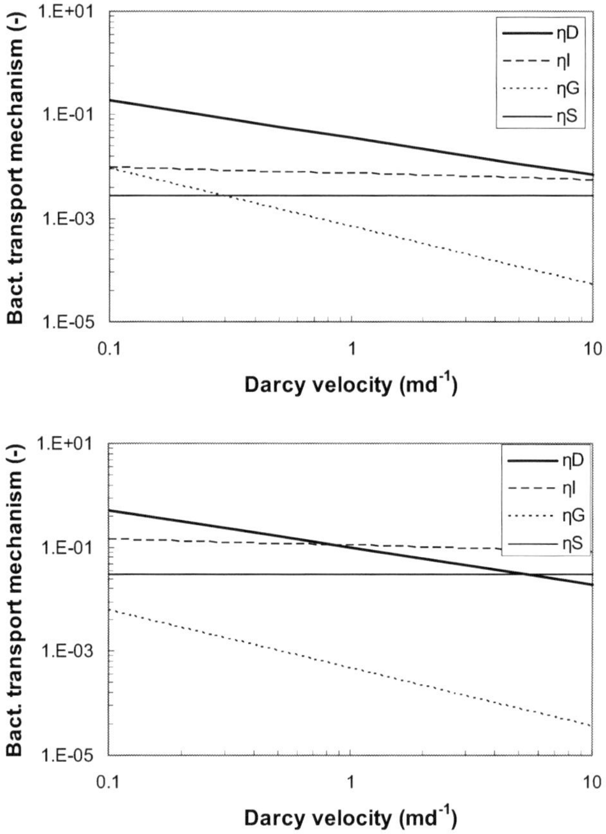

Fig 2: *Magnitude of the single bacterial transport mechanisms as a function of Darcy flow velocity. Upper graph: a_c = 1 mm; lower graph: a_c = 0.02 mm. Other parameters (for both graphs): T = 283 K, a_p = 1 μm, θ = 0.35, $H = 6.5x10^{-21}$ J; ρ_p = 1050 kgm^{-3}, ρ_{fl} = 1000 kgm^{-3}, μ = $1.3x10^{-3}$ $kgsm^{-1}$*

2.7 The filter coefficient (β)

The filter coefficient, β (m^{-1}), of an aquifer is the filter effect with respect to a certain flow length (Matthess et al., 1988; Rijnaarts et al, 1996). The initial concentration of a microbial suspension C_0 decreases along a flow length, x, to the observed concentration C according to (Iwasaki, 1937; Matthess et al., 1988)

$$C = C_0 e^{-\beta x} \tag{20}$$

The filter coefficient is constant only at the beginning of the filter process during the clean bed collision phase, when deposited particles do not yet influence the transport of particles (Matthess et al., 1991a; Rijnaarts et al., 1996). In general, the filter coefficient, the single collector removal efficiency, and the collision efficiency are related according to (Rijnaarts et al., 1996)

$$\beta = \frac{3(1-\theta)}{2a_c}\eta_0\alpha \tag{21a}$$

and, when straining has to be included, the filter coefficient becomes

$$\beta = \frac{3(1-\theta)}{2a_c}(\eta_0\alpha + \eta_S\alpha_{str}) \tag{21b}$$

Combining eqs. (12), (19) and (21b) yields

$$k_a + k_{str} = \frac{\theta a_c}{6(1-\theta)}v\beta \tag{22}$$

If the rate coefficients need to be expressed in d^{-1}, then the right-hand side of eq. (22) has to be multiplied by the specific surface area $f_s = \dfrac{6(1-\theta)}{\theta a_c}$.

2.8 Factors affecting the contact efficiencies

2.8.1 Effect of velocity

According to eqs. (15) and (18), a number of parameters affect the SCCE and the straining contact efficiency: bacteria size, grain size, density of the fluid and the bacteria, temperature, velocity, and porosity. Matthess et al. (1991 a and b) carried out an extensive set of laboratory pulse source column experiments with pure quartz sand (unfavorable attachment) and *E. coli* ATCC11229 to test the effect of grain size, grain size uniformity, and velocity with constant bacteria size (~3 μm), constant density of fluid and bacteria (1000 and 1050 kgm^{-3}), and constant temperature (283 K). The experiments were all carried out with artificial groundwater containing 526 mgL^{-1} $CaCl_2{\cdot}2H_2O$ and 184 mgL^{-1} $MgSO_4{\cdot}7H_2O$ (ionic strength = 0.014 mol(kg)$^{-1}$) in order to exclude variations in bacteria attachment resulting from variations in solution chemistry. The variations in sediment chemistry between experiments were also negligible, since only quartz sand was used. The aim of the experiments was to evaluate the applicability of the filtration theory by comparing calculated and measured filter coefficients. The former was calculated with eqs. (15), (18), and (21b) and the latter was calculated with eq. (20), with $\dfrac{C}{C_0}$ being determined from pulse source column breakthrough curves of known length x. The results of more than 100 (triplicate!) column experiments are presented in Fig. 3. In order to arrive at the calculated filter coefficient, for all experiments we "calibrated" the same single collision efficiency, $\alpha = 0.43$: this value gave the best fit between observed and calculated filter coefficients.

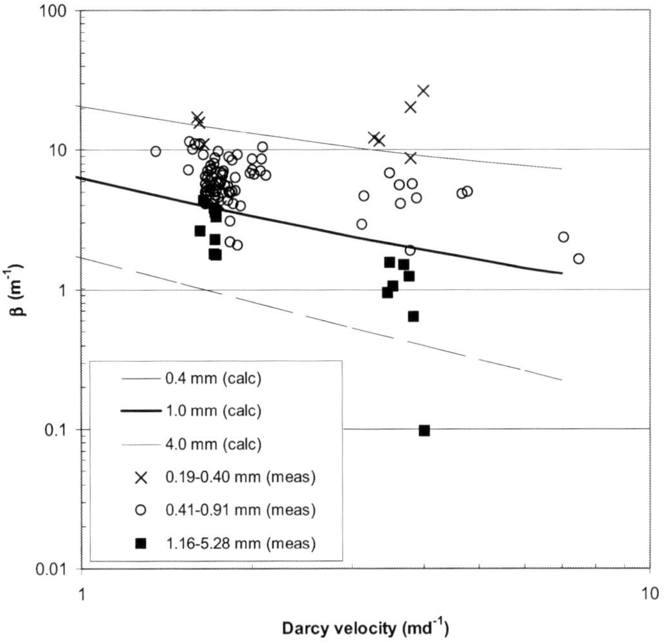

Fig. 3: *Comparison of measured and calculated filter coefficients as a function of Darcy flow velocity (md^{-1}) and median of the grain size number distribution (m). The calculated filter coefficient was determined with the SCCE according to the TE correlation equation (Tufenkji and Elimelech, 2004) and a collision efficiency, α = 0.43*

Matthess et al. (1991b) used a sticking efficiency value of 0.4. He arrived at this value by using other equations for the diffusion, interception, and sedimentation terms in eq. (15) (based on Yao et al., 1971). Because the grain sizes that Matthess and co-workers used were mainly between 0.3 and 3 mm at a Darcy flow velocity range of 1-4 md^{-1}, the straining contact efficiency (eq. 18) was almost negligible. The fit was indeed reasonable, but not perfect. For instance, in the 0.4-1.0 mm grain size range, measured filter coefficients were scattered without a clear relationship between filter coefficient and velocity. The same was true for the 1.0-4.0 mm range. The reason for the scatter was that grain size uniformity fluctuated (see next section). Upon increasing grain size uniformity, the filter coefficient decreased somewhat. In general, the filter coefficient depended on groundwater flow velocity and on the grain size of the sediment. Also, there was reasonable agreement between the calculated and measured filter coefficients for various combinations of grain size and Darcy flow velocity.

2.8.2 Effect of grain size

Fig. 4 shows the calculated and measured filter coefficients plotted against a_c, the median of the grain size number distribution. Again, measured and calculated filter coefficients agreed reasonable well, but most of the measured filter coefficients for experiments with a grain size median around 0.45 mm were rather low, especially within the 1.0-3.0 md^{-1} Darcy velocity range. This was because the grain size of this group of experiments was more uniform (U>2) than in the other experiments (U<2), bringing about a minor decrease in filter coefficient (see next section). Nevertheless, in this case too, the dependency of the filter coefficient on grain size and velocity was demonstrated, and there was reasonable agreement between the calculated and measured filter coefficients for various combinations of grain size and Darcy flow velocity.

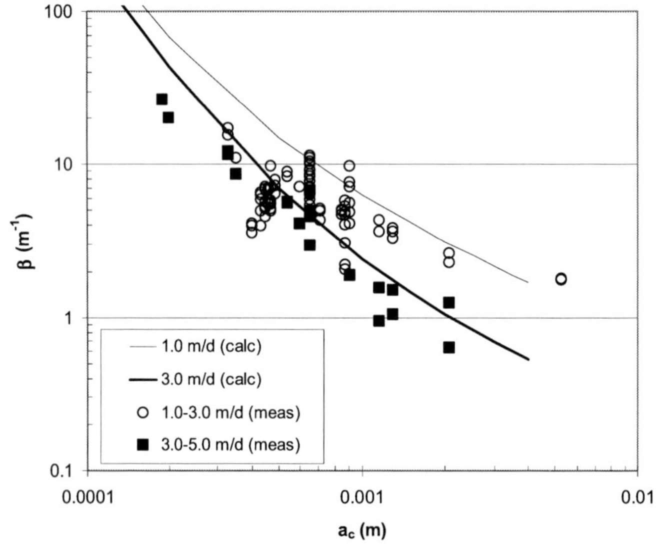

Fig. 4: Comparison of measured and calculated filter coefficients as a function of the median of the grain size number distribution (m) and Darcy flow velocity (md^{-1}). The calculated filter coefficient was determined with the SCCE according to the TE correlation equation (Tufenkji and Elimelech, 2004) and a collision efficiency, $\alpha = 0.43$

2.8.3 Effect of grain size uniformity of sediment

The uniformity of the sediment is defined as (Matthess et al., 1991a)

$$U = \frac{a_{c,60}}{a_{c,10}} \tag{23}$$

where the numbers 10 and 60 refer to the 10th and 60th percentile values of the log-normal grain size number distribution curve. Matthess et al. (1991a) prepared a number of sediment mixtures in such a way that the median of the grain size number distribution curve (used in eq. (21)) did not vary (it was taken at 0.45 mm). The results demonstrated that the uniformity, U, did not significantly influence the filter coefficient of *E. coli*. While U increased from 1.71 to 9.98, the filter coefficient hardly changed.

We agree with the conclusions reached by Matthess et al. (1991a and b) after their experiments, that measured filter coefficients and calculated filter coefficients determined with the SCCE (determined with the TE correlation equation), neglecting straining and one sticking efficiency value of 0.43, were in reasonable agreement. This demonstrates the usefulness of colloid filtration theory for predicting the transport of bacteria in porous media under saturated conditions.

2.9 Factors affecting the collision efficiency

2.9.1 Effect of zeta potential and solution chemistry

The attachment of *E. coli* is governed not only by diffusion, sedimentation, interception, and straining, but also by the collision or sticking efficiency. As mentioned above, the empirical dimensionless sticking efficiency, α , describes the fraction of collisions with filter grains that result in attachment, and according to Gilbert et al. (1991), one of the main factors in the attachment process is the zeta potential. In addition to the strain of the bacterium, the main factor affecting the zeta potential – and therefore the collision efficiency – is the solution chemistry (see also Fig. 1). In a study by Sharma et al. (1985), the zeta potential of gram-positive (*Bacillus subtillis*) and gram-negative (*Pseudomonas fluorescens*) bacteria in solutions consisting of various chemicals (heparin, polyacrylic acid, lignosulfonate, sodium pyrophosphate) varied between –20 and –80 mV, depending on the chemical used. Sharma and co-workers found an excellent correlation between the surface charge of these bacteria and transportability (or lack of attachment) in columns filled with Ottawa sand (zeta potential between –30 and –80 mV), which clearly suggested that electrostatic interactions between bacteria and sand grains were a dominant factor in their retention (Sharma et al., 1985). In a number of column experiments with *E. coli* ATCC 25922 and sand carried out by Powelson and Mills (2001), bacterial breakthrough was strongly dependent on the suspending solution used (see Fig. 5; the geochemical composition of the sand was not mentioned, but the breakthrough curves suggest that quartz sand was used = unfavorable conditions). In an attempt to explain these variations, in our experiments we determined the zeta potential of the bacteria in these solutions. To do so, solutions with *E. coli* ATCC 25922 similar to the ones made by Powelson and Mills (2001) were prepared and the zeta potential was determined in the same way as described in section 2.2. Like Sharma et al. (1985), we also found an excellent correlation between surface charge (ranging from –17 to –46 mV; see Fig. 5) and bacterial breakthrough. More generally, in a number of column experiments with quartz sand (92.35%) with some carbonates (2.4%) and iron (Fe_2O_3+Al_2O_3 was 0.77%), Goldshmid et al. (1973) found good agreement between the filter coefficient, the ionic strength of the electrolyte of the suspending solution, and the charge of the electrolyte used – which is in qualitative accordance with the Schulze-Hardy rule (see also section 2.2).

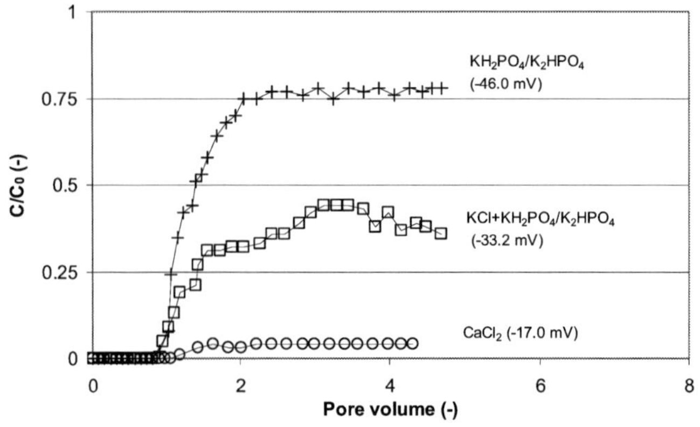

Fig. 5: *Zeta potential values determined with the Helmholtz-Smoluchowksi equation and breakthrough curves of E. coli ATCC 25922 suspended in various solutions ($CaCl_2$; KCl and dilute phosphate buffered saline; phosphate buffered saline; C/C_0 data are from Powelson and Mills, 2001)*

Note that on the negatively charged quartz sand in Goldshmid's experiments there were patches with a positive charge. At least the carbonates and possibly some of the iron were present as positively charged iron oxyhydroxide. We determined collision efficiencies from the filter coefficients and relative breakthrough concentrations determined by Goldshmid et al. (1973) and Powelson and Mills (2001) by using eq. (20) or (21); they are plotted in Fig. 6. Though the α-values were α_{total}-values in the case of Goldshmid et al. (1973) and unfavorable α_u-values in the case of Powelson and Mills (2001), they can be compared, if it is assumed that the sediment fraction available for favorable attachment was limited. In all cases the α_{str}-value was negligible given the grain sizes used (ranging between 0.12-0.71 mm). Fig. 6 clearly demonstrates the Schulze-Hardy effect: similar collision efficiencies are found for monovalent (Na, K), divalent (Mg, Ca), and trivalent (Fe(III)) electrolytes for concentration ratios of about 1 : 0.2 : 0.0002. The same effect is also achieved by changing the pH. An increase in filtration efficiency was measured when the pH was lowered (from 9.3 to 3.9; Goldshmid et al., 1973). Interestingly, an increase in conductivity caused an increase in filtration efficiency at pH 9.3, and a decrease of efficiency at low pH. At high pH, the concentration of hydronium ions was so low that addition of sodium cations decreased the zeta potential slightly and increased the filtration efficiency. At low pH the effect was different and is attributable to the pH at the bacterial surface, pH_s, which differs from that in the bulk solution pH_b, particularly at low ionic strengths according to (Hartley and Roe, 1940, in: James, 1957)

$$pH_s = pH_b + 0.217EM \tag{24}$$

where EM is the electrophoretic mobility ((μm/s)/(V/cm)). Since the EM of *E. coli* in low ionic strength solutions is around 2, the difference between both pH values is in the order of 0.4 pH unit. As the salt concentration increases, the pH of the surface decreases towards that of the medium and the filtration efficiency increases. A similar phenomenon was observed by Scholl and Harvey (1992) in determining sorption of indigenous bacteria using contaminated and uncontaminated aquifer material from Cape Cod, MA.

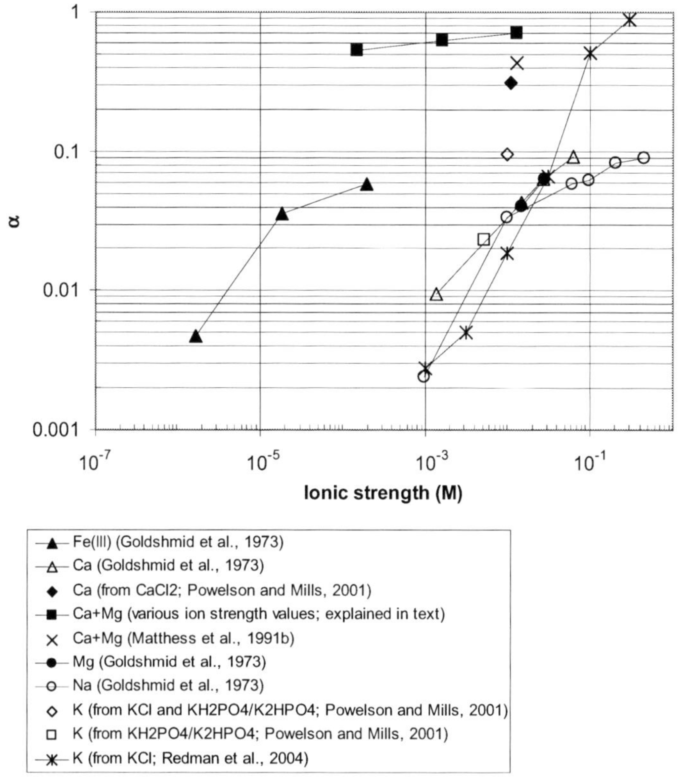

Fig. 6: *Collision efficiency of E. coli as a function of the ionic strength of the solution and the valence of the electrolytes in solution*

2.9.2 Effect of ionic strength

It is also clear from Fig. 6 that the collision efficiency increases with increasing ionic strength. This is the effect of ionic strength, which is usually explained by double-layer compression or a reduction in electrostatic force, which might cause the attractive van der Waals' force to predominate, resulting in more attachment. This is the basis of the so-called DLVO theory (after Derjaguin, Landau, Verwey and Overbeek; e.g. Schijven, 2001; Ryan and Elimelech, 1996; Sawyer et al., 1994). For *E. coli* and for the type of water (a mixture of Ca, Mg, Cl, and SO_4 ions) used in the experiments with unfavorable quartz sand, Matthess et al. (1991a and b) found a relationship between filter coefficient and ionic strength, *I*:

$$\beta = 16.03 + 1.68 \log I \tag{25}$$

From this and by using eq. (21), we calculated the relationship between unfavorable collision efficiency, α_u, and solution molarity (M):

$$\alpha_u = 0.09 \log(3.17M - 6.34 \cdot 10^{-8}) + 0.868 \tag{26}$$

This relationship is plotted in Fig. 6 ("Ca+Mg various ionic strength values"). Note that at an ionic strength of 0.01, the α_u deviates from the value 0.43 (see Fig. 6: "Ca+Mg"), which was used to determine the calculated filter coefficient (solid and dashed lines in Figs. 3 and 4). We attribute this to eqs. (25) and (26) being based on a limited dataset, whereas the α-value of 0.43 was based on more than 100 column experiments. Matthess et al. (1991b) concluded that when the ionic strength was greater than 0.01, the unfavorable collision efficiency hardly varied with ionic strength. However, the data presented in Fig. 6 suggest otherwise: for solution ionic strength values between 10^{-3} and 1 M, the total collision efficiency (α_{total}) for the bivalent ions Ca and Mg and the monovalent ion Na at least, showed a considerable increase.

2.9.3 Effect of lipopolysaccharide composition in the outer membrane

Walker et al. (2004) demonstrated that the composition of LPS in the outer membrane of *E. coli* can have a significant effect on collision efficiency (Fig. 7). Type D21g had a core oligosaccharide containing negatively charged phosphate groups, which resulted in little attachment (low sticking efficiency) in column experiments with negatively charged quartz sand. In contrast, LPS of strain JM109g consisted of an uncharged O-antigen, which appeared to be able to shield the phosphate charge, resulting in a marked increase of the sticking efficiency (more attachment).

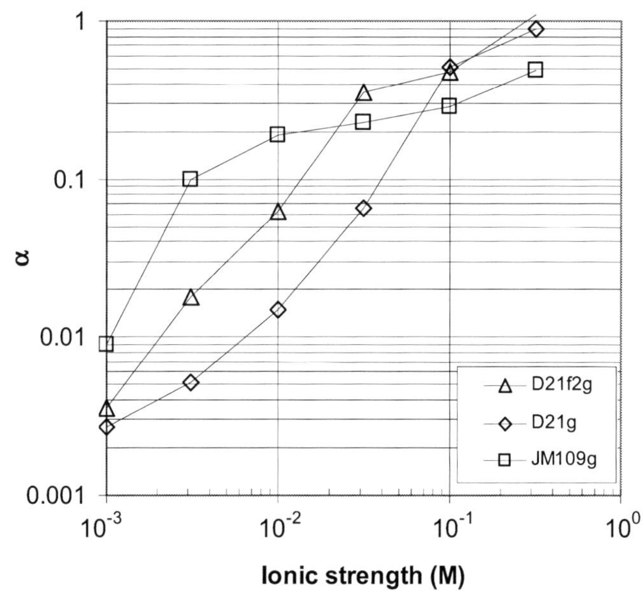

Fig. 7: *Collision efficiency of three mutants of E. coli K12 with distinct portions of the LPS molecule (Walker et al., 2004) in column experiments containing quartz sediment. LPS of D21f2g consists of lipid A and KDO; LPS of D21g has lipid A, KDO and core oligosaccharide containing charged phosphate groups; LPS of JM109g consists of lipid A, KDO, core oligosaccharide and O-antigen. JM109g mutant "shields" the charge of the phosphate groups, resulting in a higher sticking efficiency and more attachment*

2.9.4 Effect of geochemical heterogeneity

Johnson and Elimelech (1995) and Johnson et al. (1996) studied the effect of geochemical heterogeneity on the attachment of negatively charged artificial colloids during transport. They introduced the heterogeneity parameter, λ, in order to account for enhanced attachment due to the addition of positively charged iron-coated sand. Based on their results, the bacterial transport model in charged heterogeneous porous media (eq. (13)) was developed. Foppen and Schijven (2005; Chapter 5) determined the sticking efficiency of *E. coli* ATCC 25922 in a series of column experiments with sediments consisting of 0.18-0.50 mm quartz sand, goethite-coated grains, calcite grains or grains of activated carbon (AC), in varying fractions (λ = 0, 0.05, 0.1, 0.2, 0.4, 0.7, 1.0) and all of similar diameter to the quartz sand. The weighted sum of favorable and unfavorable sticking efficiencies (α_{total}) showed that upon increasing the fraction of favorable mineral grains (λ) there was an initial rapid increase, which then slowed down (Fig. 8). This was most pronounced in the AC experiments, followed by the calcite experiments, and then the goethite experiments. Foppen and Schijven (2005; Chapter 5) attributed this non-linear relation to the surface charge and hydrophobic heterogeneity of the *E. coli* population. In a number of batch experiments, Scholl et al. (1990) investigated the rate and extent of attachment of Lula-D, an *Arthrobacter sp.*, to chips of quartz, muscovite, limestone, and iron-coated quartz and muscovite, and found that the degree of attachment correlated with the sign of the surface charge of the minerals. In column experiments, maximum bacterial breakthrough in the case of iron-coated quartz was 2 log units lower than in the case of clean quartz. Also, a distinct "tailing" (or detachment) was observed.

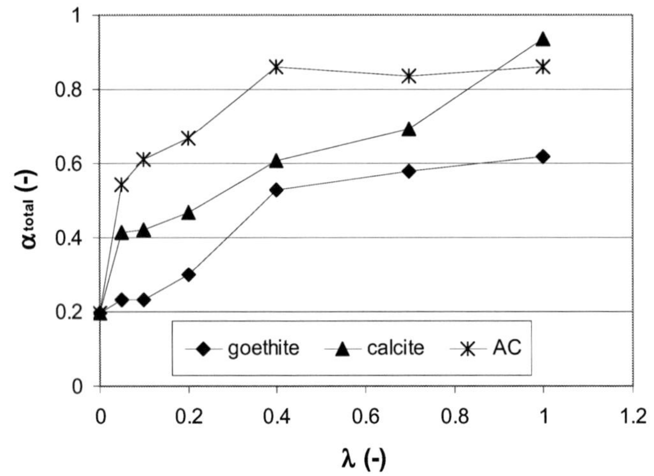

Fig. 8: *Relation between calculated total sticking efficiency (α_{total}) and geochemical heterogeneity parameter (λ) for experiments carried out with quartz sand and goethite coated sand, quartz sand and calcite grains, and quartz sand and granular activated carbon (from: Foppen and Schijven, 2005, Chapter 5). Fractions of goethite, calcite, and granular activated carbon varied from 5% to 100%*

In a series of batch and column experiments with indigenous bacteria using contaminated and uncontaminated aquifer material, Scholl and Harvey (1992) found that bacterial sorption was affected both by groundwater pH and by the presence of iron oxyhydroxide coatings around grains. The rate of attachment of *E. coli* to serpentine sand (with a high content of Mg minerals) increased concomitantly with the amount of sand added to the batch (Merkli, 1975). Also, the amount of adsorbed *E. coli* at equilibrium (after 2-8 hours) increased upon increased amount of serpentine sand. The amount of *E. coli* attached to plagioclase was much smaller, while the amount attached to quartz was negligible. This underlines the importance of the mineral surface charge for the attachment of *E. coli* (see also section 2.2).

2.9.5 Effect of grain surface roughness

Bedbur (1989) and Gimbel and Sontheimer (1980) found that grain surface roughness influenced the collision efficiency to a certain extent. Photos taken with an electron microscope clearly showed that polystyrene colloids attached to edges, pinnacles, ridges, and other elements on the surface of grains. Also, concentrations of colloids were found in holes and excavations on the grain surface, probably due to a reduction of the flow velocity in the vicinity of these surface elements. And the attachment of colloids to flatter parts of the grain surface was markedly less. For larger colloids, like *Cryptosporidium*, Tufenkji et al. (2004) have reported that the shape of the individual grains can contribute significantly to the porous medium straining potential, as it may give rise to a very wide pore size distribution. However, the effect of grain surface roughness on the sticking efficiency of *E. coli* has not been studied, and no data are available.

2.10 Evidence for bimodal sticking efficiencies

Although data on *E. coli* are lacking, recent evidence suggests that filtration of microbial particles may not be consistent with the classical colloid filtration theory. Direct evidence for the non-exponential removal of microorganisms can be found by examining microbial deposition patterns in laboratory scale columns of uniform collector grains (Tufenkji et al., 2003). Redman et al. (2001) observed that the profile of retained viruses decayed according to a power-law. Li et al. (2004) utilized a log-normal distribution of deposition rate coefficients among the colloid population in order to simulate the effluent curves and retained profiles simultaneously. Simoni et al. (1998), by determining effluent concentrations of columns of various lengths, used two subpopulations of *Pseudomonas* strain B13, each with different collision efficiency. Tufenkji and Elimelech (2004a, 2005 and 2005a) also employed a similar type of dual deposition mode model in order to explain their polystyrene latex colloids and *Cryptosporidium parvum* oocysts retention profiles. In all of the abovementioned studies, the retained particle concentration profiles on semi-log scale resulted in an initial steep slope followed by a shallower slope (classical colloid filtration theory predicts one slope). However, the explanations for this non log-linear behavior offered by these authors differ somewhat. Li et al. (2004) attributed their decrease in deposition rate coefficient with transport distance to 1) distributed deposition rate coefficients, 2) straining at the influent end of the column, and 3) depletion of colloid concentration in the perimeter of the pores adjacent to grain surfaces due to deposition from solution. Li et al. (2004) calculated that the non log-linear retained particle concentrations could be explained if surface potentials determined for the polystyrene latex microspheres would show a variation of less than 0.5% and believed that this was a viable mechanism that could yield apparent decreases in deposition rate coefficients. Simoni et al. (1998) assumed two subpopulations of cells and attributed their non log-linear profiles to two distinct sticking efficiencies (α_{fast} and α_{slow}). Tufenkji and Elimelech (2004a, 2005, 2005a) also gave 3 explanations for their non log-linear profiles: 1) based on Hahn's model (Hahn and O'Melia, 2004), a fraction of the particles has sufficient kinetic energy to escape from the secondary energy well and to remain in the fluid, 2) a fraction of the particle population overcomes the secondary energy barrier and reaches the primary energy well (resulting in distributed particle deposition rates), and 3) surface charge heterogeneities provide sites for fast and slow deposition and the concurrent existence of both favorable and unfavorable colloidal interactions in an otherwise homogeneous system can be described by considering a bimodal distribution.

Since the types of experiments described above have not been carried out with *E. coli*, it remains unknown whether non log-linear retention profiles also apply to *E. coli*. More future research on deposition patterns is required.

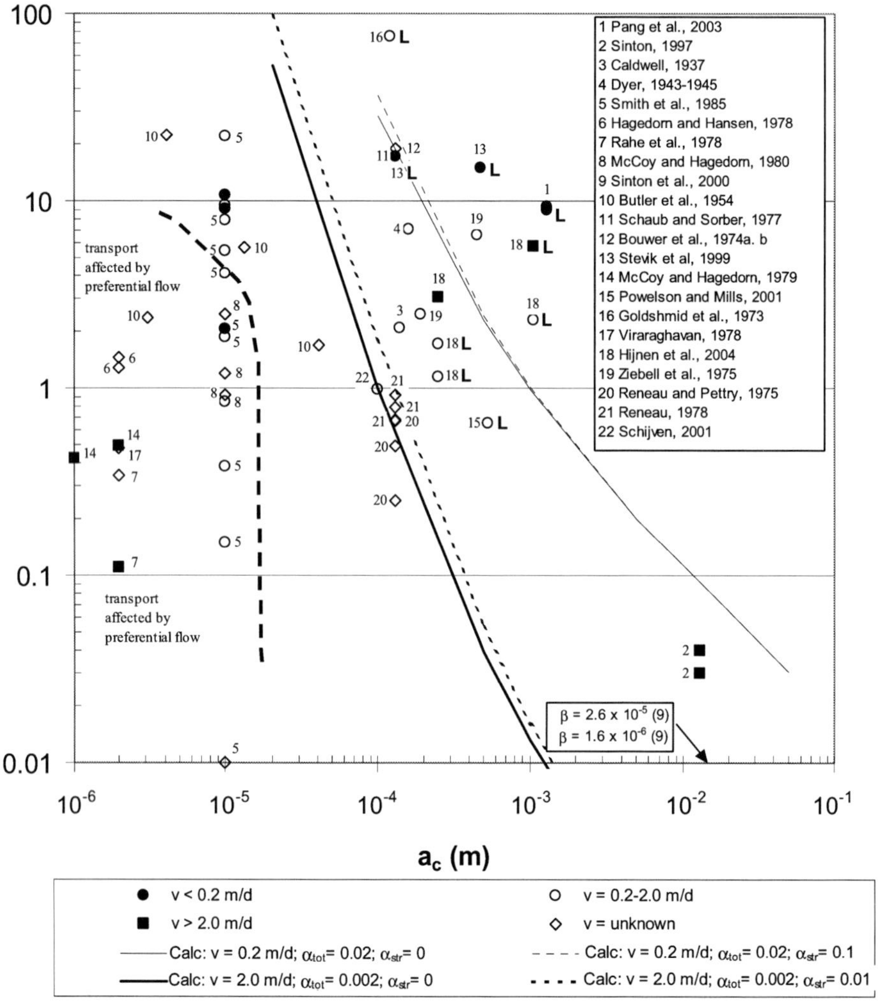

Fig. 9: *Comparison of filter coefficients determined from field and laboratory studies and calculated filter coefficients. Parameters used to determine the calculated filter coefficients were* $\alpha = 0.02\text{-}0.002$, $T = 283$ *K*, $a_p = 2$ μm, *and* $v = 0.2\text{-}2.0$ md^{-1}*). See text for explanation. Parameters used to calculate filter coefficients were* $T = 283$ *K*, $a_p = 2$ μm, $\theta = 0.3$, $H = 6.5x10^{-21}$ *J*; $\rho_p = 1050$ kgm^{-3}, $\rho_{fl} = 1000$ kgm^{-3}, $\mu = 1.3x10^{-3}$ $kgsm^{-1}$; *area left of bold dashed line: transport of E. coli and thermotolerant coliforms is influenced by preferential flow*

2.11 Filter coefficients and collision efficiencies from field and laboratory experiments

In Fig. 9 filter coefficients, grain sizes, and velocities of *E. coli* and thermotolerant coliforms from field and laboratory studies have been plotted together, with the grain size on the horizontal axis and the filter coefficient on the vertical axis. Data from field and laboratory studies are given as points (numbers refer to references given in the top right of the graph). The points were divided into four classes, based on pore water flow velocity (see legend). In a number of cases, only the type of soil (e.g. clayey loam) was mentioned in the reference. For these cases the US Soil Standards (Buol et al., 1989; Bell, 1987) were used in order to obtain an indication of the grain size. In one case, only one maximum breakthrough concentration was given (Caldwell, 1937). Also, sometimes pore water flow velocity was not measured or mentioned (see legend of Fig. 9: pore water flow velocity unknown). Finally, in none of the experiments were the die-off rate coefficients measured. In the laboratory experiments die-off was kept to a minimum or absent, and in most of the field experiments, transport was fast and therefore die-off was probably negligible. Die-off certainly played a role in longer-term field experiments (e.g. Caldwell, 1937; Dyer and Bhaskaran, 1943 and 1945a, b), but was not measured. For these reasons, the accuracy of the points in the figure is limited and they should be interpreted with care. However, the figure does illustrate a few interesting issues. Over the entire grain size and pore water flow velocity range, the filter coefficients varied between 0.1 and 20 m^{-1} only, and only in exceptional cases (Goldshmid et al., 1973; Sinton et al., 1997) did they fall outside this range. Furthermore, in some cases, low filter coefficients were found for soils with very small grain sizes (< 10 μm). In those cases, macropore flow must have taken place. Indeed, Smith et al. (1985), Hagedorn et al. (1978), Rahe et al. (1978), McCoy and Hagedorn (1979, 1980), Butler et al. (1954) all reported that macropore or pipe flow caused *E. coli* to be transported over considerable distances in short time intervals. Viraraghavan (1978) also observed fast transport, but he did not specifically refer to macropore flow. These studies were carried out in soils at shallow depth and macropore flow in soils is known to occur very frequently in the upper layers in which there is bioactivity. At the other end of the grain spectrum, similarly low filter coefficients were found at grain sizes between 0.1-1 mm. Almost all of these studies were carried out in laboratory columns (indicated with an "L" in Fig. 9) under ideal and completely controlled conditions. Between these two extremes, a variety of filter coefficients was measured – mainly in field experiments.

For grain sizes larger than 20 μm, pore water flow velocities could be roughly divided into two categories: <0.2 md^{-1} and between 0.2 and 2.0 md^{-1}. Using eq. (21), we calculated a range of collision efficiencies, straining correction factors, and corresponding calculated filter coefficients (solid and dotted lines in Fig. 9) that, by trial and error, approximately fitted within the measured field filter coefficients. For $v = 0.2\ m/d$, the resulting α_{total} was 0.02, while the straining correction factor varied between 0 and 0.1. For $v = 2.0\ m/d$, α_{total} was 0.002, while the straining correction factor varied between 0 and 0.01. From this we concluded that the range of collision efficiencies determined for geochemically heterogeneous sediments under various hydrochemical conditions was small (0.002-0.2) compared with the sticking efficiency values from Fig. 5 (0.002-0.9, but mostly 0.01-0.9) for those solutions with typical groundwater ionic strength values (ranging from a few

$mmolL^{-1}$ to several tens of $mmolL^{-1}$), especially given that divalent electrolytes (like Ca^{2+}, Mg^{2+}) are usually present in groundwater.

We think there are three possible explanations for the low sticking efficiencies that were estimated for the field studies:

- preferential flow mechanisms, or the presence of zones of high permeability in which transport of bacteria is rapid, determine to a great extent the bacteria concentration at a certain distance from a source of pollution. If such zones are not identified by those doing the research, then calculated sticking efficiencies are underestimated, and actually are inappropriate to apply;
- *E. coli* populations themselves may be heterogeneous in terms of their attachment characteristics and it might be that in each *E. coli* plume traveling through an aquifer, few will survive for a relatively long time and will not attach, giving rise to relatively low sticking efficiencies. In the laboratory, this population heterogeneity might be masked because of the relatively small scale of the experiments;
- *E. coli* or thermotolerant coliforms usually travel in a plume of wastewater, consisting of many organic and inorganic compounds, all with highly variable concentrations and possibly affecting bacterial attachment characteristics. Laboratory experiments are usually carried out with solutions that do not reflect wastewater compositions.

2.12 Kinetic desorption or detachment

To date, there have been no systematic studies of the desorption or detachment of *E. coli* or thermotolerant coliforms previously attached to sediment grains. In the experiments carried out by Sinton et al. (2000), both *E. coli* 2690 and *E. coli* J6-2 concentrations during the so-called tailing phase of the breakthrough curve (which is thought to occur as a result of slow release or detachment of attached bacteria) decreased to zero. The same rapid reduction during the tailing phase has been found in thermotolerant coliforms (Sinton et al., 1997). In field-scale fractured rock tracer experiments, Champ and Schroeter (1988) reported that *E. coli* (wild type) concentrations during the tailing phase, most likely due to detachment, were at least 30 times lower than maximum *E. coli* concentration during peak breakthrough of the experiments. The detachment of *E. coli* as a process becomes more visible when relative concentrations are plotted on a log scale, (Johnson et al., 1995, Schijven, 2001). The difference between the maximum relative concentration during seeding with *E. coli* and immediately after seeding during the tailing phase in which desorption is taking place varies from 1 log-unit (Hijnen et al., 2004) to around 2-3 log-units (Peters, 1989; Hijnen et al., 2004). From this we conclude that detachment occurs but it is relatively unimportant.

2.13 Factors affecting inactivation

2.13.1 Effect of temperature

The inactivation or die-off rate coefficients of bacteria in the fluid and bacteria attached to the sediment, k_i and k_{is}, are influenced by many factors (Murphy and Ginn, 2000; Reddy et al., 1981; Barcina et al. ,1997). One of the most important factors is the temperature. In Fig. 10 the effect of temperature on the die-off rate coefficient of various *E. coli* strains is presented. However, the data are difficult and unwieldy to compare. The data are from "water" environments without sediment (except Sjogren, 1994; Van Donsel, 1967; first graph of Bogosian et al., 1996) that were kept in the dark (except the second graph of Wang and Doyle, 1998; Korhonen and Martikainen, 1991; McFeters and Stuart, 1972) at a pH range of 6-8 (unknown for Rice et al., 1992; Nasser et al., 1993; van Donsel, 1967; McFeters and Stuart, 1972; Bogosian et al., 1996), usually in non-sterile, non-autoclaved conditions (except the first graph of Wang and Doyle, 1998; the second graph of Korhonen and Martikainen, 1991, second graph of Nasser et al., 1993; McFeters and Stuart, 1972). Despite the heterogeneous conditions during the various experiments, in the resulting data set the dependency of the die-off rate coefficients on temperature is similar: the increase in the die-off rate coefficient per degree Celsius rise is apparent and comparable in most experiments. If the experiments under sterile conditions are ignored and it is assumed that all other experimental variations are negligible, then the die-off rate coefficients are within a "band-width" of 1-log-unit. The average die-off coefficient at 10 °C is 0.15 d^{-1} and at 20 °C it is 0.50 d^{-1}.

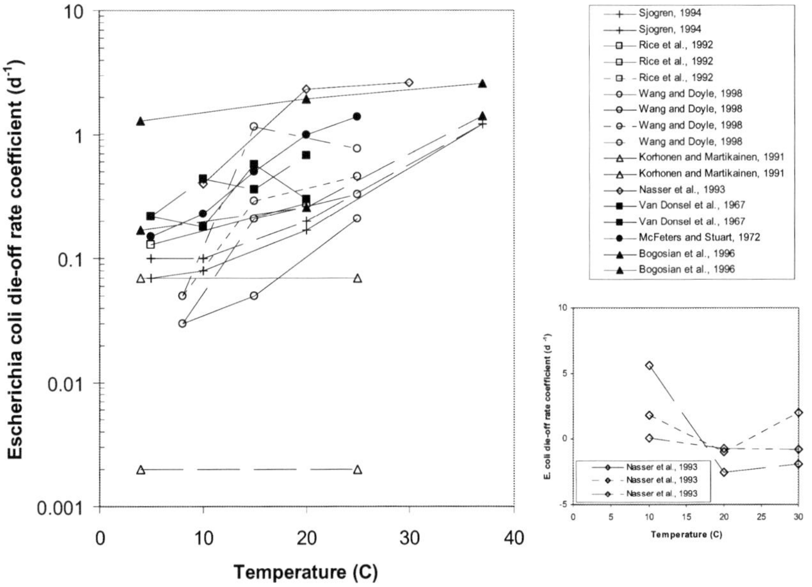

Figure 10: Effect of temperature on the E. coli die-off rate coefficient

2.13.2 Effect of protozoa and antagonists

In Fig. 11 the effect of protozoa and/or antagonists on the die-off rate coefficient of various *E. coli* strains is presented. Protozoa may ingest and digest *E. coli* (predation), while antagonists such as indigenous bacteria compete for the same source of nutrients. In the Figure each data pair is connected with a line, and the upper die-off rate coefficient represents non-autoclaved and non-sterile conditions in which predation and/or antagonism could take place, while the lower die-off rate coefficient has been determined for sterile conditions without predation/antagonism. From the data it can be concluded that due to predation/antagonism the die-off rate increases by 1 log-unit on average. Also, predation/antagonism does not seem to depend on temperature (Fig. 11).

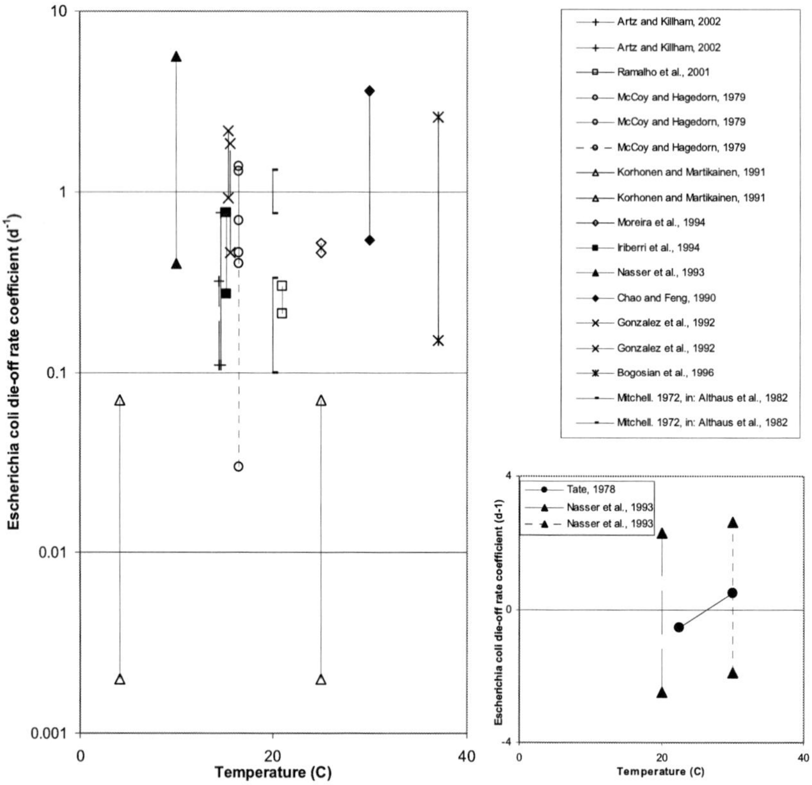

Fig. 11: Effect of autoclaving / sterilizing the medium prior to E. coli inoculation

Table 2: Effect of light, soil, pH, toxic substances, and oxygen on the die-off rate coefficient of E. coli. Lower part of the table: additional E. coli die-off rate coefficients determined in single experiments (without testing the influence of certain parameters)

Medium:	**T (°C)**	**pH**	**Strain:**	**Die-off rate coeff. (d^{-1})**	**Reference:**
Effect of light:					
Mineral water	21		*E. coli* NCTC 9001	0.58	Ramalho et al., 2001
Mineral water	21		*E. coli* NCTC 9001	0.33	idem;
Effect of soil:					
Native groundwater spiked with raw sewage	20	7	*E. coli* isolated from sewage	0.09-0.3	Pang et al., 2003
Sorbed *E. coli* (to aquifer material)	20	7	*E. coli* isolated from sewage	2.59-4.47	idem;
Groundwater				0.06	Filip et al., 1986
Groundwater + mixed sand				0.09	idem;
Sterile tap water	30		*E. coli* HB101	0.04	Chao and Feng, 1990
Soil Chung Hsing University	30		*E. coli* HB101	0.2	idem;
Wastewater Pima County	22.5		Thermotol. colif. from WWTP	0.62	Karim et al., 2004b
Sediment WWTP Pima County	22.5		Thermotol. colif. from WWTP	0.35	idem;
Effect of pH:					
Deionized water with KOH or HCl	10	2.5	*E. coli* MH3427	6.39	McFeters and Stuart, 1972
Deionized water with KOH or HCl	10	4	*E. coli* MH3427	0.58	idem;
Deionized water with KOH or HCl	10	5	*E. coli* MH3427	0.4	idem;
Deionized water with KOH or HCl	10	6	*E. coli* MH3427	0.3	idem;
Deionized water with KOH or HCl	10	7	*E. coli* MH3427	0.32	idem;
Deionized water with KOH or HCl	10	10	*E. coli* MH3427	0.71	idem;
Deionized water with KOH or HCl	10	12	*E. coli* MH3427	6.39	idem;
Seawater	20	5.2	*E. coli*	0.28	Allen et al., 1952, in: Althaus et al., 1982
Seawater	20	6.1	*E. coli*	0.42	idem;
Seawater	20	7.1	*E. coli*	0.42	idem;
Seawater	20	7.7	*E. coli*	0.33	idem;
Pompano fine sand, approx. saturated	30	6.64	*E. coli* ATCC 15597	0.59	Tate, 1978
Sterile Pahokee Muck (soil), approx. saturated	22.5	6.16	*E. coli* ATCC 15597	0.04	idem;

Table 2(cont.): Effect of light, soil, pH, toxic substances, and oxygen on the die-off rate coefficient of E. coli. Lower part of the table: additional E. coli die-off rate coefficients determined in single experiments (without testing the influence of certain parameters)

Medium:	**T (°C)**	**pH**	**Strain:**	**Die-off rate coeff. (d^{-1})**	**Reference:**
Effect of toxic substances:					
Well water 4 (Cu = 0.61 mgL^{-1})	15	7.13	O157:H7	0.77	Artz and Killham, 2002
Well water 3 (Cu = 3.91 mgL^{-1})	14.8	6.98	O157:H7	1.59	idem;
Effect of oxygen:					
0.003 M phosphate buffer (1-18% O2 saturation)		6	*E. coli*	0.58	Allen et al., 1952, in: Althaus et al., 1982
0.003 M phosphate buffer (100% O2 saturation)		6	*E. coli*	0.09	Allen et al., 1952, in: Althaus et al., 1982
Effect of water type:					
Tap water from Coimbra	25		*E. coli* ATCC 8677	0.08	Moreira et al., 1994
Mineral water (glass bottles)	25	5.64	*E. coli* ATCC 8677	0.35	idem;
Other:					
Groundwater	20		*E. coli*	0.12	Kudryavtseva, 1971
Groundwater	20		*E. coli*-408	0.16	idem;
Well water	11	7.48	Thermotol. colifom	0.77	McFeters et al., 1974
Rural domestic well 275 ft in depth	9	7.8	*E. coli* strain Hfr	0.74	Keswick et al., 1982
Gainesville water well	22	7.6	*E. coli* stock collection	0.4	Bitton et al., 1983
Infiltrated sewage effluent			Total coliform	0.03	idem;
Infiltrated sewage effluent			Thermotolerant coliform	0.06	idem;
Sterile well water	22.5		*E. coli* ED 8654	0.01	Caldwell et al., 1989
Sterile well water	22.5		*E. coli* HB 101	0.02	idem;

2.13.3 Other effects on the die-off rate coefficient

In Table 2, a variety of "other" effects on the die-off rate coefficient are given. These are the effect of light, soil, pH, toxic substances, and oxygen. The effect of light is not very relevant for groundwater conditions. Soil has the effect of increasing the die-off rate coefficient (Pang et al., 2003; Filip et al., 1986; Chao and Feng, 1990) or decreasing it (Karim et al., 2004b; Bogosian et al., 1996). The parameter "soil" incorporates several other effects (e.g. the presence of nutrients or toxic substances), and the effect of soil on the die-off rate coefficient depends very much on local conditions. The pH has a well-known effect on the die-off rate coefficient. McFeters and Stuart (1972) showed that in de-ionized water to which either potassium hydroxide or hydrochloric acid had been added, die-off rates were lowest in the pH range of 6-8. Allen et al. (1952; in: Althaus et al., 1982) demonstrated a similar phenomenon for *E. coli* in seawater. Heavy metals are known to have a toxic influence on bacteria (Althaus et al., 1982). Artz and Killham (2002) showed that the die-off rate of an *E. coli* O157:H7 strain in groundwater with a [Cu] of 3.91 mgL^{-1} was double that in groundwater with 0.61 mgL^{-1}. Finally, Allen et al. (1952; in Althaus et al., 1982) showed that the die-off of *E. coli* in water saturated with oxygen was significantly lower than in water with a low oxygen saturation.

2.14 Modeling the removal of *E. coli*

To illustrate the effect of a number of key parameters on the transport of *E. coli* in aquifers, a steady state (for the flow field) groundwater model was constructed in MODFLOW (Webtech360, 2003). The area modeled was 500 x 500 m and the model consisted of 100 columns, 100 rows, and 2 isotropic homogeneous layers (the top layer was 5 m thick and the bottom layer was 15 m thick) each with a permeability of 20 md^{-1} and a porosity of 0.25. In the middle of the model, in layer 2, an abstraction well was located with an abstraction rate of 2000 m^3d^{-1}; 100 m west of this well, in layer 1 a small recharge well was located with a recharge rate of 1 m^3d^{-1} to simulate the effect of a cesspit dug into the saturated zone. The *E. coli* concentration of the recharging water was 10^6 cells$(mL)^{-1}$, which was thought to be representative of high-strength sewage (Sawyer et al., 1994).

The first step was to calculate the pore water flow velocities in each 5x5 m cell. With this flow field, a total retention rate coefficient was calculated based on the assumption that detachment is negligible. Therefore inactivation of sorbed cells became irrelevant and only the mass balance of bacteria in the liquid phase was considered. Then, eq. (11) can be rewritten as

$$\frac{\partial S}{\partial t} = \pi a_p^2 (k_a + k_{str}) C \qquad (27)$$

and inserting in eq. (1):

$$\frac{\partial C}{\partial t} = \nabla \cdot (D \nabla C) - \nabla \cdot (vC) - k_{total} C \qquad (28)$$

in which k_{total} is the total retention rate coefficient (now expressed in d^{-1} instead of md^{-1}), which includes the effect of favorable and unfavorable attachment (eq. (13)) as well as straining and die-off or inactivation and is defined as :

$$k_{total} = \frac{f_s v \theta}{4} \{\eta_0 \alpha_{tot} + \eta_S \alpha_{str}\} + k_i \qquad (29)$$

The SCCE, η_0, was determined with eq. (15). Temperature, unfavorable collision efficiency, favorable collision efficiency, straining correction parameter, and the inactivation rate coefficient were kept constant. The k-value fields are given in Fig. 12 for an aquifer consisting of quartz grains (left-hand part) and an aquifer consisting of quartz sand with Fe-oxyhydroxides (1.5% of the total surface area;

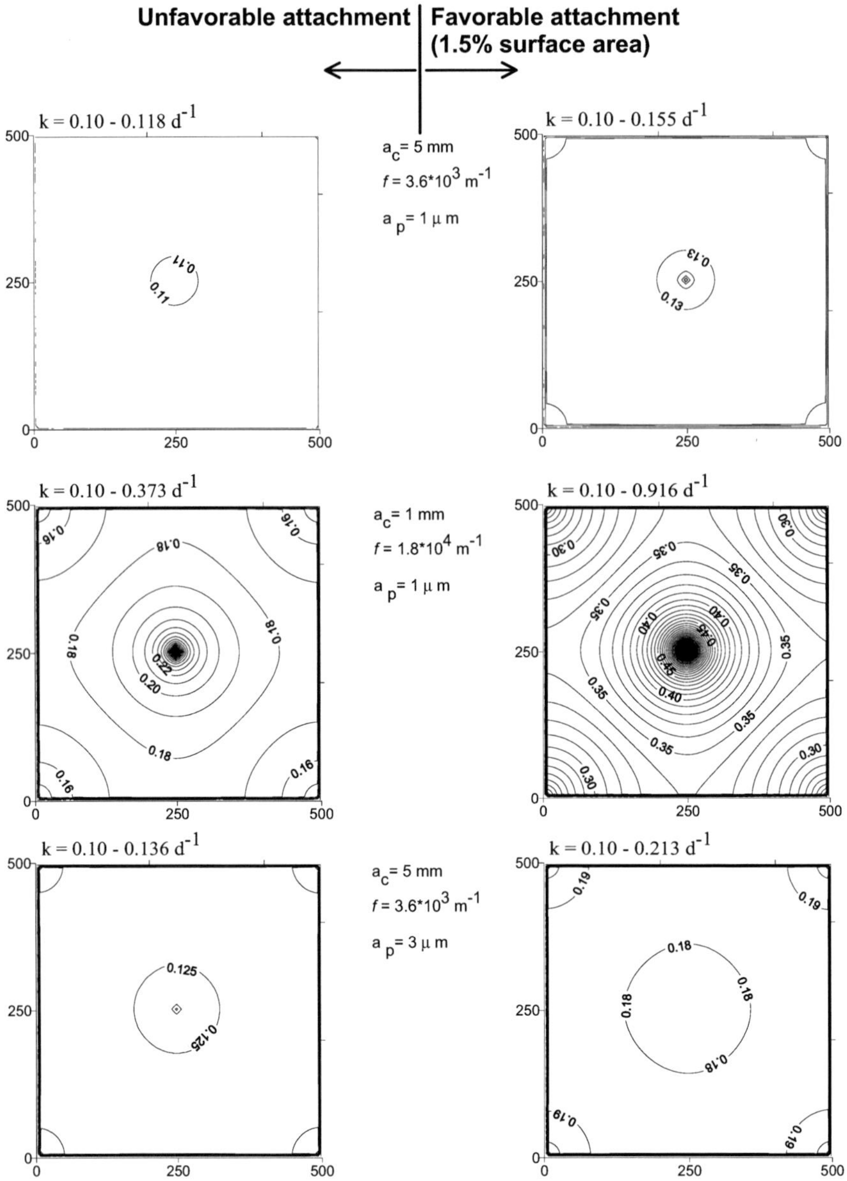

Fig. 12: Calculated total retention rate coefficient k_{total} (contour line interval: 0.01 d^{-1}). See text for explanation. Parameters used were $\alpha_f = 0.007$, $\alpha_f = 1.00$, $\alpha_{str} = 0.01$, $k_i = 0.1\ d^{-1}$, $T = 283$ K, $\theta = 0.25$, $H = 6.5 \times 10^{-21}$ J; $\rho_p = 1050\ kgm^{-3}$, $\rho_{fl} = 1000\ kgm^{-3}$, $\mu = 1.3 \times 10^{-3}\ kgsm^{-1}$; left-hand graphs: $\lambda = 0$, right-hand graphs: $\lambda = 0.015$

right-hand part of Fig. 12). Per type of aquifer, the grain size was varied (1 mm and 5 mm), as was the size of the infiltrating *E. coli* (1 μm and 3 μm). Fig. 12 shows that within these boundary conditions, the k_{total} varied from 0.1 on the boundaries of the model where flow velocities were lowest, to a maximum of 0.373 for the quartz-grained aquifer near the abstraction well where pore water flow velocities were highest. The maximum k_{total} for the Fe-coated aquifer was 0.916 d^{-1}. From the Figure, the following general observations can be made:

- In coarse-grained aquifers the total retention rate coefficient is dominated by decay (here taken at "only" 0.1 d^{-1}) while physicochemical removal processes are of secondary importance;
- When the grain size reduces (the difference between middle k -value field and the upper k -value fields), then the total retention rate coefficient increases due to an increase in the diffusion, interception, and straining components in the SCCE. Of these, the total retention rate increase due to diffusion is the most pronounced;
- When the size of the colloid increases (compare the uppermost and lowest k-value fields), then the total retention rate coefficient increases due to an increase of the sedimentation, the interception, and the straining component in the SCCE. Of these, the total retention rate increase due to sedimentation is most pronounced.

Finally, in MT3DMS (Zheng and Wang, 1999) the *E. coli* concentration in the Fe-coated aquifer was determined over time (Fig. 13), using the right-hand side k-value fields of Fig. 12. The dispersivity used was 10 m. The distribution of concentration in the second layer was at a dynamic equilibrium some 60 days after the start of infiltration of *E. coli* (the situation after 100 days is shown in Fig. 13). The maximum distance traveled from the recharge well in these 3 cases as a result of advection including dispersion on the one hand and total retention on the other is around 80 m in the coarse-grained aquifer with the small *E. coli* (1 μm). In the fine-grained aquifer, the maximum distance traveled is only a few meters, while in the coarse-grained aquifers with large *E. coli* (3 μm), it is around 60 m.

2.15 Conclusions

From this literature review, a number of important conclusions emerge:

Under fully controlled laboratory conditions, with constant chemical groundwater quality, one type of collector (quartz grains), a range of Darcy flow velocities, and a range of collector grain sizes, there was reasonable agreement between the measured filter coefficients from breakthrough curves and the filter coefficients calculated using the SCCE (determined with the TE correlation equation). Therefore, we conclude that colloid filtration theory in general and the TE correlation equation in particular provide a valuable framework for predicting the transport of *E. coli* in porous media under saturated conditions. The bacterial transport mechanism due to straining is not yet fully understood. Although theoretical relations and experimental data appear to be in agreement, the identity of the straining correction parameter remains obscure.

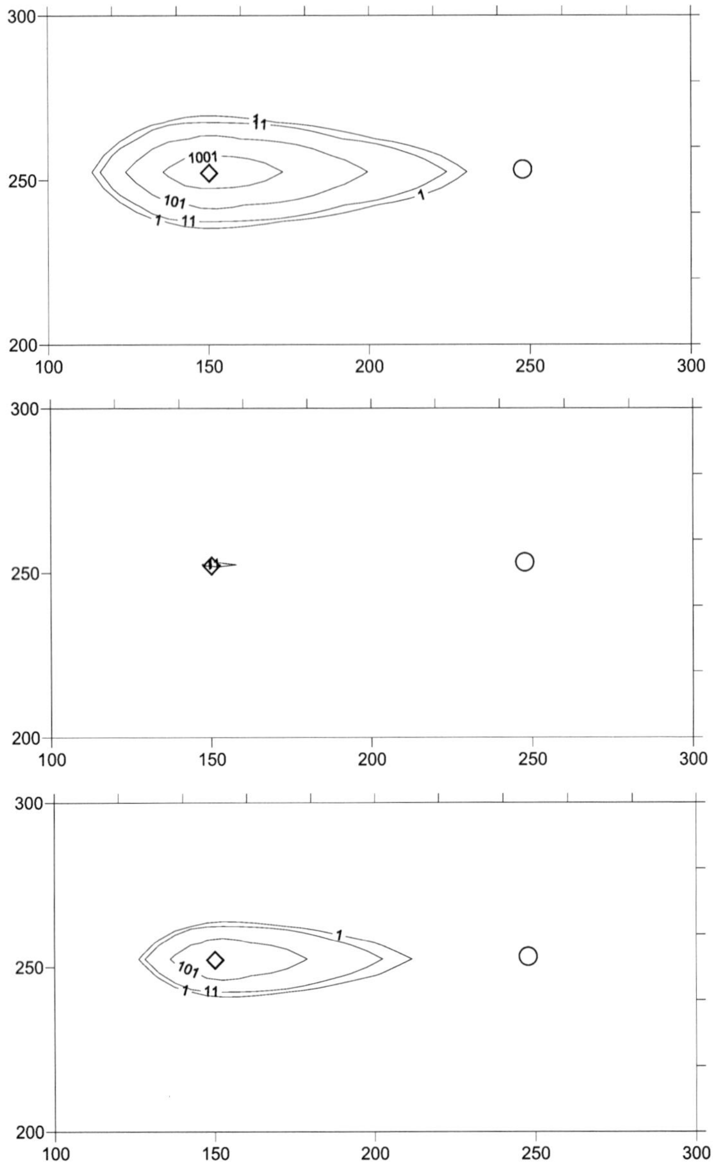

Fig. 13: *Distribution of E. coli concentration (in cells(mL)$^{-1}$) midway in layer 2 at a depth of 12.5 m after 100 days of infiltration. The abstraction well is given as a circle and the recharge well is a diamond (the recharge well is located in the middle of layer 1 at a depth of 2.5 m). Retention rate coefficients used are given in Fig. 12 (right-hand part) for the case of favorable attachment*

One of the most important factors determining the total sticking efficiency of *E. coli* in sediment is surface charge difference between collector and *E. coli*. These charge differences depend on the solution chemistry, ionic strength, geochemical heterogeneity, bacteria population heterogeneity, and the composition of lipopolysaccharides on the outer membrane of *E. coli*. In quartz-grain dominated sediments without preferential flow with an average groundwater composition consisting of monovalent and divalent ions such as Na, K, Ca, and Mg with a total concentration of < 20 $mmolL^{-1}$, the sticking efficiency values of *E. coli* (consisting of an O-antigen shielding the charged phosphate groups of the inner core oligosaccharide of the outer membrane) may vary between 0.02 and 0.9. When positively charged minerals are present as patches around sediment grains or as separate grains, there will be increased attachment. Another factor of importance in determining the total sticking efficiency of *E. coli* might be grain surface roughness. However, to date this has not been studied and so no data are available.

The range of field sticking efficiencies determined for geochemically heterogeneous sediments under various hydrochemical conditions was low (0.002-0.2) compared with laboratory sticking efficiencies (0.02-0.9). There are three possible explanations for this: (i) preferential flow mechanisms, or the presence of zones of high permeability in which transport of bacteria is rapid, may largely determine the bacterial concentration at a certain distance from a source of pollution, (ii) *E. coli* populations themselves may be heterogeneous in terms of their attachment characteristics, i.e. a fraction of the *E. coli* cells may have low sticking efficiencies, (iii) *E. coli* or thermotolerant coliforms usually travel in a plume of wastewater, consisting of many organic and inorganic compounds, all with highly variable concentrations. These compounds may block attachment sites, thereby leading to enhanced transport of *E. coli*. Laboratory experiments are usually carried out with solutions that do not reflect wastewater compositions, which might be the cause for the observed discrepancy between laboratory and field sticking efficiencies.

The detachment of *E. coli* has not been studied systematically. However, from the few cases in which deloading of *E. coli* was visible from breakthrough curves, the maximum relative *E. coli* concentrations were 1 to 3 log-units lower than during loading. This shows that detachment occurs, but as a process it is relatively unimportant.

The inactivation or die-off rate coefficients of *E. coli* and thermotolerant coliforms in solution and attached to sediment are influenced by many factors. The two major factors seem to be temperature and the presence of predators. The average die-off coefficient at 10 °C is 0.15 d^{-1} and increases to 0.50 d^{-1} at 20 °C. However, these rate coefficients are extremely variable. Predation causes the die-off rate coefficient to increase by 1 log-unit on average.

In our opinion, an important area of research on the transport of *E. coli* in aquifers is the effect of typical wastewater characteristics on the transport of *E. coli* and thermotolerant coliforms. These characteristics include high ionic strength values, high concentrations of organic compounds (e.g. aromatic and aliphatic carbohydrates and humic substances), and the presence of high concentrations of other microorganisms. How do *E. coli* and thermotolerant coliform organisms behave under these conditions in terms of sticking efficiency, attachment rate, and detachment rate coefficients, survival, and, more generally, in terms of the ability of these pathogens to be transported through aquifers? Little is known about the transport behavior of *E. coli* under those conditions, yet leakage of wastewater is one of the most important sources of *E. coli* and pathogens found in groundwater.

Finally, the upscaling of data obtained from controlled experiments in the laboratory to typical field conditions needs due attention. If preferential flow occurs, then which factors are most important in determining *E. coli* transport behavior, how should those characteristics be incorporated into the conceptualization of an *E. coli* transport model, and which of the existing models describing preferential flow (e.g. dual porosity, dual permeability, multi-porosity, and/or multi-permeability; Simunek et al., 2003) is relevant? In addition to the conceptualization of an *E. coli* transport model under field conditions, attention needs to be given to the aspect of population heterogeneity of *E. coli* and thermotolerant coliforms perhaps resulting in varying (decreasing) sticking efficiencies along the flow path in order to arrive at a better framework for predicting fecal indicator organism concentrations in aquifers.

CHAPTER 3:

DETERMINING STRAINING OF *ESCHERICHIA COLI* FROM BREAKTHROUGH CURVES

Foppen, J.W.A., A. Mporokoso, and J.F. Schijven, 2005. Determining straining of *Escherichia coli* from breakthrough curves. J. Contam. Hydrol. 76 (2005), 191-210.

Abstract

Though coliform bacteria are used world wide as an indication of faecal pollution, the parameters determining the transport of *E. coli* in aquifers are relatively unknown, especially for the period after the clean bed collision phase brought about by prolonged infiltration of waste water. In this research, the breakthrough curves of *E. coli* after total flushing of 50-200 pore volumes were studied for various influent concentrations in various sediments at different pore water flow velocities. The results indicated that straining in Dead End Pores (DEPs) was an important process that dominated bacteria breakthrough in fine-grained sediment (0.06-0.2 mm). The filling of the DEP space with bacteria took 5-65 pore volumes and was dependent on concentration. Column breakthrough curves were modelled and from this the DEP volumes were determined. These volumes (0.21-0.35 % of total column volume) corresponded well with values calculated with a formula based on purely geometrical considerations and also with values calculated with a pore size density function. For this function the so-called Van Genuchten parameters of the sediments used in the experiments were determined. The results indicate that straining might be a dominant process affecting colloid transport in the natural environment and therefore it is concluded that proper knowledge of the pore size distribution is crucial to an understanding of the retention of bacteria.

3.1 Introduction

In some developing countries in arid regions, the main source of drinking water is groundwater abstracted from dug or drilled wells. As wastewater collection systems are uncommon in these countries, most households dispose of their solid and liquid waste via soakaways (or pit latrines). In some cases, the distance from pit latrine to abstraction well is small (less than 200 m) and there is then a real risk of the abstracted water being contaminated with pathogens (Chapter 10).

Total Coliforms (TC) and Faecal Coliforms (FC) are commonly used as indicators of pathogens related to faecal contamination of groundwater as they are simple and relatively cheap to determine and give a good indication of the microbiological contamination. The main bacterial strain in TC and FC is *Escherichia coli*, which is rod shaped with an average length of 2-4 μm and an average diameter of 1 μm (Matthess et al., 1991b). In general, *E. coli* is hydrophilic (Van Loosdrecht et al., 1987b) and its zeta potential, which is a measure of the charge near the surface of the bacterium, varies between -10 to -30 mV (Van Loosdrecht, 1987a).

When it comes to assessing the risk of microbiological pollution or predicting the distance a wastewater plume with its microbiological load can travel in aquifers it seems appropriate to assess first the behaviour of *E. coli* in terms of its potential to travel various distances in aquifers. Various studies have focused on the transport behaviour of *E. coli*. The most well known field experiments were carried out by Caldwell and Parr (1937), who followed the breakthrough of "*B. coli*" (and "*B. aerogenes*", *B = Bacterium*) in time and space by taking daily samples of groundwater in more than 100 observation wells at regular distances downstream from a pit latrine. Lewis et al. (1982) summarised a number of field studies focusing on the distance TC and/or FC bacteria travelled in aquifers. They concluded that bacteria can travel several hundred metres in aquifers; the actual distance travelled depends on groundwater flow velocity, survival rate, initial concentration, dilution and dispersion of groundwater, and the sensitivity of the method used to detect bacteria. A more recent field study by Sinton et al. (1997) on wastewater infiltration and wastewater injection experiments supports this conclusion. Matthess et al. (1991a, b) used filtration theory (Yao et al., 1971) to describe the transport of *E. coli* ATCC 11229 in laboratory columns. The columns with a fixed inner diameter of 9.9 cm and various heights (from 0.1 to 0.6 m) contained sediment of various grain sizes and were seeded with *E. coli* for a total duration of 1-5 pore volumes at various pore water flow velocities. The results indicated two types of *E. coli* breakthrough. At low pore water velocities of around 0.7-1.5 m/d, breakthrough reached a plateau phase at low C/C0 value (type 1); at higher pore water flow velocities (2-7.5 m/d) the breakthrough was faster and C/C0 values were higher (type 2). In both cases, retardation seemed to be insignificant, which points to the lack of equilibrium sorption. According to Matthess et al. (1988), transport of *E. coli* in laboratory columns under conditions that rule out decay or growth of bacteria can largely be explained by using a first order filter factor that can be incorporated into the advection–dispersion equation. The filter factor is mainly dependent on pore water flow velocity and grain size of the sediment.

Since the 1970s much research has been done on the transport of bacteria in aquifers (e.g. Ryan and Elimelech, 1996; Murphy and Ginn, 2000), and from this it appears that the most important physico-chemical factors determining transport of colloids in aquifers are straining and physical filtration as well as hydrodynamic dispersion. Straining is the trapping of microbes in pore throats that are too small to allow passage. It is exclusively a result of pore geometry. Matthess and Pekdeger (1988) defined a so-called geometrical suffusion security δ, or straining, as

$$\delta = \frac{a_p}{0.12 a_{c,10}} \qquad (1)$$

where a_p is the diameter of the bacterium (m) and $a_{c,10}$ is the 10th percentile of the log-normal grain size number distribution (m). Those authors hypothesized that straining occurs only if $\delta \geq 1.5$, i.e. when the bacteria size is greater than 18% of d_{10}, but not below this critical value. However, Bradford et al. (2002, 2003) reported that straining was more pronounced than predictions based upon eq. (1). A less empirical approach was followed by Herzig et al. (1970), who calculated the volume of spherical colloids that could be retained based on purely geometrical considerations. They concluded that this type of retention in so-called "crevice sites" could be an important mechanism if the colloid diameter is more than 5% of the grain diameter of the porous medium. This value has also been reported in other studies (Corapcioglu and Haridas, 1985; McDowell-Boyer, 1986).

Physical filtration is the removal of particle mass from solution via collision with and deposition on the surface of the grains of the porous medium. The term includes both sedimentation and kinetic or equilibrium attachment. Several studies (Ryan and Elimelech, 1996; Johnson and Elimelech, 1995; Johnson et al., 1996; Elimelech and O'Melia, 1990; Sun et al., 2001) have indicated that geochemically heterogeneous aquifer sediment can have a profound influence on the kinetic attachment and detachment of colloids. It is thought that favourable attachment of bacteria with a negative surface charge can occur on patches of positively charged surfaces (e.g. iron oxyhydroxides) present as patchy coatings around grains in aquifers. The intensity of attachment depends not only on the opposed surface charge of bacterium and "collector" but also on the available patch surface area of the aquifer sediment (Sun et al., 2001). Finally, unfavourable attachment occurs when negatively charged bacteria attach to negatively charged surfaces (Sun et al., 2001; Bhattacharjee et al., 2002). This type of attachment is usually explained with DLVO theory (e.g. Ryan and Elimelech, 1996). Although bacterium and collector surfaces are both negatively charged and therefore repulsive, it is possible for the colloids approaching a collector surface to overcome this force, because when the distance between colloid and collector is small enough (less than a few nm), the attractive Van der Waals force becomes dominant. The colloids are then able to attach to collector surfaces. In contrast to favourable attachment, unfavourable attachment involves both attachment and detachment.

In the research described in this paper, the general objective was to elucidate the transport of indicator organisms in aquifers. More specifically, the objective was to understand under which conditions straining of *E. coli* occurs under controlled laboratory conditions without the interference of biological factors. In particular, we were interested to investigate theoretically and experimentally the dynamics of straining and the possibility that straining will cease after the smaller pores that are responsible for straining are filled. In contrast to the experiments carried out by Matthess et al. (1991a, b), the length of the experiments was 50-200 pore volumes in order to simulate the effect of prolonged wastewater infiltration. In the following account, an overview of the materials and methods is given and then the results of the experiments are analysed together with the modelling results. Finally, the most important implications are discussed.

3.2 Theory

The trapping of bacteria solely as a result of pore geometry is thought to be determined by the diameter of the pore throat, which is defined as the beginning or end of a pore, and by the diameter and length of the pore itself, which determines the volume of bacteria that can be retained. In a regular rhombohedral packing of identical spheres the largest spherical particle that can pass through without being trapped has a diameter of 7.7% of the grain diameter (Graton and Fraser, 1935; Herzig et al., 1970; Dullien, 1991). In general, however, the pore throat diameter distribution of interconnected pores in sediments is random (Dullien, 1991; Taylor and Jaffe, 1990). Also, pore throat diameters are a few orders of magnitude smaller than the pore radii (Dullien, 1991).

The volume of bacteria that could be retained was determined on purely geometric considerations by Herzig et al. (1970) for the case in which the exit pore throat diameter (where bacteria leave the pore) equals zero (a so called 'Dead End Pore' or 'DEP'). Assuming spherical colloids and grains and assuming that the available volume for straining in one single layer of colloids between two grains is fully

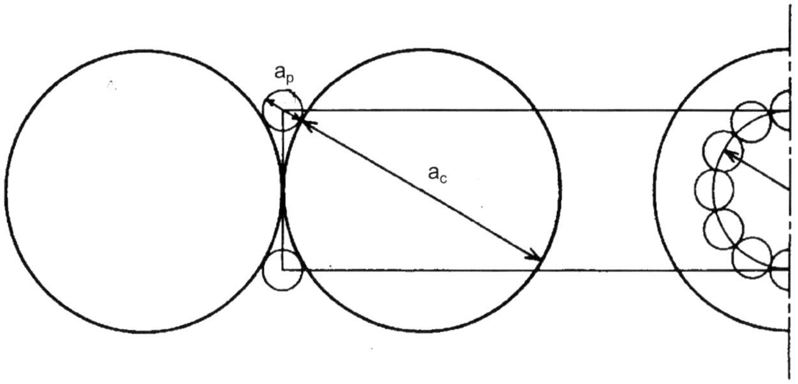

Fig. 1: Straining of colloids in pores with no exit pore throat (from Herzig et al., 1970)

occupied (see Figure 1), then the volume of the bacteria retained per unit of sediment volume, σ (-), can be calculated as

$$\sigma = \frac{1-\theta}{2}\pi N\left(\frac{a_p}{a_c}\right)^2\sqrt{\left(1+\frac{a_p}{a_c}\right)^2-1} \qquad (2)$$

where

θ porosity of the sediment (-)

N co-ordination number or the number of contact points between grains (-),

a_p diameter of the colloid (m)

a_c diameter of the grain (m)

For instance, for $\theta = 0.40$ and N = 7 (corresponding to an orthorhombic packing of spheres; e.g. Taylor and Jaffe, 1990), the σ values are given in Table 1 (upper part). Based on these values, Herzig et al. (1970) concluded that straining in this type of pores becomes an important process if $\frac{a_p}{a_c} \geq 0.05$. However, we hypothesize that the importance of straining in saturated porous media with a relatively small portion of DEPs is only a temporary phenomenon if the colloid flushing occurs over a sustained period of time: after the volume of dead end pores is filled with colloids, the water flow rate through the DEP is drastically reduced. Because the pore is effectively plugged and does not significantly participate in the flow process, it also ceases its straining function. Instead, flow is redirected through non-straining pores. If only a small percentage of pores are DEP, the plugging may not result in significant changes in the hydraulic conductivity and flow rate. The time t necessary to plug the DEP volume, σ, effectively depends on the influent colloid concentration, C_0 (in g/mL), in the fluid:

$$\frac{t}{t_0} = \frac{\sigma \rho_p}{\theta C_0} \qquad (3)$$

where t_0 is the time needed to flush one pore volume from a unit volume, $\frac{t}{t_0}$ represents the number of pore volumes to plug the DEP volume, and ρ_p is the bacterial density (g/mL). As an example, Table 1 lists the pore volumes (ranging between 0.8 and 48.1*10^6) needed to plug the DEP volume as a function of C_0 and relative colloid size, $\frac{a_p}{a_c}$, for a hypothetical flushing experiment in a 5 cm long column with a diameter of 5 cm and filled with a porous medium with $\theta = 0.4$ and $N = 7$. Eq. (3) provides a simple measure for the time needed to fill the DEP volume once the pore volume distribution is known.

Table 1: Volume of retained bacteria for various values of $\frac{a_p}{a_c}$ ($\theta = 0.40$ and $N = 7$) and the number of pore volume required to fill this volume if all colloids flowing into the column are strained in the case of a column of 5 cm diameter and height. A colloid was assumed to have a height of 2 μm, a diameter of 1 μm, and a density ρ_p of 1 g/mL

a_p/a_c	0.02	0.05	0.08	0.1
σ	0.053%	0.53%	1.72%	3.02%
Number of PVrequired ($C_0 = 10^3$ cells/mL)	0.8*10^6	8.4*10^6	27.4*10^6	48.1*10^6
Number of PVrequired ($C_0 = 10^9$ cells/mL)	0.8	8.4	27.4	48.1

3.2.1 Pore size density function

Since sediment usually consists of an array of grain sizes, the relation between grain size, colloid size and the magnitude of straining is not as straightforward as in the simple model (Eq. (2)) proposed by Herzig et al. (1970). Instead, the actual pore-size distribution function should be used to determine the relative DEP volume, σ. We propose that σ of a non-uniform porous medium with a pore size density function $f(r)$, where r is the pore-radius (m), is

$$\sigma = \int_0^{2d} f(r)dr \qquad (4)$$

The pore size distribution can be derived from the soil-water retention curve, which in turn can be measured in the laboratory. Wise et al. (1994) used Van Genuchten's (1980) soil–water retention function to arrive at a pore size density function *f*:

$$f(r) = m_v n_v A r^{-(n_v+1)}[1 + A r^{-n_v}]^{-(m_v+1)} \tag{5a}$$

and

$$A = \left(\frac{\alpha_v 2\delta_{aw} \cos\beta_{aws}}{\rho_{fl} g}\right)^{n_v} \tag{5b}$$

where m_v, n_v, α_v are the so-called Van Genuchten parameters ($m_v = 1-1/n_v$), whose values depend upon soil properties. The parameter α_v is a measure of the first moment of the pore size density function (in m^{-1}) and n_v is an inverse measure of the second moment of the pore size density function (dim. less). Furthermore, δ_{aw} is the interfacial tension between air and water phases (taken here as 0.0725 Nm^{-1}; from Wise et al., 1994), β_{aws} is the contact angle formed in the air–water–solid system (here taken as zero), ρ_{fl} is the density of water (here taken as 1000 kgm^{-3}), and g is the gravitational constant (9.81 $m\ s^{-2}$).

3.3.3 Bacteria transport model

The macroscopic mass balance equation for mobile bacteria suspended in the aqueous phase without the interference of biological factors such as growth and decay can be expressed as (Cameron and Klute, 1977; Corapcioglu and Haridas, 1984, 1985; Kim and Corapcioglu, 1996; Bengtsson and Lindqvist, 1995; Bengtsson and Ekere, 2001)

$$\frac{\partial C}{\partial t} = D\frac{\partial^2 C}{\partial x^2} - v\frac{\partial C}{\partial x} - \frac{\rho_{bulk}}{\theta}\frac{\partial S}{\partial t} \tag{6}$$

where C is the mass concentration of suspended bacteria in the aqueous phase (g/mL), D is the hydrodynamic dispersion coefficient (m^2/d) and includes the effects of random motility and chemotaxis, v is the pore water flow velocity (m/d), ρ_{bulk} is the bulk density of the porous medium (g/mL), θ is the volume occupied by the fluid per unit total volume (-), S is the total retained bacteria concentration (g/g) and x is the distance travelled (m). For the bacteria retained by the solid matrix, it was assumed that a fraction was available for DEP straining and a fraction was available for temporary or reversible retention. Because quartz sand was used, the temporary or reversible retention was believed to be mainly the result of unfavourable attachment and detachment. The mass balance equation for retained bacteria can be expressed as

$$\frac{\partial S}{\partial t} = \frac{\partial S_{DEP}}{\partial t} + \frac{\partial S_{TEMP}}{\partial t} \tag{7}$$

in which

$$\frac{\partial S_{DEP}}{\partial t} = k_{DEP} \frac{\theta}{\rho_{bulk}} C \tag{7a}$$

and

$$\frac{\partial S_{TEMP}}{\partial t} = k_{ret} \frac{\theta}{\rho_{bulk}} C - k_r S_{TEMP} \tag{7b}$$

where S_{DEP} represents the bacteria concentrations due to straining in dead end pores and S_{TEMP} represents the bacteria concentrations due to temporary retention (both in g/g). Furthermore, k_{DEP} and k_{ret} represent the rate coefficients for DEP straining and for temporary retention (both in 1/d), while k_r is the release rate coefficient (1/d) which determines the rate of detachment. For DEP straining we assumed that during flushing the pore volume available for DEP would fill up with bacteria at a constant rate dependent on bacteria concentration, until a maximum was reached. Once the maximum pore volume values were reached, the rate parameter k_{DEP} reduced to zero. In formula:

$$k_{DEP} = 0 \qquad \text{if} \qquad S_{DEP} \geq S_{MAX,DEP} \tag{8}$$

where $S_{MAX,DEP}$ (g/g) is the maximum bacteria concentration that can be retained due to DEP straining. Finally, the relation between $S_{MAX,DEP}$ and the volume of the bacteria retained per unit of sediment volume, σ, was defined as:

$$S_{MAX,DEP} = \sigma \frac{\rho_{bact}}{\rho_{bulk}} \tag{9}$$

3.3 Materials and methods

3.3.1 Preparation of bacteria suspensions

Bacteria suspensions were prepared by taking a needle scoop of *E. coli* ATCC25922 and mixing it with 50 mL of nutrient broth. The mixture was shaken for 24 hours in a temperature controlled room at 37 °C after which a bacteria concentration of around 10^9 cells/mL was obtained (determined by plate counting on Chromocult). The suspension was centrifuged for 30 minutes at 3000 rpm and the supernatant was removed and replaced by sterile tap water with a temperature of around 4 °C. The resulting bacteria suspension was used for the column experiments. Bacteria suspensions with different concentrations were prepared by diluting with sterile tap water. Concentrations in each of these dilutions were determined by plate counting. In the experiments, lithium was used as a tracer to determine pore water velocity and dispersivity. To do so 100 mL of demineralised water in which 152 mg of LiCl was dissolved was added to each litre of bacteria suspension. The resulting [Li^+] was around 22.7 mg/L. The [Li^+] was measured using an Atomic Emission Spectroscope. A series of column experiments was carried out with bacteria suspensions prepared 1 to 5 days prior to loading the columns. The average pH of the bacteria suspensions was around 7.0, while the EC values ranged between 750-780 µS/cm.

3.3.2 Column experiments

Two quartz sand fractions (Filcom Industries) with a grain size of 60-200 µm and 200-500 µm were used. Prior to the experiments the sand was sterilised for 24 hours in an autoclave at 105 °C. Total porosity of each sand fraction was determined by taking an oven-dried sample and measuring its weight and volume. The porosity was calculated by adding a known volume of water and measuring the resulting total volume. Two perspex columns (with an inside ∅ of 5 cm and a height of 5 cm) were incrementally packed with one of the two sand fractions. Prior to the packing, the columns were sterilised with alcohol. A stainless steel mesh overlain by a nylon mesh with an aperture opening of 50 µm was placed at the bottom of the column in order to retain the sand. A 2.5 mm diameter teflon tube connected the columns to a HPLC pump. The bacteria suspension was kept in 5-litre containers connected to the columns with teflon tubes and rubber stoppers. The containers were positioned about 50 cm above the columns in order to maintain a hydrostatic pressure on the column. All experiments were conducted at 4° C in order to prevent growth and decay of *E. coli*. Prior to each experiment, the columns were drained for 24 hours in order to remove any colloid particles present in the column.

In all, 4 experiments (EXP1, EXP2, EXP3 and EXP 4 in Table 2) were carried out at a pump speed of 0.5 mL/min on average (values ranged between 0.41 and 0.64 mL/min). Each experiment consisted of 3 runs (see Table 2). The average pore water flow velocity (0.7 m/d) was considered to be within the range of natural groundwater flow velocities. The exact flow velocity in each column was determined by measuring the amount of effluent at given time intervals. Columns were loaded with an *E. coli* suspension with varying concentrations: around 1.8 x 10^8 cells/mL for EXP1 (fine sand) and EXP3 (coarse sand) and around 18 x 10^8 cells/mL for EXP2 (fine sand) and EXP4 (coarse sand). An additional experiment (EXP1a) was carried out, with a low bacteria concentration (0.8 x 10^8 cells/mL) and two pump rates (0.5 mL/min and 2.0 mL/min). The range of concentrations was used to assess a possible concentration effect on straining (the filling of DEP space) and attachment, whereas the range of flow rates was applied to assess the effect of pore water flow velocity on these processes. We expected that straining would be little affected by a change in pore water flow velocity (Table 2).

The length of each experiment was determined by the shape of the breakthrough curve. When the fraction C/C_0 (C = concentration in the effluent and C_0 = the bacteria concentration in the container) became constant, the container containing the bacteria suspension was replaced by a container of sterile tap water. In general, the total duration of each experiment (loading with *E. coli* and deloading with sterile tap water) ranged between 50 and 200 pore volumes. Bacteria concentrations in the effluent samples were determined by measuring the optical density (OD) with a spectrophotometer at 410 nm. A calibration curve was prepared in order to translate

OD values into bacteria concentrations. The maximum permissible OD value was 1.5. Below this value, there is an almost linear relationship between OD and bacteria concentration. After the experiments, a regression analysis was carried out for all measured bacteria concentrations versus their OD values.

Table 2: Overview of parameters and their values per experimental run

Experiment	Grain size (mm)	C_0 (cells * 10^8 /mL)	Pump speed (mL/min)	Disp. Coeff (m^2/d).	k_{ret} (1/d)	k_r	k_{DEP} (1/d)	$S_{MAX, DEP}$ (-)
		1.8	0.55	4E-4 (R^2 = 0.99)	5.0	0.5	38	2.0E-3
EXP1	0.06-0.2	1.9	0.46	3E-4 (R^2 = 0.96)	5.0	0.2	17	2.2E-3
		1.8	0.44	3E-4 (R^2 = 0.93)	5.0	0.1	30	1.3E-3
EXP1a	0.06-0.2	0.8	0.5	5E-4 (assumed)	5.0	0.3	38	1.3E-3
		0.8	2.0	5E-4 (assumed)	20.0	1.0	152	1.3E-3
		16	0.55	3E-4 (R^2 = 0.99)	3.0	0.2	38	2.0E-3
EXP2	0.06-0.2	18	0.49	5E-4 (R^2 = 0.92)	1.5	0.3	34	2.0E-3
		19	0.41	4E-4 (R^2 = 0.97)	4.0	0.1	35	2.0E-3
		1.8	0.64	4E-4 (R^2 = 0.99)	3.9	0.5	-	-
EXP3	0.2-0.5	1.9	0.44	1E-3 (R^2 = 0.97)	5.5	0.1	-	-
		1.8	0.45	5E-4 (R^2 = 0.92)	3.4	0.3	-	-
		16	0.63	3E-4 (R^2 = 0.97)	11.8	8.0	-	-
EXP4	0.2-0.5	18	0.48	1E-3 (R^2 = 0.93)	3.6	3.4	-	-
		19	0.41	6E-4 (R^2 = 0.97)	2.6	1.3	-	-

3.3.3 Determining the Van Genuchten parameters

Starting from saturation, water potential was measured in a laboratory column (height 5 cm, inner diameter 10 cm) connected to a hanging column (Jin et al., 2000). This was done using a micro-tensiometer with a ceramic cup (diameter 0.5 cm) that was attached to the wall of the column. The tensiometer was connected to electronic pressure transducers; electrical voltage readings were collected intermittently. Voltage readings were tranferred to matric potential values using linear calibration curves obtained separately with the hanging columns. Soil water content was determined by weighing. The pressure head range that could be determined with the micro-tensiometer varied between 0 cm (saturation) and 125 cm suction. Using RETC, Van Genuchten's soil water retention function was fitted to the soil water content and pressure head data in order to determine the Van Genuchten parameters , n_v *and* α_v, used in eq. (5a).

3.3.4 Numerical model implementation

Equations (6), (7) and (8) were solved numerically with an explicit finite difference scheme that is forward in time, central in space for dispersion and upwind for advective transport. With each time step, first advective transport is calculated, then rate reactions and finally dispersive transport. Rates must be integrated over a time interval, which involves calculating the changes in bacteria concentrations both in the fluid and on the solid matrix. The scheme was prepared in a spreadsheet and is the same as that used in PHREEQC version 2 (Parkhurst and Appelo, 1999). Part of the modelling was performed in a spreadsheet, as this has the advantage of flexibility. Prior to this, test runs were carried out in order to compare the spreadsheet results with results obtained with PHREEQC and CXTFIT version 2.1 (Toride et al., 1999). In the experiments that appeared to involve straining, the spreadsheet was used and all parameters were determined by fitting curves visually. In the experiments that did not involve straining, curves were fitted with CXTFIT version 2.1 (Toride et al., 1999). CXTFIT is a computer code for estimating solute transport parameters from observed concentrations or for predicting solute concentrations using the convection–dispersion equation as the transport model. CXTFIT, PHREEQC and RETC (for estimating the pore size density function) are public domain codes and can be downloaded from the Internet.

3.4 Results

3.4.1 Tracer breakthrough

The measured total porosity was 0.40 for the fine sand and 0.38 for the coarse sand. For all experiments, the recovery of the tracer for both sand fractions was around 0.5 after flushing of 1 pore volume and around 1.0 after flushing of about 3 pore volumes (graphs not given). The dispersion coefficients were calculated by fitting the observed tracer data within the first 4 pore volumes of flushing with the analytical equation for conservative transport and by taking into account the measured pump discharge (see Table 2). The dispersion coefficients for the fine sand ranged between 5E-4 and 3E-4 m^2/d and for the coarse sand between 3E-4 and 10E-4 m^2/d. These variations within one type of sand originated from differences in packing, and were considered to be within acceptable ranges. Dispersion coefficients for the coarse sand were somewhat higher than for the fine sand. For all experiments, the coefficient of determination, R^2, ranged between 0.92 and 0.99. In all, it was concluded that the columns were properly set up. Front movement through the column was according to the average pore water flow velocity and there seemed to be no preferential flow in the columns or along the column walls.

3.4.2 *E. coli* breakthrough

From the regression analysis of all measured bacteria concentrations versus their OD values we obtained the relationship

$$C = 1.17 * 10^8 * OD^{1.089} \qquad (10)$$

The fit between observed and calculated data was good (N = 96; $R^2 = 0.96$) and the relationship was used to calculate bacteria concentrations from their respective OD value. Selected *E. coli* breakthrough curves (BTC) for all experiments are given in Figure 2a-f. In EXP1 (fine sand, $C_0=1.8*10^8$ cells/mL) there was initially rapid breakthrough and then at C/C_0 values of 0.1 - 0.2 a temporary plateau developed. This plateau was also present in EXP2 (fine sand, $C_0=18*10^8$ cells/mL), but was

much less pronounced (Figure 2c). EXP1A (fine sand) with the lowest concentrations (C_0=0.8*10^8 cells/mL), showed a plateau too (Figure 2d). EXP1A also clearly demonstrated that this plateau was not affected by pump rate variations: breakthrough at 0.5 mL/min was similar to breakthrough at 2.0 mL/min. Hence, this plateau was determined much more by a difference in C_0 than by a difference in flow rate. The various runs for EXP1 and EXP2 showed the same levelling-off, but sometimes not so clearly. It seems that there may be variations between column experiments and that these confound the effects of C_0. Nevertheless, the duration of the initial plateaus seemed to strongly depend on C_0, while the small variations in flow rate did not seem to be important. Moreover, plateaus did not develop in the BTCs of the coarse sand columns, EXP3 and EXP4 (Figure 2e and 2f).
This behaviour (dependent on concentration and pore size, but independent of pore water flow velocity), plus the fact that both quartz sediment and bacteria were negatively charged (at the neutral pH of the bacteria suspension), showed that differences in sorption were probably not the reason why these plateaus occurred only in the fine sand. A more likely reason for these plateaus is the presence of pores too small for the bacteria to pass through. These dead end pores are more likely to occur in fine sand. It would be reasonable to assume that the total volume of such pores was less than the total volume of bacteria flowing into the column. Obviously, if C_0 values were high, these pores filled rapidly, but if C_0 values were lower they took longer to fill. The concentration dependence (R^2 = 0.96) is also shown in Figure 3, in which the number of pore volumes required to fill the DEP space for all fine sand experiments (EXP1, EXP1A and EXP2) has been plotted against the influent bacteria concentration. Filling the DEP space took 5 to 65 pore volumes; the exact number depended on the influent concentration. The logarithmic linearity of this dependence certainly reinforced the hypothesis that the fine sand had a more or less constant volume of pores with radii that were too small for bacteria to pass through. Such type of relationship was also implicitly suggested in the example given in Table 1 (lower part).

As long as there was still an excess of available DEPs, the deposition or trapping of bacteria in the DEPs could be described as a first order sink. As soon as the fine sand DEPs were filled, the C/C_0 values increased to 0.75-0.9 and then in all the fine sand experiments increased only slightly. This slow but steady increase was particularly visible in one run of EXP1 (data not shown), in which loading with bacteria was continued for some 150 pore volumes. The same slow increase was also visible in the coarse sand experiment with low C_0 (EXP3). For the fine sand this process seemed to be independent from either pump rate or C_0. In all cases, this behaviour pointed towards a process of retaining bacteria on the one hand and releasing a small portion of the retained bacteria on the other hand. This seemed to be temporary retention, in which the process of unfavourable attachment and detachment was thought to be taking place. The BTCs of EXP4 (high concentration) reached C/C_0 values of 1 after only few pore volumes (almost tracer-like), whereas the C/C_0 values for EXP2 (high concentration) remained below 1 during the entire period of flushing. From this we conclude that the sum of pore volume available for temporary retention was less for the coarse sand than for the fine sand.

Finally, in all experiments, the release of bacteria during deloading was small. The C/C_0 values during the deloading of the high C_0 experiments were slightly lower than those of the low C_0 experiments. This behaviour suggests that the number of bacteria detaching from the grains was not fully determined by the influent bacteria

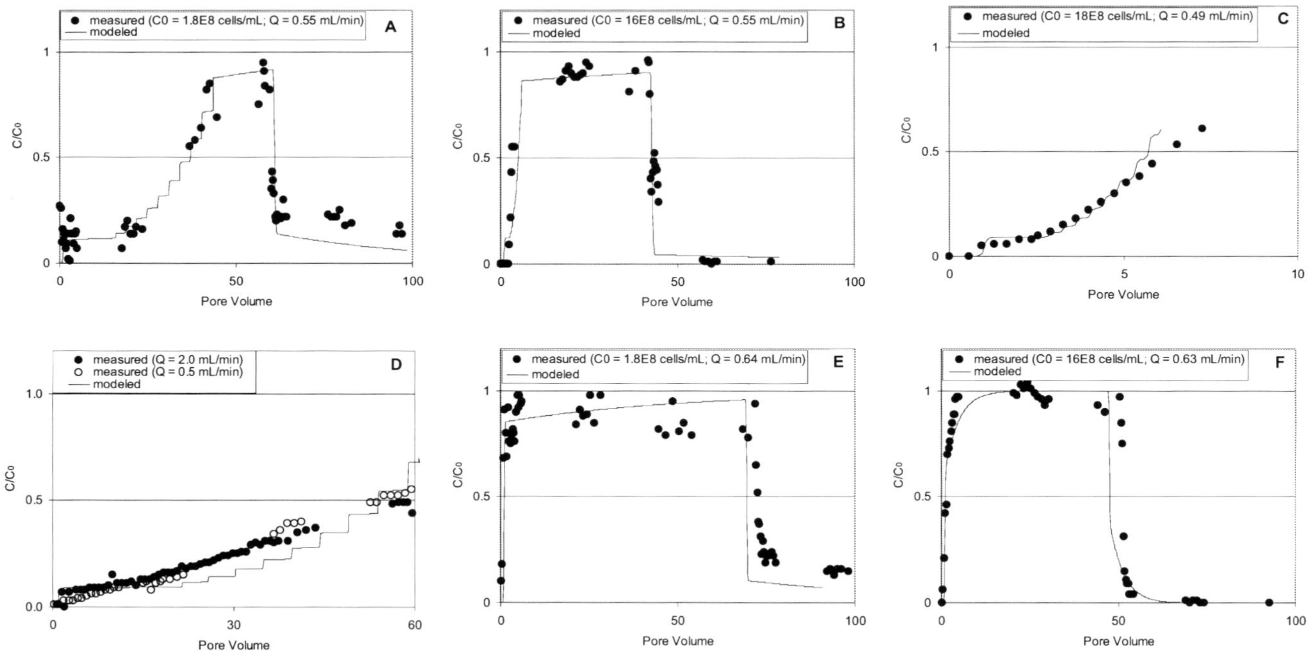

Fig. 2a-f: E. coli breakthrough curves: (a) fine sand and low concentration (EXP1), (b) fine sand and high concentration (EXP2), (c) detail of breakthrough during the first 7 pore volumes of EXP2 (second run, see Table 2), (d) fine sand, very low concentration ($0.8*10^8$ cells/mL) at two pump rates (0.5 and 2.0 mL/min), (e) coarse sand (EXP3) and (f) coarse sand (EXP4). Measured values are given in dots; modelled breakthrough is given as a line

concentration alone. In other words, the release was not entirely a first order kinetic process, but a process between first and zero order.

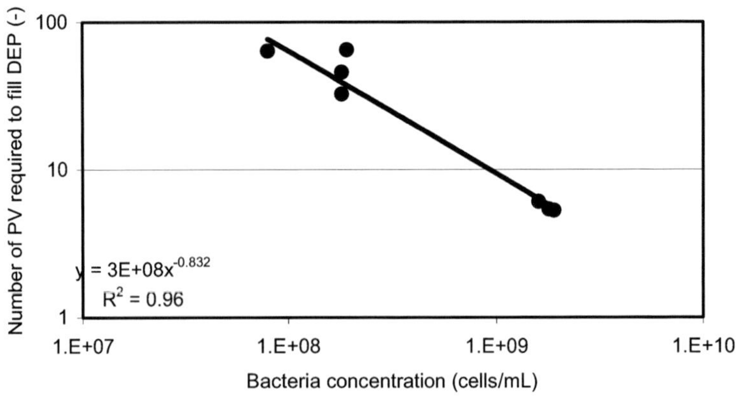

Fig. 3: Number of pore volumes required to fill the DEP space versus concentration of the bacteria suspension for all fine sand experiments (EXP1, EXP2 and EXP1a)

3.4.3 Modelling of breakthrough curves

The values for the model parameters are given in Table 2. For the fine sand, the values for $S_{MAX,DEP}$ were all within a narrow range of 1.3E-3 to 2.2E-3 gram bacteria per gram sediment, indicating that a small but almost constant value of the pore volume was available for DEP straining. Expressed as a percentage of the total volume of the column, these values ranged from 0.21 - 0.35%. The rate coefficient due to DEP straining, k_{DEP}, was linearly dependent on pore water flow velocity (Table 2, EXP1A), which indicates that the DEP rate coefficient as modelled could also be regarded as the product of pore water flow velocity and a rate coefficient that depends on the distance travelled (with dimension m^{-1} instead of d^{-1}). Such a relationship supports the interpretation that the plateaus observed in the BTCs of the fine sand were caused by straining instead of kinetic sorption. It will be recalled that k_{DEP} varied somewhat and that we attributed this to variations between similar sand column experiments, which confounded the effects of C_0. The modelled rate coefficients k_{ret} and k_r for the fine sand experiments were low and did not show a very clear relation with either velocity or influent concentration. However, the k_{ret} values for EXP1were somewhat higher than the k_{ret} values for EXP2. This result again indicated that temporary retention was not entirely a first order kinetic process.

Because of the observed lack of straining, breakthrough in the coarse sand was modelled without k_{DEP} and $S_{MAX,DEP}$, but with k_{ret} and k_r attachment and detachment coefficients (using CXTFIT). In the case of EXP3, k_{ret} varied between 3.4 and 5.5, while k_r varied between 0.1 and 0.5. The almost tracer-like breakthrough curves of EXP4 were simulated best with a relatively high k_r, indicating that attachment and detachment were of the same order of magnitude.

3.4.4 Volume available for straining

Was $S_{MAX,DEP}$ indeed found to be related to a volume of pores with a radius less than 1-3 μm (the average size of an *E. coli*; see Introduction)? The measured pressure heads and soil water content values for the fine and coarse sand (values not given) were used to fit Van Genuchten's soil-water retention function with RETC, which resulted in values for the parameters n_v and α_v (see Fig. 4a). The resulting pore size density functions are also given in Fig. 4a, while the cumulative values as part of the total available pore volume are given in Fig. 4b. The cumulative pore volume is similar to the so-called "effective saturation" (Van Genuchten, 1980; Wise, 1992; Wise et al., 1994) and values range from 0 to 1 (in Figure 4b only the interval from 0-1% is given). Pore radii for the fine sand ranged from 1 to 40 μm with a median around 10μm. Pore radii for the coarse sand ranged from 6 to 228 μm with a median around 62 μm. The cumulative pore volume for the fine sand (Figure 4b) showed that 0.7-0.9% of the total available pore volume was within the pore radii range of 1-3 μm, while there was practically no such volume (< 0.003%) present in the coarse sand. This amount (0.7-0.9%) may not be much, but, as long as the DEP space was not filled or saturated, it was responsible for retaining 80-90% of the influent bacteria.

Table 3 summarises the volumes of bacteria retained per unit volume of sediment, according to eq. (2), the cumulative pore volume according to eq. (5) (now expressed per unit volume of sediment) and the pore volume for straining according to the modelled values of $S_{MAX,DEP}$. We concluded that the ranges of volume available for straining calculated with the Herzig and Wise formula (eq. (2) and eq. (5) resp.) were similar and that the modelled S values of 0.21-0.35% were well within these ranges. Therefore, we infer that $S_{MAX,DEP}$ was related to the pore volume according to eq. (9) and the modelled $S_{MAX,DEP}$ values closely corresponded to this pore volume.

Table 3: Volume of the bacteria retained per unit of sediment volume (σ) according to eq. (2), the volume of pores with a radius less than 2 μm according to eq. (5), and the pore volume according to the modelled values of $S_{MAX,DEP}$

	Fine sand (0.06-0.2 mm)	Coarse sand (0.2-0.5 mm)
σ (acc. to Herzig et al., 1970) (θ= 0.4; N = 7; a_p = 1-3 μm)	0.002 - 0.53%	0.0002 - 0.027%
Volume of pores with a radius of 1-3 μm (acc. to Wise et al., 1994)	0.019 - 0.36%	< 0.001%
Pore volume belonging to $S_{MAX,DEP}$= $1.3x10^{-3}$-$2.2x10^{-3}$ (ρ_{bulk} of fine sand = 1.60 g/mL; (ρ_p = 1.00 g/mL (assumed))	0.21 – 0.35%	-

3.5 Discussion

The long duration of flushing (50-200 pore volumes) in our research yielded information crucial for the interpretation of the breakthrough curves in terms of the processes (DEP straining) taking place in the column. Matthess et al. (1991a, b), who also flushed with *E. coli,* obtained their results after only 1-5 pore volumes of flushing, so they never observed the filling of the DEP volume and associated increase in C/C_0 value. Other researchers probably could have observed the process but did not, perhaps because of limited flushing times.

In Table 4, data from a number of experiments from the literature have been used to determine the volume of bacteria retained per unit of sediment volume, σ , according to eq. (2). Based on the minimum and maximum grain size of the sediments used and the minimum and maximum bacteria size used, a range for σ was determined. In the last column of Table 4 the minimum number of pore volumes required to fill the Dead End Pore space is shown, calculated from the column dimensions, porosity, influent bacteria concentration and shape and size of the bacteria/colloids (all given in the references used). In the fine sand (0-63 μm) column experiment of Bengtsson and Ekere (2001; see Table 4), a recovery of around 0.3 for a duration of 22 pore volumes was observed, which is proportional to 22*(1-0.3) = 15.4 pore volumes of complete bacteria removal. The minimum number of pore volumes calculated with Herzig's formula (13.7, right-hand column of Table 4) was close to this value, indicating that the DEP space was almost completely filled and recovery values might have increased if the experiment had continued for longer. The same applies to the experiments done by Bradford et al. (2003). Although their experiments were specifically aimed at identifying straining processes, it would have been worthwhile to continue the experiments for 5-10 pore volumes in order to be able to calculate the Dead End Pore space. In the case of Brown et al. (2002), a plateau was visible between 1 and 3 pore volumes before recovery values increased to 0.8-1.0. Brown and his colleagues opine that this might have been caused by complex multi-layer deposition and that this needs further study. Based on the calculated σ , we postulate that these plateaus could very well have been caused by straining.

Another example shows that it might be crucial to calculate σ and to relate it to influent concentrations. For instance, Tan et al. (1994) discarded straining as a possible process (they called it "pore plugging" in their Discussion and Conclusions). Given the number of pore volumes of their experiments, and given the influent concentrations, only in the case of the highest concentrations (10^9 cells/mL) does straining seem irrelevant. It might be that Tan's threshold retention capacity in the low concentration experiment (2×10^7 cells/mL) was mainly caused by straining. Also, in Lindqvist et al. (1994), the low concentration experiment (2.2×10^7 cells/mL of IS1 bacteria in the Lincoln sand) might at least have been influenced by straining.

Table 4: The potential importance of straining for a number of experiments from literature. The volume of bacteria retained per unit of sediment volume, σ, was calculated according to eq. (2). Based on σ and the influent bacteria concentrations, C_0, the minimum number of pore volumes required to fill the Dead End Pore volume could be calculated; this was then compared with the number of pore volumes the columns were flushed with in the actual experiments

Reference	θ (-)	a_p/a_c (-)		σ (%)		C_0 (cells/mL)	Number of pore volumes of flushing bacteria in experiment	Minimum number of pore volumes required to fill DEP
		min.	max.	min.	max.			
Bengtsson and Ekere (2001)	0.6		0.0318	0	0.113	1.60E+08	22	13.7
Bradford et al. (2003)	0.35	0.0018	0.009	0	0.008	4.24E+08	2-3	7.5
Bradford et al. (2003)	0.35	0.0029	0.0067	0	0.004	3.86E+07	2-3	3.8
Bradford et al. (2003)	0.35	0.0044	0.0057	0	0.002	4.85E+06	2-3	1.9
Bradford et al. (2003)	0.35	0.0035	0.0046	0	0.001	1.18E+06	2-3	1.1
Brown et al. (2002)	0.49	0.0017	0.0133	0	0.016	1.60E+08	4-7	5.2
Tan et al. (1994)	0.38	0.0005	0.004	0	0.001	2.00E+07	1-2	1.7
Tan et al. (1994)	0.38	0.0005	0.004	0	0.001	1.50E+08	1-2	0.2
Tan et al. (1994)	0.38	0.0005	0.004	0	0.001	1.00E+09	1-2	0
Lindqvist et al. (1994)	0.36	0.004	0.01	0	0.01	2.20E+07	50	16.1
Lindqvist et al. (1994)	0.36	0.004	0.01	0	0.01	2.00E+08	50	1.8
Lindqvist et al. (1994)	0.4	0.002	0.004	0	0.001	2.50E+07	50	1.3
Lindqvist et al. (1994)	0.34	0.0008	0.001	0	3E-05	1.20E+09	50	0

It should be noted that eq. (2) has some shortcomings. The most important ones are:

- A uniform porous medium grain size is assumed. As already stated, this is certainly unrealistic for sediments in their natural environment;
- Pores with an exit pore throat radius between zero and the diameter of the colloid are available for straining but are neglected in eq. (2);
- It might not be very realistic to assume that colloids completely fill the space between two grains. This might depend on the orientation of that space to the direction of flow (perpendicular or parallel);
- A Dead End Pore may contain more than one layer of colloids.

Despite these shortcomings, the results of the modeling justify concluding that eq. (2) still appears to offer a valuable framework for assessing the importance of straining and for calculating the available straining volume.

So far we have not discussed the origin and value of k_{DEP}. Bradford et al. (2003) established a power function correlation between their k_{DEP} and the ratio $\frac{a_p}{a_c}$, in which a_c was taken as the median of the porous medium grain diameter according to

$$k_{DEP} = a(\frac{a_p}{a_c})^b \qquad (11a)$$

and a and b were fitting parameters. Since we used colloids of one size only, whereas Bradford used various colloid sizes and porous media with various grain diameters, it is not appropriate to evaluate Bradford's findings here. However, as well as observing that k_{DEP} reduces to zero once $S_{MAX,DEP}$ is exceeded (see eq. (8)), on the basis of EXP1A (k_{DEP} = 38 d^{-1} for the low flow velocity experiment and 152 d^{-1} for the high flow velocity experiment; see Table 2), we can refine eq. (11a) by including pore water flow velocity:

$$k_{DEP} = a(\frac{a_p}{a_c})^b v \qquad (11b)$$

When the pore water flow velocity is taken as 0.9 m/d, then two points can be examined with Bradford's power function correlation. For the fine sand, $\frac{a_p}{a_c}$ (= 2/130) = 0.015, while k_{DEP} is 38 day^{-1} and for the coarse sand, $\frac{a_p}{a_c}$ (= 2/350) = 0.006, while in this case k_{DEP} is zero. Changing the unit from day^{-1} to min^{-1} means that the k_{DEP} for the fine sand is roughly around 1.4 min^{-1} (established via visual comparison of BTCs and then multiplied by a factor of 4.6 to adjust for pore water flow differences). This value is double the k_{DEP} value (0.69 min^{-1}) corresponding to a $\frac{a_p}{a_c}$ ratio of 0.015 determined with Bradford's power function correlation. As already stated by Bradford et al. (2003), straining probably also depends on factors not yet included in eq. (11b), such as soil grain size uniformity, colloid distribution, water content and experimental scale. Therefore, additional experimental studies are needed to assess the influence of these factors on straining.

Outbreaks of groundwater-borne diseases may be associated with processes like the reduction of attachment sites (causing bacteria to move freely through the aquifer) or a sudden growth of pathogens in wells. Acknowledging the existence of a Dead End Pore space means that another process can be added to theoretical and practical

Table 5: Average values for the Van Genuchten parameters for 6 major soil textural groups (according to Carsel and Parrish, in: Van Genuchten et al., 1991) and the cumulative volume of pores with a radius less than 2.5 μm, determined with eq. (5)

Texture	α_v (1/m)	n_v (-)	Pores with radius ≤ 2.5 μm (% of total pore volume)
Sand	14.5	2.68	0.05
Loamy sand	12.4	2.28	0.43
Sandy loam	7.5	1.89	41.1
Loam	3.6	1.56	99.9
Silt	1.6	1.37	100.0
Clay	0.8	1.09	100.0

applications. In fine porous sediments the filling of Dead End Pores can result in breakthrough of bacteria in the same way that a reduction of attachment sites causes bacteria to move freely. The easiest way of determining whether straining can occur is to use Herzig's formula. However, since sediments in their natural environment have certain grain size ranges, it might be more useful to consider the sediment pore size distribution, which can be determined indirectly via soil water retention functions. In Table 5 the results of applying the formula of Wise et al. (1994) to a number of major soil textural groups are presented. From the Table it can be concluded that straining must be significant in virtually all soil groups. Even in the "sand" group the process of straining can be significant (0.05%), especially when colloid concentrations are low. This demonstrates once again the importance of straining during transport of colloids in the natural environment and it also demonstrates that even in sandy soils phenomena like macro-pore flow and/or flow in fissures and cracks (Powell et al., 2003) are probably important in the breakthrough of colloids.

CHAPTER 4:

MEASURING AND MODELING STRAINING OF *ESCHERICHIA COLI* IN SATURATED POROUS MEDIA

Foppen, J.W.A., M. van Herwerden, and J.F. Schijven, 2007a. Measuring and modeling straining of *Escherichia coli* in saturated porous media. J. Contam. Hydrol. DOI: 10.1016/j.jconhyd.2007.03.001.

Abstract

Though coliform bacteria are used worldwide to indicate fecal pollution of groundwater, the parameters determining the transport of *Escherichia coli* in aquifers are relatively unknown. We evaluated the occurrence of both straining and attachment of *E. coli* ATCC25922 in columns of ultra-pure, angular, saturated quartz sand. The column experiments were conducted over a wide range of porous medium sizes, column heights, input concentrations, and pore water flow velocities. Straining and attachment were examined by modeling the breakthrough curves (with HYDRUS 1D). In addition, model output was compared with measured strained and attached bacteria via column extrusion experiments (in which sand was extruded from the column and placed in excess water) and flow reversal experiments (in which the pore water flow direction was reversed, thereby dislodging strained bacteria). Our model consisted of an attachment rate coefficient and a straining rate coefficient; both these decreased with transport distance. The straining rate coefficient also decreased in a Langmuirian way, in response to the filling of available pore space, which in turn depended on influent bacteria concentration, quartz grain diameter, and transport distance. The maximum strained fraction was 25-30% of total bacteria mass applied to the column; the maximum attached fraction was 30-35%. The fit between modelled and measured (strained and attached) bacteria masses was acceptable, as was the sensitivity of the model output to fitted parameter values. Our results lead to a new description for the time-dependent mass balance of strained bacteria, which entails using three fitting parameters. The results also imply that column experiments in combination with retention profiles (or various column lengths) are not enough to explain the retention processes in a column. Column extrusion and flow reversal experiments provide vital additional information on the occurrence and magnitude of straining. Our straining model could be of assistance in evaluating the importance of straining and in incorporating the straining process in bacteria transport modeling.

4.1 Introduction

In some developing countries in arid regions, the main source of drinking water is groundwater abstracted from dug or drilled wells. As wastewater collection systems are uncommon in these countries, most households dispose of their solid and liquid waste via soakaways (or pit latrines). In some cases, the distance from pit latrine to abstraction well is small (less than 200 m) and there is then a real risk of the abstracted water being contaminated with pathogens (Chapter 11).

Total Coliform (TC) and Fecal Coliform (FC) concentrations are commonly used as indicators of pathogens related to fecal contamination of groundwater as they are simple and relatively cheap to determine and give a good indication of the microbiological contamination. The main bacterial strain in TC and FC is *Escherichia coli.* When it comes to assessing the risk of microbiological pollution or predicting the distance a wastewater plume with its microbiological load can travel in aquifers, it seems appropriate to assess first the behavior of *E. coli* in terms of its potential to travel in aquifers. The classical colloid filtration theory [*Yao et al.*, 1971; *Rajagopalan and Tien*, 1976; *Tobiason and O'Melia*, 1988; *Logan et al.*, 1995; *Ryan and Elimelech*, 1996; *Tufenkji and Elimelech,* 2004a] that is commonly used to predict the removal of colloids and biocolloids does not account for straining [*Bradford et al.*, 2002, 2003, 2005; *Bradford and Bettahar*, 2005, 2006]. Straining is the trapping of colloid particles in down-gradient pore throats that are too small to allow the particles to pass through [*Herzig et al.*, 1970; *Corapcioglu and Haridas*, 1984, 1985]. *Matthess and Pekdeger* [1985] concluded that straining occurs when the bacteria size is greater than 18% of the 10th percentile of the cumulative grain size distribution, but not when it is below this critical value. For this reason most previous studies on colloid transport have neglected straining as a mechanism for retention. Recently, however, *Bradford et al.* [2002, 2003, 2005], and *Bradford and Bettahar* [2005, 2006] clearly demonstrated that straining becomes a significant process when the ratio of colloid to median grain size is greater than 0.5%. *Foppen et al.* [2005; Chapter 3] found considerable masses of bacteria due to straining when the colloid to median grain size was 1.2% and *Li et al.* [2004] and *Tufenkji et al.* [2004] also found evidence for straining at colloid to median grain size ratios less than 18%. Furthermore, the results of *Bradford et al.* [2002, 2003, 2005], and *Bradford and Bettahar* [2005, 2006] indicated that retained particle concentrations decayed in an extremely non log-linear or hyper-exponential manner with distance, and that colloid mass retention by straining occurs primarily at the column inlet or textural interface, because of colloids being retained in dead end pores and/or at grain junctions smaller than some critical size. The apparent number of dead end pores was hypothesized to decrease with increasing distance, since advection, dispersion, and size exclusion tend to keep mobile colloids within the larger pore networks, thus bypassing smaller pores.

Bradford et al. [2002, 2003, 2005] and *Bradford and Bettahar* [2005, 2006] considered the sticking efficiency, and therefore the attachment rate coefficient, to be constant in time and space throughout the duration of an experiment. In column experiments, *Simoni et al.* [1998] (*Pseudomonas*), *Baygents et al.* [1998] (groundwater isolates A1264 and CD1), *Redman et al.* [2001] (recombinant Norwalk virus), *Li et al.* [2004] (latex colloids of different surface charge densities) and *Tufenkji and Elimelech* [2004a, 2005, 2005a] (latex colloids of different sizes and *Cryptosporidium* oocysts) have clearly demonstrated that the attachment rate coefficient decreased with distance from the column inlet. Obviously, in many cases, straining and attachment occur simultaneously and they are difficult to distinguish from each other. Moreover, they may not be independent and both processes may explain the depth dependence of retained particles or colloids. The purpose of the current study was to distinguish straining and attachment of *E. coli* ATCC 25922 in columns of ultra-pure quartz sand. Since attachment and straining are a strong function of porous medium size and system dynamics, column experiments, flow reversal experiments, and column extrusion experiments were conducted over a wide range of porous medium sizes, column lengths, input concentrations, and pore water flow velocities. By modeling the transient mass balance equations using breakthrough data, model parameters that determine straining and attachment were investigated.

4.2 Theory

4.2.1 General model description

The transport of mass in porous media may be generally described by the advection-dispersion-sorption (ADS) equation (*De Marsily* [1986]), and various expressions of the ADS equation have been used for the transport of *E. coli* (*Pang et al.* [2003]; *Powelson and Mills* [2001]; *Matthess and Pekdeger* [1981] and [1985]; *Matthess et al.* [1985], [1988]; Chapter 3 of this study). The one-dimensional macroscopic mass balance equation for *E. coli* without the interference of biological factors such as growth and decay can be expressed as [e.g. *Herzig et al.* 1970]:

$$\frac{\partial C}{\partial t} = D\frac{\partial^2 C}{\partial x^2} - v\frac{\partial C}{\partial x} - \frac{\rho_{bulk}}{\theta}\frac{\partial S}{\partial t} \tag{1}$$

where C is the mass concentration of suspended bacteria in the aqueous phase (# of cells/mL), t is time (s), D is the hydrodynamic dispersion coefficient (cm^2/s) and includes the effect of random motility, v is the pore water flow velocity (cm/s), ρ_{bulk} is the bulk density of the porous medium (g/mL), θ is the volume occupied by the fluid per unit total volume (-), S is the total retained bacteria concentration (# of cells/gram sediment) and x is the distance traveled (cm). The retained bacteria fraction can be expressed by a first-order kinetic term (*Bradford et al.* [2003], *Bradford and Bettahar* [2006]):

$$\frac{\partial S}{\partial t} = \frac{\partial S_{att}}{\partial t} + \frac{\partial S_{str}}{\partial t} \tag{2}$$

whereby

$$\frac{\partial S_{att}}{\partial t} = \frac{\theta}{\rho_{bulk}} k_a C - k_r S_{att} \tag{3}$$

$$\frac{\partial S_{str}}{\partial t} = \frac{\theta}{\rho_{bulk}} k_{str}(1 - BS_{str})C \tag{4}$$

where S_{att} is the retained bacteria concentration due to kinetic attachment (# of cells/gram sediment), S_{str} is the bacteria concentration (# of cells/gram sediment) retained due to straining, k_a is the attachment rate coefficient (s^{-1}), k_r is the detachment rate coefficient (s^{-1}), k_{str} is the straining rate coefficient (s^{-1}), and B is a fitting parameter that may vary depending on the conditions of the experiment (pore water flow velocity, influent bacteria concentration, type of collector material, etc.). The attachment rate coefficient, $\mathrm{k_a}$, can be determined by (e.g. *Yao et al.*, 1971]):

$$k_a = \frac{3(1-\theta)v\eta_0\alpha}{2a_c} \tag{5}$$

where a_c is the grain diameter (cm), η_0 is the dimensionless single collector contact efficiency, determined from physical considerations [e.g. *Tufenkji and Elimelech*, 2004], and α is the dimensionless sticking efficiency. We determined η_0 with the Tufenkji-Elimelech correlation equation [*Tufenkji and Elimelech*,

2004]. Because of population heterogeneity of micro-organisms, the sticking efficiency, α, may vary (*Simoni et al.* [1998], *Baygents et al.* [1998], *Redman et al.* [2001], *Li et al.* [2004] and *Tufenkji and Elimelech* [2004a, 2005, 2005a]).

4.3 Materials and methods

4.3.1 Bacterial suspensions

Bacterial suspensions were prepared in a similar way as described in Chapter 5. Bacteria were suspended in demineralized water to concentrations ranging between 10^3 and 10^9 cells/mL. The EC value of the suspension was 1-3 μS/cm. In two experiments, bacteria were suspended in 1 and 10 mmol/L NaCl to assess the effect of ion strength on breakthrough.

4.3.2 Porous media

Ultra-pure quartz sand (J.T. Baker, Phillipsburg, NJ) with a diameter of 1 mm was ground with a mortar and sieved into 4 size fractions (38-45, 45-53, 75-90, and 180-212 μm). To remove impurities on the surface of the grains that might be the cause of variability in bacteria attachment, the sand was combusted at 800°C for 4 h. Then, the sand was acid-washed by soaking in 12 N HCl for 24 h, and rinsed with deionized water until the electrical conductivity of the rinse water was less than 1 μS/cm.

4.3.3 Column experiments and reversed flow experiments

Column experiments were conducted in borosilicate glass columns with an inner diameter of 2.5 cm (Omnifit, Cambridge, U.K.) with polyethylene frits (25 μm pore diameter) and one adjustable endpiece. The columns were packed wet with one of the quartz size fractions with vibration to minimize any layering or air entrapment. Column sediment length varied between 0.25 and 5 cm. Prior to each experiment, the packed column was equilibrated by pumping (Watson-Marlow 101U/R) 10 pore volumes of demineralized water through the column at a constant discharge rate of 6-7 mL/min. Usually, a suspension of *E. coli* with a concentration of ~ 4×10^7 cells/mL was flushed through the column for 10 minutes (loading phase) at a constant pore water flow velocity (~ 0.05 cm/s) followed by an 8-minute flush of *E. coli*-free demineralized water (deloading phase). Then, the flow direction was reversed and *E. coli*-free demineralized water was flushed through the column for 10 minutes and the effluent at the top end of the column was collected. The flow reversal experiments are based on the assumption that bacteria strained in a column are not attached to the collector and can therefore be easily dislodged. By reversing the flow at the end of a column experiment, we expected strained *E. coli* cells to be (partly) mobilized, which was a measure of the amount of strained *E. coli* cells. We

Experimental parameters								Mass balance		Modeled (HYDRUS 1D)							Calculated	
# of exp.	grain size	Duration (min)			C_0 (x10^7)	v	col. length	Effluent	Retained	R^2	k_a	Standard	k_{str}	Standard	B	Standard	η_0	α
	(μm)	load.	deload.	backfl.	(cells/mL)	(cm/s)	(cm)				(s^{-1})	error	(s^{-1})	error	(gram sed/cells)	error	(-)	(-)
1	180-212	10	8	10	3.86	0.0543	5	0.92	0.08	0.996	9.28E-04	6.97E-05	7.69E-03	7.76E-04	6.53E-07	6.93E-08	0.0031	0.130
2	180-212	10	8	10	4.67	0.0464	2	0.90	0.10	0.999	2.46E-03	3.82E-05	2.52E-02	7.05E-04	3.01E-07	9.18E-09	0.0032	0.381
3	180-212	10	8	10	4.61	0.0541	1	0.88	0.12	0.998	2.08E-03	1.50E-04	3.87E-02	4.91E-03	1.18E-07	5.23E-09	0.0031	0.293
4	180-212	10	8	10	17.80	0.0541	1	0.90	0.10	0.975	6.21E-03	4.84E-04	***9.94E-02***	2.82E-01	***2.58E-08***	2.25E-08	0.0031	0.875
5	180-212	10	8	10	120.24	0.0541	1	0.95	0.05	0.998	3.56E-03	2.53E-04	***7.60E-02***	6.76E-02	3.99E-09	1.04E-09	0.0031	0.502
6	180-212	3	3	10	4.74	0.0529	1	0.91	0.09									
7	180-212	10	8	-	4.18	0.0415	1											
8	180-212	10	8	-	4.25	0.0387	1											
9	75-90	10	8	10	4.32	0.0520	5	0.84	0.16	0.996	1.89E-03	6.22E-05	6.95E-03	6.15E-04	5.55E-07	5.72E-08	0.0089	0.042
10	75-90	10	8	10	4.54	0.0512	2	0.88	0.12	0.996	3.99E-03	1.08E-04	3.34E-02	1.72E-03	1.85E-07	9.80E-09	0.0089	0.090
11	75-90	10	8	10	6.45	0.0454	1	0.96	0.04	0.999	2.94E-03	3.85E-05	6.31E-02	8.88E-03	7.48E-08	3.26E-09	0.0092	0.072
12	75-90	10	8	10	13.87	0.0512	1	0.93	0.07	0.994	5.11E-03	2.28E-04	***6.60E-02***	1.56E-02	2.68E-08	1.44E-09	0.0089	0.115
13	75-90	10	8	10	58.95	0.0506	1	0.97	0.03	0.964	***4.47E-03***	4.16E-03	***5.79E-01***	5.33E+02	***3.19E-09***	2.38E-07	0.0079	0.115
14	75-90	3	3	10	4.69	0.0498	1	0.97	0.03									
15	75-90	10	8	-	4.20	0.0443	1											
16	75-90	10	8	-	4.21	0.0401	1											
17	45-53	10	8	10	4.22	0.0484	5	0.71	0.29	0.963	2.82E-03	1.07E-04	4.39E-03	8.69E-04	8.42E-08	5.59E-08	0.0170	0.022
18	45-53	10	8	10	4.63	0.0484	2	0.78	0.22	0.985	6.68E-03	2.07E-04	2.00E-02	1.80E-03	1.22E-07	1.17E-08	0.0170	0.052
19	45-53	10	8	10	4.34	0.0429	1	0.79	0.21	0.993	1.03E-02	3.38E-04	2.93E-02	2.16E-03	3.57E-08	2.46E-09	0.0175	0.088
20	45-53	10	8	10	4.83	0.0326	5	0.67	0.33	0.965	2.43E-03	3.81E-04	6.46E-03	8.37E-04	1.00E-07	2.54E-08	0.0187	0.026
21	45-53	10	8	10	4.46	0.0214	5	0.65	0.35	0.934	***2.05E-03***	1.10E-03	***3.57E-03***	8.37E-04	***1.29E-07***	1.28E-07	0.0208	0.029
22	45-53	10	8	10	0.0109	0.0415	5	0.66	0.34	0.782	***9.33E-04***	2.16E-04	***3.80E-03***	2.86E-03	***8.29E-05***	1.90E-04	0.0177	0.008
23	45-53	10	8	10	0.0011	0.0429	5	0.56	0.44	0.673			1.77E-02	1.98E-03	1.95E-05	1.01E-05	0.0175	
24	45-53	10	8	10	0.0001	0.0415	5	0.60	0.40	0.922			1.26E-02	7.89E-04	1.62E-04	6.38E-05	0.0177	
25	45-53	10	8	10	12.15	0.0505	1	0.86	0.14	0.993	8.09E-03	2.74E-04	3.85E-02	3.73E-03	1.94E-08	1.18E-09	0.0162	0.064
26	45-53	10	8	10	366.53	0.0456	1	0.89	0.11	0.733	***5.56E-03***	2.07E-03	***3.11E-02***	2.15E-02	6.09E-10	3.09E-10	0.0166	0.047
27	45-53	10	8	10	1.77	0.0456	0.25	0.92	0.08	0.947	1.80E-02	1.03E-03	1.76E-01	2.13E-02	8.16E-08	8.24E-09	0.0166	0.153
28	45-53	3	3	10	4.27	0.0461	1	0.84	0.16									
29	45-53	10	8	-	4.09	0.0429	1											
30	45-53	10	8	-	4.22	0.0429	1											
31	38-45	10	8	10	4.41	0.0462	5	0.44	0.56	0.984	5.02E-03	6.25E-04	1.36E-02	7.47E-04	3.87E-08	6.67E-09	0.0212	0.028
32	38-45	10	8	10	3.76	0.0475	2	0.75	0.25	0.980	7.54E-03	3.45E-04	1.99E-02	1.81E-03	6.16E-08	6.14E-09	0.0211	0.042
33	38-45	10	8	10	3.77	0.0435	1	0.73	0.27	0.995	1.43E-02	2.87E-04	4.09E-02	2.06E-03	3.28E-08	1.42E-09	0.0215	0.085
34	38-45	10	8	10	4.95	0.0312	5	0.58	0.42	0.973	3.49E-03	3.92E-04	8.89E-03	8.28E-04	6.67E-08	1.23E-08	0.0231	0.027
35	38-45	10	8	10	4.89	0.0190	5	0.46	0.54	0.985	***2.16E-03***	4.95E-04	6.80E-03	3.90E-04	7.22E-08	1.86E-08	0.0262	0.024
36	38-45	10	8	10	0.0130	0.0367	5	0.31	0.69	0.713			2.07E-02	1.30E-03	8.83E-07	5.48E-07	0.0223	
37	38-45	10	8	10	0.0001	0.0367	5	0.40	0.60	0.576			2.26E-02	1.94E-03	1.68E-04	7.20E-05	0.0223	
38	38-45	10	8	10	15.57	0.0475	1	0.81	0.19	0.995	1.18E-02	2.42E-04	4.32E-02	4.50E-03	1.79E-08	9.91E-10	0.0211	0.065
39	38-45	10	8	10	90.12	0.0478	1	0.83	0.17	0.988	1.05E-02	2.96E-04	***5.23E-02***	1.23E-02	4.17E-09	3.09E-10	0.0210	0.058
40	38-45	10	8	10	4.80	0.0448	0.25	0.86	0.14	0.960	2.28E-02	3.09E-03	1.89E-01	2.13E-02	2.09E-08	4.73E-09	0.0213	0.132
41	38-45	3	3	10	4.38	0.0462	1	0.69	0.31									
42	38-45	3	6	10	4.64	0.0446	1											
43	38-45	3	15	10	4.54	0.0412	1											
44	38-45	3	30	10	4.61	0.0394	1											
45	38-45	10	8	-	4.06	0.0435	1											
46	38-45	10	8	-	4.13	0.0435	1											

Table 1. *Overview of experimental conditions, modelled parameters, and calculated parameters. For the calculation of* η_0*, the following parameter values were assumed: temperature* $T = 293$ *K, bacteria diameter* $a_p = 1.43$ *μm, Hamaker constant* $H = 6.5x10^{-21}$ *J, particle density* $\rho_p = 1055$ kgm^{-3}*, fluid density* $\rho_{fl} = 1000$ kgm^{-3}*, and fluid viscosity* $\mu = 1.3x10^{-3}$ $kgsm^{-1}$*. Measured porosities were 0.439, 0.464, 0.491, and 0.500 (from coarse to fine fraction). Model output was insensitive to parameters (*k_a*,* k_{str}*, and* B*) given in bold and italics (see section "Sensitivity to model results" for explanation)*

therefore hypothesized that the elution of bacteria following flow reversal would provide information on the spatial distribution of previously strained cells.

In order to assess the effect of the duration of loading and deloading on the amount of strained bacteria mass, the durations were varied in a number of cases (experiments #6, 14, 28, 41-46 in Table 1). To assess the effect of concentration on the retained bacteria mass, the concentration of the influent bacteria suspension was varied between 10^3 and 10^9 cells/mL (exps. #4, 5, 12, 13, 22-26, and 36-39). In order to assess the effect of transport distance, column lengths were varied between 0.25 and 5.0 cm (exps. #1-3, 9-11, 17-19, 27, 31-33, 40). In order to assess the effect of velocity, pore water flow velocities were varied between 0.01 and 0.05 cm/s (exps. #17, 20-21, 31, 34-35). The influence of bacteria concentration and of pore water velocity was studied using the smallest grain size fractions, in which straining was expected to be most pronounced. The *E. coli* concentration was determined using optical density measurements (at 220 nm) with a UV-visible spectrophotometer (Cecil 1021, Cecil Instruments Inc., Cambridge, England). For the breakthrough curves, a 1 cm flow-through quartz cuvette was used, while the reversed flow effluent was collected in 12.5 mL polyethylene containers and then measured using a 1 cm quartz cuvette. Absolute cell numbers were deduced after calibration with plate counts. *E. coli* concentrations below 10^6 cells/mL were determined with plate counts (Chromocult™ agar; Merck, Whitehouse Station, NJ).

4.3.4 Column extrusion experiments

For a number of cases (in duplicate; exps. #7-8, 15-16, 29-30, and 45-46), flow was not reversed after the experiment, but packed beds were extruded from the column to release strained bacteria. Conditions for these experiments were similar to exps. #3, 11, 19, 33 (1 cm packed bed, $C_0 \sim 4\text{x}10^7$ cells/mL, $v \sim 0.05$ cm/s, all grain sizes). Immediately after the column experiment (= loading and deloading of *E. coli* cells), the bottom end piece of the column was removed without disturbing the packed bed, and the quartz grains were extruded. The packed bed remained saturated during the entire column extrusion process so as not to cause release of retained bacteria. The column was placed in a polyethylene container filled with 50 mL demineralized water. Then, for a 120-minute period, the container was occasionally firmly shaken by hand, and every 30 minutes (at t = 0, 30, 60, 90, 120 min) samples were taken and plated (4 plates per sample) on Chromocult™ agar. To account for the possibility of decay of the bacteria, the *E. coli* influent suspension used to load the column was also plated in quadruplicate every 30 minutes for 120 minutes. The amount of cells immediately released in the container was a measure for the number of cells strained in the column, while the development of the bacteria concentration in time was a measure for the net attachment and detachment of the bacteria cells.

4.3.5 Electrokinetic characterization of *E. coli* and cell size

A zeta-meter similar to the one made by *Neihof* [1969] was used. Movement of bacteria was visible on a video screen attached to a camera mounted on top of a light microscope (Olympus EHT) in phase contrast mode. Particle mobility values were obtained from measurements on at least 50 particles. Velocity measurements were used to calculate the zeta potential with the Smoluchowski-Helmholtz equation.

4.3.6 Model fitting and numerical modeling tools

In order to fit the model to the *E. coli* breakthrough data and to determine the parameters k_a, k_{str}, and S_{max}, we used the HYDRUS 1D computer code [*Simunek et al.*, 1998]. In HYDRUS, we employed two first-order deposition coefficients, one for attachment and one for straining. To simulate reductions in straining due to the filling of pores, we used the Langmuirian dynamics equation offered in HYDRUS. Because HYDRUS did not provide detailed information on retained bacteria concentrations as a function of transport distance and as a result from straining or attachment only, the combined mass balances of bacteria in the fluid (eq. 1) and retained by the sediment (eq. 2) were solved numerically with an explicit finite difference scheme that is forward in time, central in space for dispersion and upwind for advective transport. With each time step, first advective transport is calculated, then rate reactions and finally dispersive transport. Rates must be integrated over a time interval, which involves calculating the changes in bacteria concentrations both in the fluid and on the solid matrix. The scheme was prepared in a spreadsheet and is the same as that used in PHREEQC version 2 (*Parkhurst and Appelo* [1999]). The scheme was extensively tested, compared with analytical solutions, and then used by *Foppen et al.* [2005; Chapter 3]. Because of the flexibility of the spreadsheet, newly formulated mass balances for strained bacteria, which are not available in other model codes, could be easily implemented.

4.4 Results

4.4.1 Electrokinetic characterization of *E. coli*, cell size and stability of the *E. coli* suspension

The zeta potential of the *E. coli* in demineralized water was -165.5 mV (standard deviation 21.7 mV). The zeta potential of the quartz sand in demineralized water was -78.3 mV (standard deviation 6.0 mV). The average *E. coli* length was 2.27 μm and the average diameter was 0.90 μm. This corresponded to an equivalent spherical diameter [*Rijnaarts et al.*, 1993] of 1.43 μm. Both bacteria length and diameter did not change for the duration of the experiments.

4.4.2 Breakthrough curves

Some selected breakthrough curves are shown in Fig. 1. Breakthrough was rapid and without retardation. For the smaller grain sizes, breakthrough increased gradually and the retained bacteria mass increased with decreasing grain size. For the smallest two quartz grain sizes, we observed a sudden C/C_0-increase during initial deloading with demineralized water (e.g. Fig. 1a between pore volumes 6 and 7). Although we used demineralized water, the electrical conductivity of the bacteria suspension increased to 1-3 μS/cm during the experiments. When we increased the ionic strength of the bacteria suspension to 1 and 10 mmol/L NaCl, not only did the attachment increase, but the peaks at the end of the loading phase to the beginning of the deloading phase also disappeared (Fig. 1c). Presumably, at low ionic strength (1-3 μS/cm) attachment took place, but a minor decrease in ionic strength during deloading (the electrical conductivity of the demineralized water was 0.4 μS/cm)

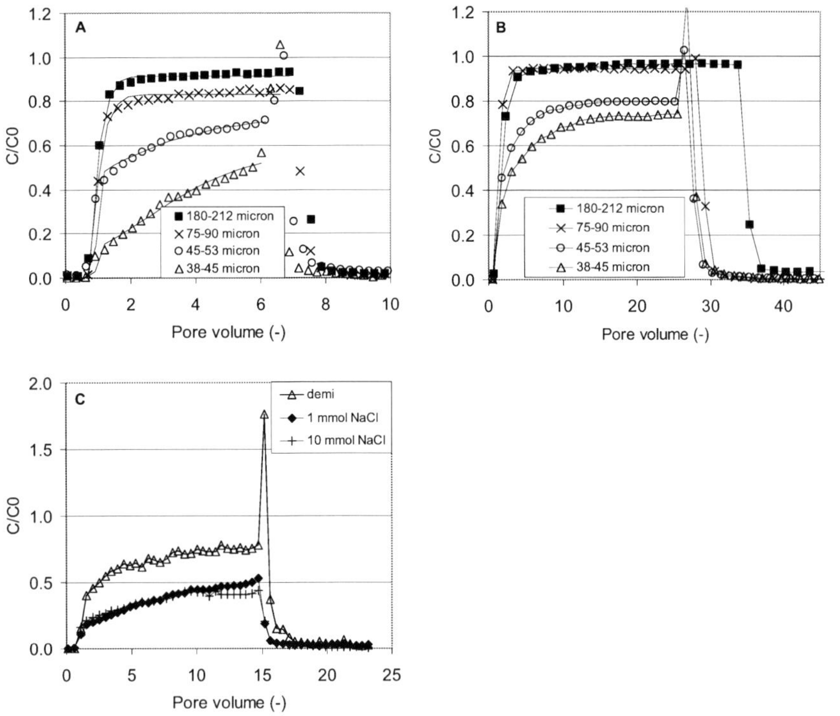

Fig. 1: Breakthrough curves for the four quartz sand fractions for 5-cm columns (a. exps. #1, 9, 17, and 31; see Table 1) and 1-cm columns (b. exps. #3, 11, 19, and 33). Lines in (a) are curves fitted with HYDRUS. Figure 1c: Breakthrough curves for 2-cm columns of 38-45 μm quartz sand for bacteria suspensions (~$4x10^7$ cells/mL; pore water flow velocity: ~ 0.05 cm/s) of demi water, 1 mmol/L NaCl and 10 mmol/L NaCl.

was enough to detach some of these cells. This ionic strength decrease became completely insignificant at higher ionic strengths (1-10 mmol/L NaCl) and therefore the peaks with associated detachment disappeared. Total retained bacteria mass ranged from 0.03 to 0.16 (fraction of total mass applied to the column) for the two largest grain size fractions (see Table 1; Mass balance section). For the two smallest fractions, retained masses were as high as 0.69 of the total mass applied to the column, depending on column length, pore water flow velocity, and influent bacteria concentration. Finally, after deloading, *E. coli* concentrations decreased rapidly and from this lack of "tailing" we concluded that detachment was not important.

4.4.3 Nature and occurrence of straining, attachment and detachment

From those columns that were extruded and placed in 50 mL demineralized water, there was an immediate release of bacteria that remained constant for the entire duration of the experiment (Fig. 2). We considered these immediately released cells as having been strained; they were present in the column without being attached. The concentration of the effluent suspension did not decrease (diamonds in Fig. 2a and b), and so there was no decay of bacteria cells. Therefore, decay in the container

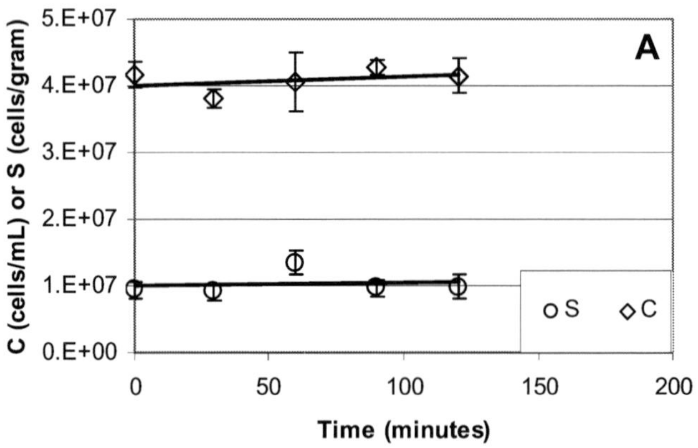

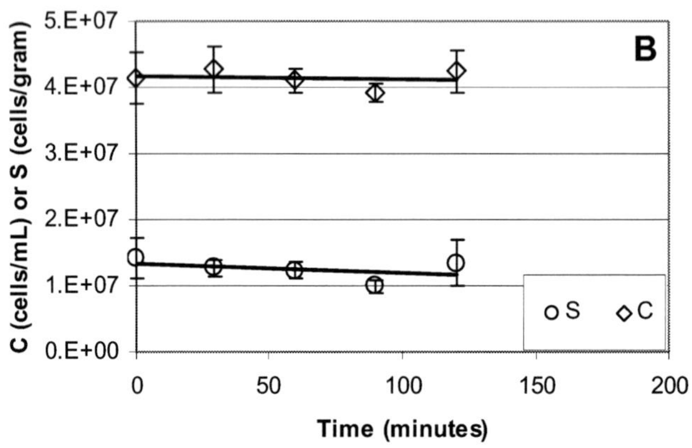

Fig. 2: Results of the extrusion experiments (a. exp. #29; b. exp. #45). S *was calculated by dividing the total number of cells in the container with demineralized water by the dry weight of the sediment in the container. The concentrations of the E. coli influent suspension used to load the column (* C *) are also given*

with water and sediment was also likely to be insignificant and there was also hardly any net detachment or attachment. This lack of detachment confirmed the relative unimportance of the detachment process observed from the lack of tailing in the breakthrough curves (see previous section). The lack of attachment contrasted with the 15-21% of the total bacteria mass applied to the columns that had attached to the quartz grains while being flushed through the columns. This 15-21% was determined from the fraction not recovered during the extrusion experiments (Table 2, right column). Apparently, conditions for attachment to the quartz sand in the column differed from those in the container.

Table 2. Mass balance of bacteria during the column extrusion experiments. Effluent = bacteria mass in the effluent (as fraction of total bacteria mass applied to the column); released = fraction of total bacteria mass released in polyethylene container; retained = not recovered fraction = 1 - effluent – released.

					Fraction	
Exp. #	grain size	C_0 (x10^7)	col. length	Effluent	Released	Retained
(from Table 1)	(μm)	(cells/mL)	(cm)	-	-	1-Effl.-Rel.
29	45-53	4.09	1	0.79	0.04	0.18
30	45-53	4.22	1	0.82	0.03	0.15
45	38-45	4.06	1	0.74	0.05	0.21
46	38-45	4.13	1	0.79	0.04	0.17

It was also remarkable that, despite the rather high negatively charged surfaces of both quartz and *E. coli* (very low zeta-potential values; see previous sections), a considerable amount of bacteria still appeared to be able to attach.
As soon as the flow was reversed, bacteria began to be released; the release ceased within 1-1.5 pore volumes (Fig. 3). Because the release process always lasted around 1 pore volume for all experiments, and because detachment was negligible, we concluded that these released cells were strained and must have been present throughout the column, not just at the column inlet. However, the total amount of cells released from flow reversal did not equal the total amount of strained cells. We believe this is because straining and attachment very probably also took place when the flow was reversed, thereby reducing the total amount of cells released. The reduction in amount must have been least for the cells released immediately during reverse-flushing, and most for the cells that had to travel back through the entire column. Despite this pore-volume-dependent reduction of backwashed and released cells, the decrease of the normalized concentrations during the first 1-1.5 pore volume of flow reversal was still linear on a logarithmic scale (Fig. 4a and b). This finding justified neglecting a depth-dependent power law in our model approach, because such a power law may describe straining as a process that occurs primarily at the column inlet and then decreases hyper-exponentially [*Bradford et al.*, 2003], instead of decreasing log-linearly as a function of transport distance. The log-linear decrease we observed could not have resulted from strained cells distributed in the column according to a depth-dependent power law.
When the duration of deloading was increased from 3 to 30 minutes (exps. #43-46; Fig. 5a), the ratio of the total amount of bacteria in the reversed flow fraction over the total amount of bacteria applied to the column decreased from 0.12 to 0.07. When data from experiments of 10 minutes loading and 8 minutes deloading were compared with data from 3 minutes loading and 3 minutes deloading for the various quartz grain sizes (Fig. 5b), the amount of bacteria in the reversed flow fraction over the total amount of bacteria applied to the column was 1.89 times higher for the shorter experiments. Since we assumed that the amount of bacteria in the reversed flow fraction was a measure of the amount of strained bacteria (see above), we concluded that the amount of strained bacteria was indeed influenced by the duration of the experiment. Apparently, bacteria that were initially retained in the column due to straining rapidly started to attach to the surface of the quartz grains.

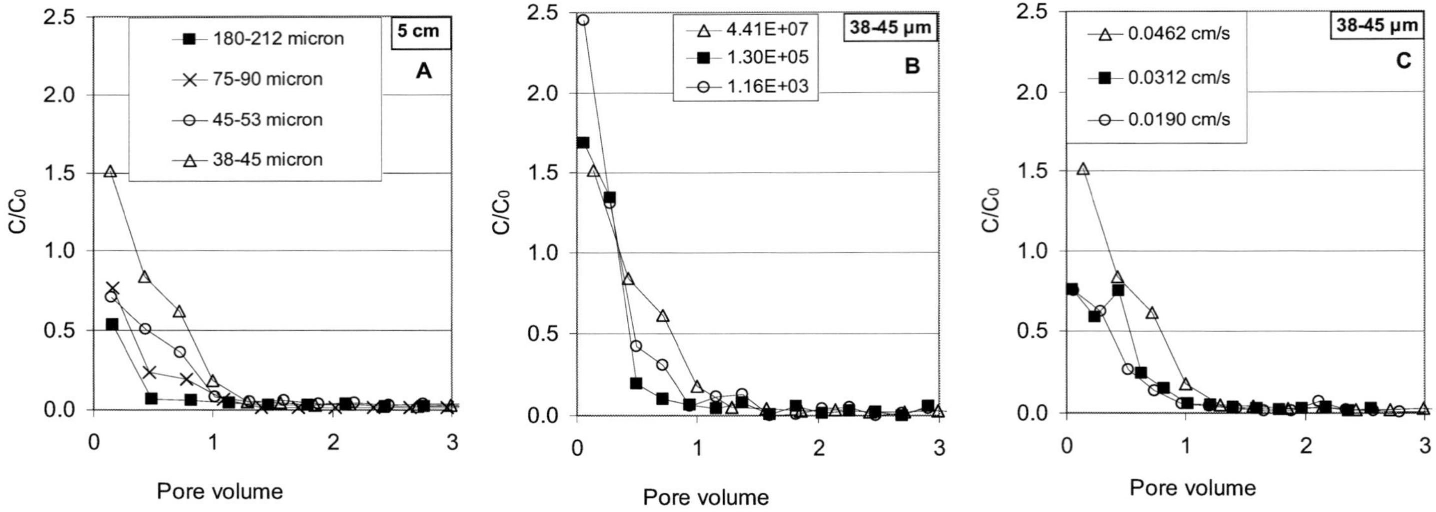

Fig. 3: E. coli released (expressed as fraction of the influent bacteria mass) as a result of reversing the flow direction for the 5-cm columns for the various quartz grain size fractions (a), for the 5-cm columns of the 38-45 µm quartz grains for various influent concentrations (b), and pore water flow velocities (c).

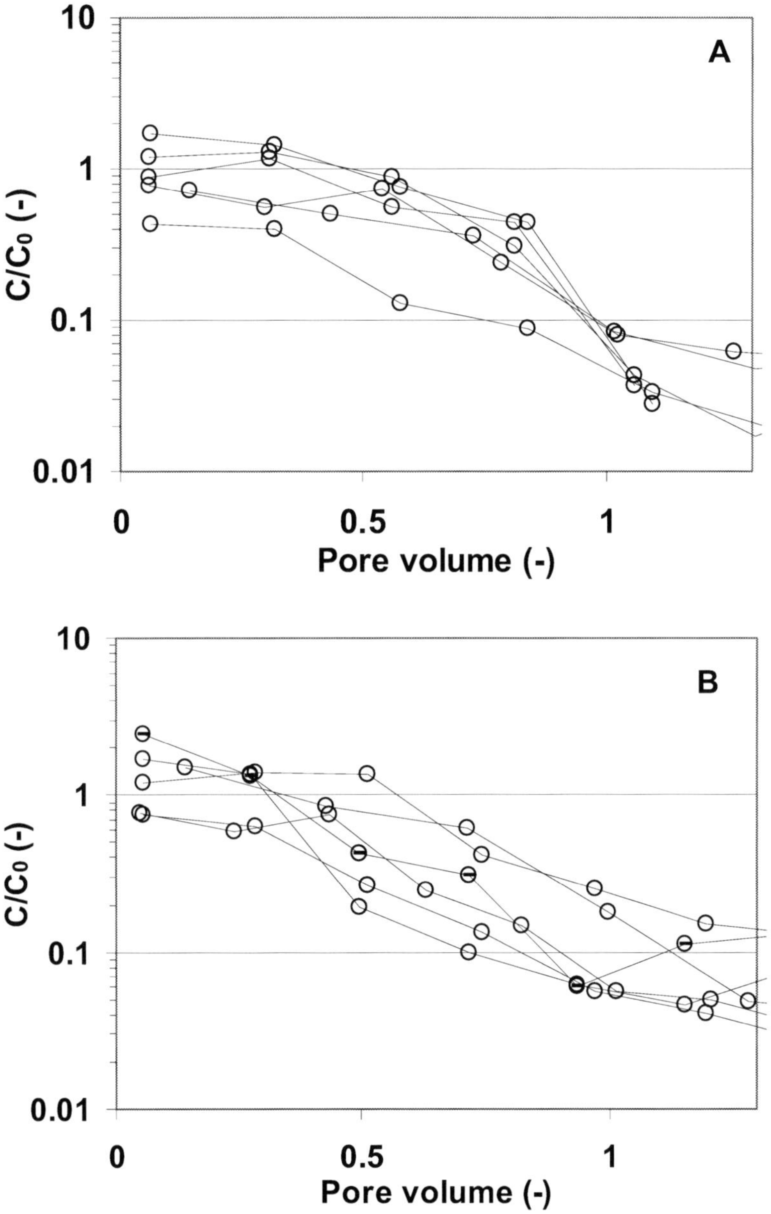

Fig. 4: Normalized concentrations on logarithmic scale of the reverse flow experiments (a: 45-53 μm; b: 38-45 μm) during the first 1.3 pore volume of backflushing

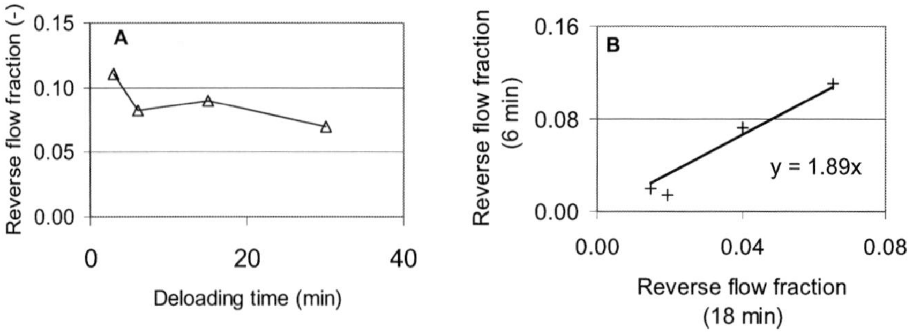

Fig. 5: Reversed flow fraction (= mass of bacteria as fraction of the total mass applied to the column) as a function of deloading time in the 38-45 μm quartz sand (a. exps. #41-44), and the relation between the reversed flow fraction (= mass of bacteria as fraction of the total mass applied to the column) of the 6-minute experiments and the reversed flow fraction of the 18-minute experiments (b. exp. #3 vs. #6, #11 vs. #14, #19 vs. #28, #33 vs. #41).

4.4.4 Model fitting

An example of the increase of cumulative amounts of bacteria (flowing into the column at the top, out of the column at the bottom, retained due to attachment and due to straining) versus time, calculated with HYDRUS 1D, is given in Fig. 6. The figure clearly shows the leveling off of the cumulative amount of strained bacteria, due to the reduction of k_{str}. This reduction of k_{str} to a very low value was characteristic for many experiments, especially for the two coarsest quartz grain fractions. Results of the fitting process with HYDRUS-1D are summarized in Fig. 7a-i. Since detachment was negligible, we assumed k_r to be zero, and we only considered the rising limb of the breakthrough curves. The fits between modelled and measured breakthrough data were high in most cases (R^2>0.9; see Table 1; a number of fitted curves are given in Fig. 1a.). Less good fits (R^2=0.576-0.922) were obtained for the low concentration experiments (< 10^6 cells/mL; exps. #22-24, 36, and 37 in Table 1). Due to the method of determining these low *E. coli* concentrations (by plating), breakthrough data were more scattered than for the high concentration experiments in which the flow-through cuvette in the spectrophotometer was used. For some of these experiments (#23, 24, 36, and 37), k_a was zero, which was probably related to the scatter in the breakthrough data. Apparently, minimization of the objective function in HYDRUS-1D did not necessitate the use of two rate coefficients: one (k_{str}) was enough.

B increased with increasing grain size and was 3-4 times higher in the 38-45 μm fraction than in the 180-212 μm fraction (Fig. 7a). We expected less strained bacteria upon increasing grain size, nevertheless, the amount of strained cells in the 180-212 μm fraction was still substantial. In addition, B increased when column lengths increased: on average, B was 3 times higher for the 5-cm columns than for the 1 and 2-cm columns. Furthermore, B was almost linearly inversely dependent on the influent concentration (Fig. 7b), ranging from 10^3 to more than 10^9 cells /mL.

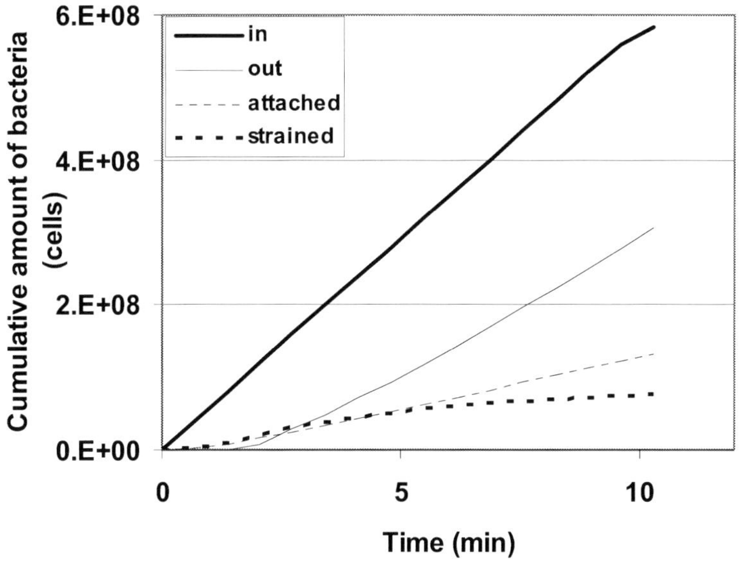

Fig. 6: Increase of cumulative amounts of bacteria (flowing into the column at the top, out of the column at the bottom, retained due to attachment and to straining) versus time, calculated with HYDRUS 1D for exp #17 (Table 1).

Compared to grain size and pore water flow velocity, the dependency of B on influent concentration seemed the most important relation. When pore water flow velocity increased 2-3 fold, B remained constant (Fig. 7c). Standard errors of estimate (error bars) were relatively high for the fitted B values in Fig. 7c, which indicated that model output in those cases was not very sensitive to variations in B (see also next paragraph).

The value of k_a depended not only on grain size (Fig. 7d), but also varied with column length. The calculated sticking efficiencies (eq. (5)) increased with decreasing column length (Fig. 8) from 0.022 for the 5-cm columns to 0.153 for the 0.25-cm columns. These sticking efficiency reductions with increasing column length were a clear indication of the non log-linear attachment of the *E. coli* we used, as was also observed by *Simoni et al.* [1998], *Baygents et al.* [1998], *Redman et al.* [2001], *Li et al.* [2004] and *Tufenkji and Elimelech* [2004a, 2005, 2005a]. At the same time, k_{str} reduced with increasing column length (Fig. 7g), but here the reductions were more pronounced than for k_a, while the variations with grain size were not as obvious as for k_a. Especially for the 2- and 5-cm columns, k_{str} hardly varied with grain size. Upon increasing influent concentration, k_a remained very low (Fig. 7e) and suffered from the scatter in breakthrough data (see above), while k_{str} decreased from 0.02 to 0.01 (Fig. 7h). Finally, both k_a and k_{str} increased with increasing pore water flow velocity (Fig. 7f and 7i), although the increase was minor.

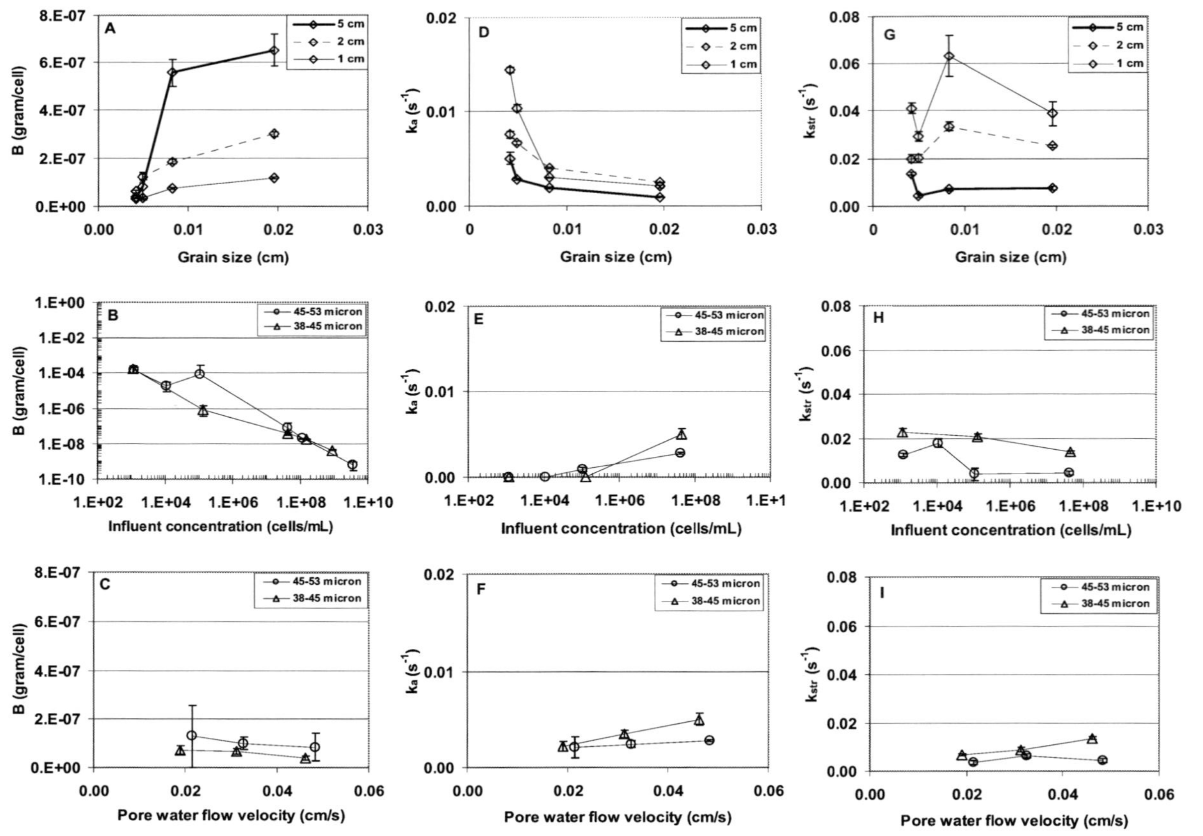

Fig. 7: Relation between modelled parameters B (a, b, and c), k_a (d, e, and f) and k_{str} (g, h, and i) as a function of grain size (upper graphs), influent bacteria concentration, and pore water flow velocity (lower graphs). One error bar indicates the standard error of estimate

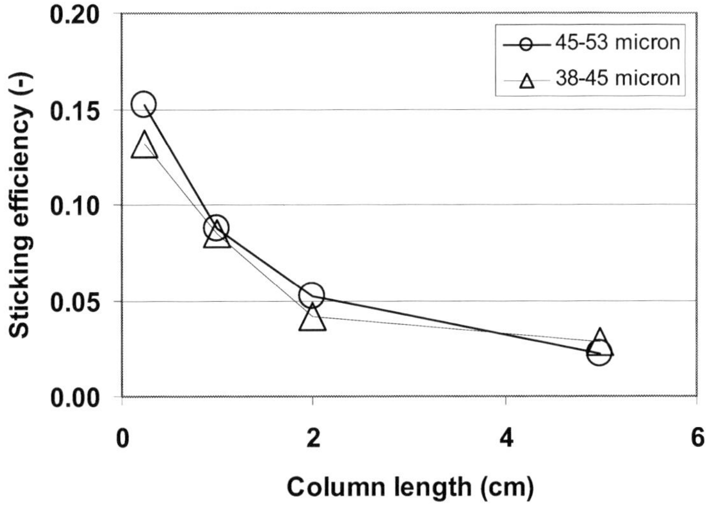

Fig. 8: Relation between transport distance and sticking efficiency, α

4.4.5 Sensitivity of the model results to parameter values

In order to obtain more information about the sensitivity of the model results to parameter values, we determined the standard error of estimate ($s_{Y'}$), and the T-value (*Daniel and Wood* [1971]; *Simunek et al.* [1998]). In general, the $s_{Y'}$ values, given as error bars in Fig. 7a-i (values are given in Table 1), were small. This indicates that the model results were sensitive to variations in the parameters modelled. The T-value is obtained from the mean (ψ) and standard error of estimate using the equation (*Simunek et al.* [1998])

$$T-value = \frac{\psi}{s_{Y'}} \tag{6}$$

T-values provide an absolute measure of the deviations around the mean. Higher values indicate that the scatter around the mean is relatively small, and that modelled values are more sensitive to the fitted parameter value than lower T-values. In general, our T-values ranged from 1-100 for the 3 fitted parameters (Fig. 9a and b), which indicated that the standard error of estimate ranged from relatively low (high T-value, sensitive model results) to relatively high (low T-value, insensitive model results). The T-values of the modelled k_a were highest, followed by those of k_{str}, and then B. If the T-values of k_a and k_{str} were larger than or equal to 5, we

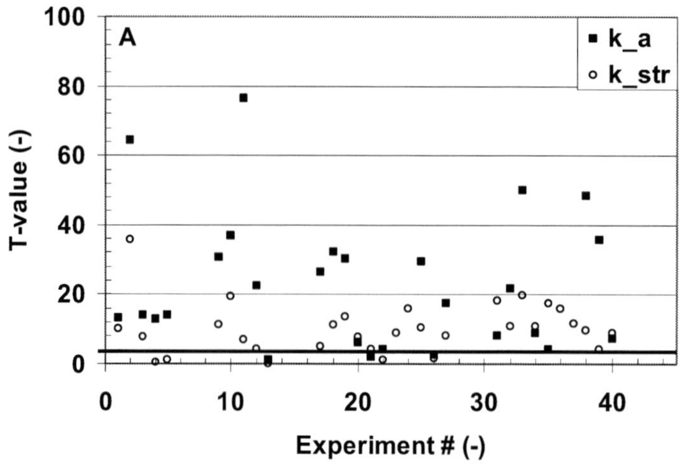

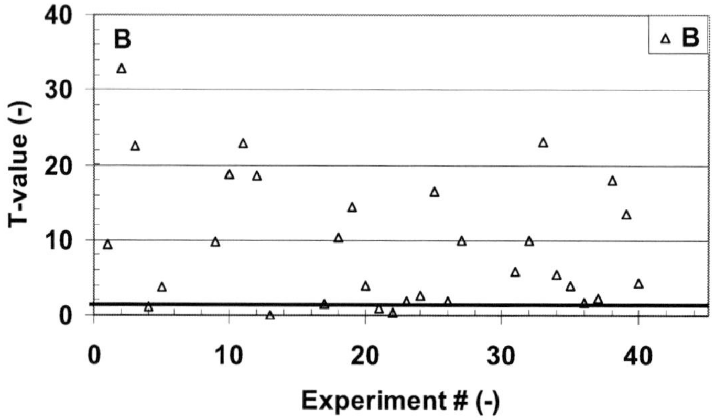

Figure 9. The T-value statistic to evaluate the sensitivity of the model output for parameter values determined with HYDRUS (k_a , k_{str} , and B). Experiment numbers on the x-axis correspond with Table 1. Bold lines indicate the lowest acceptable T-values.

considered the model results to be sensitive to the parameter value. If the T-values of B were larger than or equal to 1.5, we considered the model results to be sensitive to this fitted parameter value. We selected a lower cut-off T-value for B , because the B values ranged over more than 6 log units, and therefore could still yield a model that was significantly sensitive to the fitted parameter value, despite relatively high standard errors of estimate. As a result of these lower T-value cut-offs (bold horizontal lines in Fig. 9a and b), a number of experiments yielded models that were not sensitive to the fitted parameter values. Table 1 shows these parameter values in bold and italics. From the Table, it can be concluded that the insensitivities were mainly related to the straining rate parameter in the experiments with very high concentrations of *E. coli* in the influent ($C_0 > 10^8$ cells/mL; exp. # 4, 5, 12, 13, 26, and 39). It seems that these experiments produced breakthrough curves that were mainly determined by attachment, while straining was hardly visible. In addition, other insensitive model outputs were found to be correlated with attachment rate coefficient values of the lowest velocity experiments (# 21 and 35) and one low concentration experiment (Exp. # 22). In light of the insensitive model output, in the following sections we have not considered the calculated parameter values of experiments 4, 5, 12, 13, 21, 22, 26, 35, and 39.

4.4.6 Comparing modelled and measured values

Modelled S_{str} - values from the entire 1 cm column experiments (exp. # 3, 11, 19, and 33) were compared with measured S_{str} -values from the extrusion experiments, which were multiplied with a factor of 1.89 to account for the duration of the experiment (see above and Fig. 5). Measured S_{str} -values (open circles in Fig. 10) depended on grain size, similar to the modelled S_{str} -values. However, measured values were 10-50% lower than the modelled S_{str} -values. We attributed this to the extra time (a few minutes) that was required to extrude the quartz grains out of the column, and in which strained bacteria were able to attach. In addition, due to the extrusion of the quartz grains, the packed bed was slightly compressed, despite efforts to minimize this. Such compression might have resulted in additional attachment of strained bacteria.

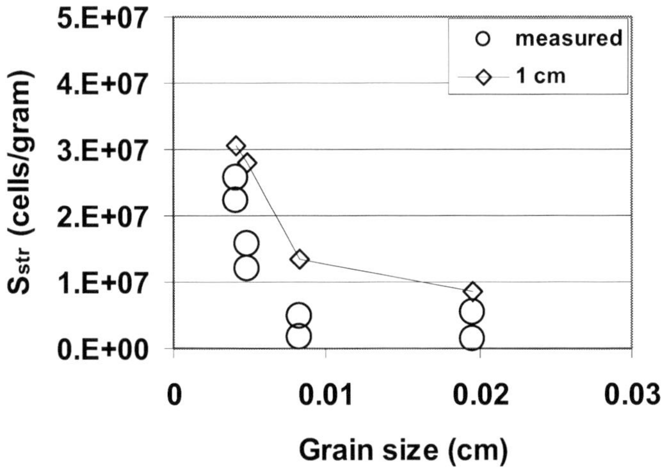

Fig. 10: Modelled S_{str} values versus S_{str} values determined from the extrusion experiments

Modelled S_{str} -values were also compared with measured S_{str} -values from the flow reversal experiments. The modelled S_{str} -values were determined with the spreadsheet model (see "Model fitting and numerical modeling tools" in the Methods section), because HYDRUS did not provide detailed information on retained bacteria concentrations as a function of transport distance and as a result from straining only. Therefore, parameter values determined from fitting with HYDRUS, were used in the spreadsheet model, and then modelled strained bacteria concentrations in the upper 1 cm of the column were determined and expressed per gram sediment. The measured S_{str}-values were determined by:

$$S_{str} \approx C_{reverse,initial} * \frac{\theta}{\rho_{bulk}} \tag{7}$$

where $C_{reverse,initial}$ was the initial measured concentration in the reversed flow effluent (cells/mL). By taking only the concentration initially measured during flow reversal, we minimized the effects of straining and attachment, which most likely also took place upon reversing the flow. In addition, we only used measured data from the 2- and 5-cm columns, because the pore volume of the 1-cm columns was less than the sampling volume, which caused dilution of the released bacteria concentration, thereby reducing the measured S_{str}-values. Values were also multiplied with 1.89 to account for the duration effect (see above and Fig. 5) and then log-transformed to equally include low concentration experiments.

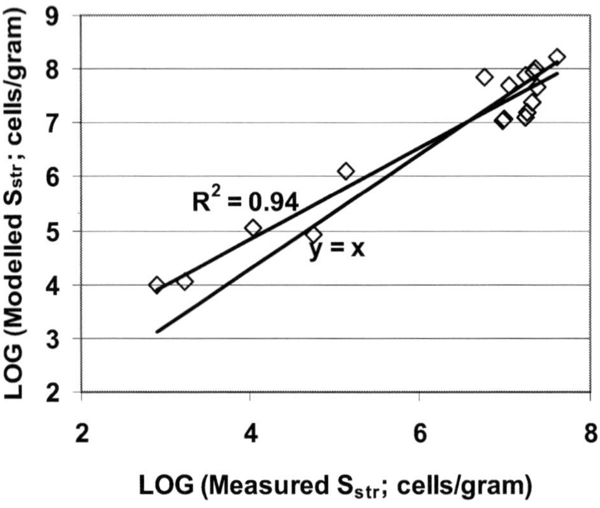

Fig. 11: Modelled S_2 values versus S_2 values determined from the flow reversal experiments

The best fit line ($R^2 = 0.94$; Fig. 11) deviated from the ideal line (y = x), especially for the low concentrations, for which modelled S_{str} values were overestimated 0.5-1.0 log-unit compared to measured values. We attributed this to the method of determining these low *E. coli* concentrations (by plating) that caused more scattering in the breakthrough data than for the high concentration experiments, and that resulted in very low values for the modelled attachment rate coefficient, k_a. As a consequence, the modelled S_{str}-values for these experiments appeared to be high compared to the measured S_{str}-values.

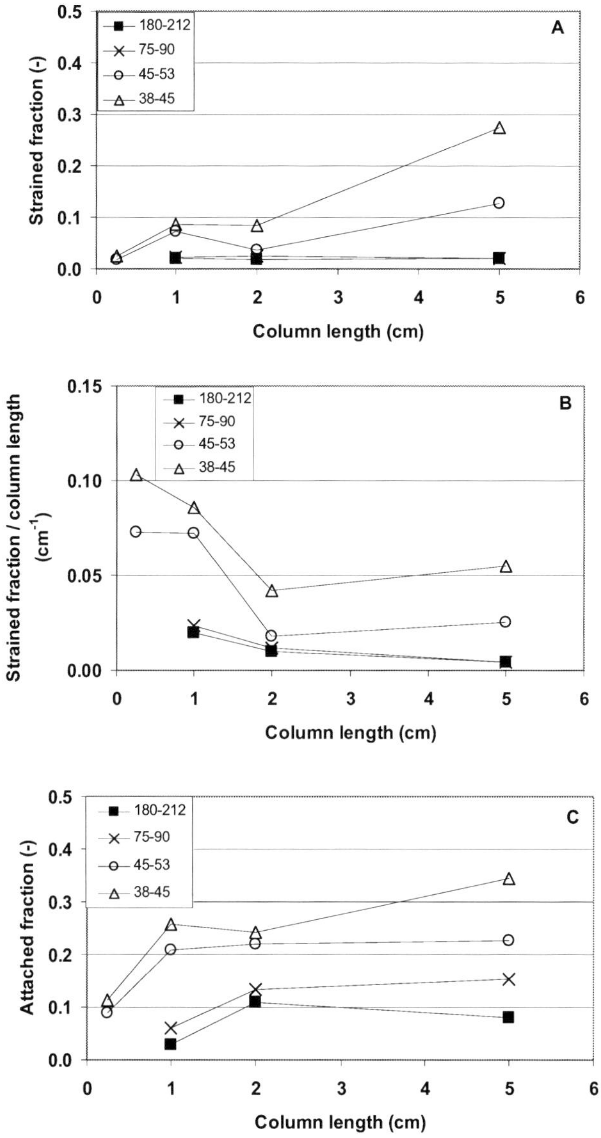

Figure 12. Strained bacteria, as a fraction of total cells applied to the column (a), strained fraction, expressed per cm column length (b) and attached fraction (c), as a function of column length. Amounts of strained and attached bacteria were determined with HYDRUS 1D.

4.4.7 Effect of transport distance

Total amounts of strained and attached bacteria obtained from modeling for the various column lengths were expressed as a fraction of the total bacteria mass applied to the column (Fig. 12a, b and c). The strained fraction was roughly linearly dependent on transport distance. When the strained fraction was divided by the column length (Fig. 12b), it was clear that the relative contribution to straining in the first cm of the column was higher than at greater distances from the column inlet. A similar pattern was observed for the attached fraction (Fig. 12c). Attachment was maximal in the first 1-2 cm of the column and then, at greater distances from the column inlet, attachment dropped almost to zero, since attached fractions in the figure are seen to have remained constant.

4.4.8 The fitting parameter B and the mass balance of strained bacteria

In our experiments, k_{str} was linearly dependent on the inverse of the column length, $\frac{1}{x}$, with a constant ratio of 0.045 ($R^2 = 0.98$; Fig. 13a). We did not attempt to distinguish between the four grain size fractions, although the ratio for the smaller grain size fractions was a fraction higher than the ratio for the larger fractions. Furthermore, B was inversely dependent on C_0, and exponentially dependent on the column length, L, and on a_c. The product of these three parameters is given on the x-axis of Fig. 13b (values have been log-transformed to be able to compare high and low concentration experiments)to equally include lower concentration experiments). When a_c was raised to the power of -1.20, we obtained the best fit ($R^2 = 0.93$) between B and the product $C_0 x^{-1} a_c^{-1.20}$; the slope of the fitted line was -0.717. Therefore, B was best described by:

$$B = [C_0 L^{-1} a_c^{-1.20}]^{-0.717} \tag{8}$$

Inserting eq. (8) in the mass balance equation of strained bacteria (eq. (4)), whereby B was expressed as a function of (C,x) instead of C_0 and L, and k_{str} was expressed as a function of the variable x instead of the constant L, yielded:

$$\frac{\partial S_2}{\partial t} = \frac{\theta}{\rho_{bulk}} \frac{0.045}{x} (C - C^{1-a} x^a a_c^{ab} S_{str}) \tag{9}$$

whereby in our case $a = 0.717$ and $b = 1.20$. So, instead of the Langmuirian type of site saturation given in eq. (4), the newly derived mass balance for strained bacteria, eq. (9), was dependent on the transport distance, because the straining rate coefficient decreased with increasing column length, and because B increased with increasing column length. In addition, the filling of sites was dependent on C^{1-a} instead of C as in eq. (4). So, for a set of experiments with varying concentrations and constant term a_c^{ab}, $\frac{\partial S_{str}}{\partial t}$ was smaller for the C^{1-a} case than for the C case, and therefore S_{str} increased more slowly than would be the case if the site saturation occurred in a typical Langmuirian way. Thus, the straining that

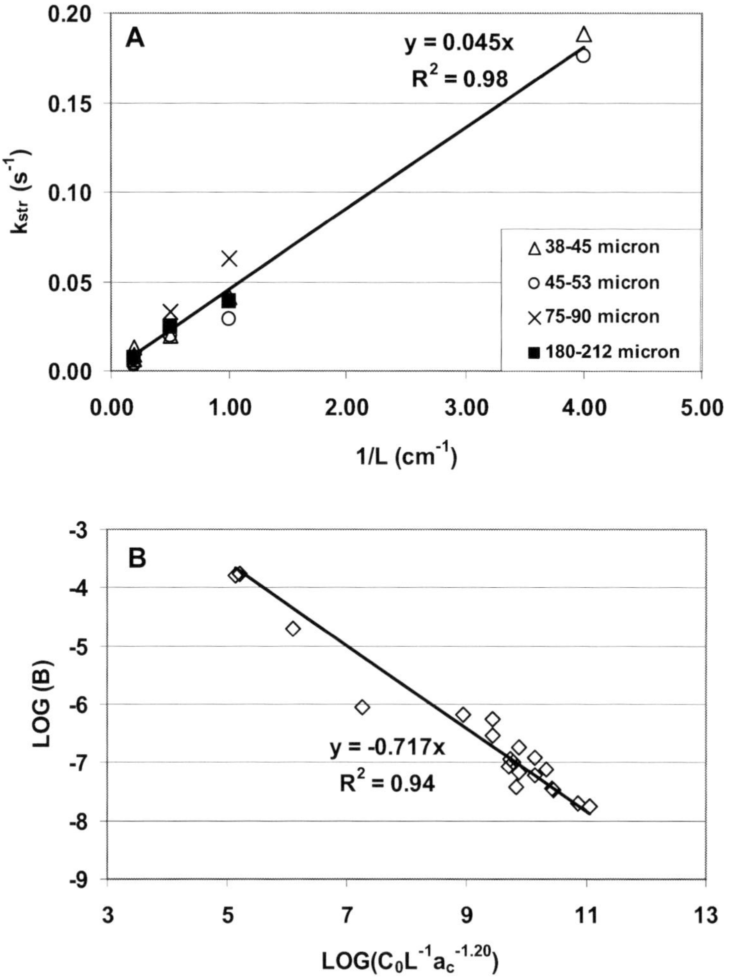

Fig. 13: Straining rate coefficient, k_{str}*, as a function of* $\frac{1}{x}$ *(a), and* B *as a function of the product* $C_0L^{-1}a_c^{-1.20}$ *(b).*

occurred in our experiments was more important than straining with a Langmuirian way of site saturation. To demonstrate the effects of this modified Langmuir type of site saturation of eq. (9) on the breakthrough of *E. coli*, using the spreadsheet, we numerically solved the combined mass balances of bacteria in the fluid (eq. 1) and retained by the sediment (eq. 2, whereby $\frac{\partial S_{str}}{\partial t}$ was given by eq. (9)). For the example, attachment was assumed to be zero. For comparison, the typical Langmuir model, similar to eq. (4), and written as:

$$\frac{\partial S_2}{\partial t} = \frac{\theta}{\rho_{bulk}} \frac{0.045}{L} (C - CL^a a_c^{ab} S_{str}) \qquad (10)$$

was modelled separately, also with the spreadsheet. For each mass balance (eq. (9) or (10)), the influent concentration was varied between 10^3 and 10^9 cells/mL, while the column length was kept constant at 5 cm. Then, the influent concentration was kept at 10^3 cells/mL, while the transport distance was varied between 5 and 500 cm. The parameters used in eq. (9) are given in Table 3 for the case of $C_{inf} = 10^3\ cells / mL$ (Fig. 14a; C_inf= 1E3 cells/mL).

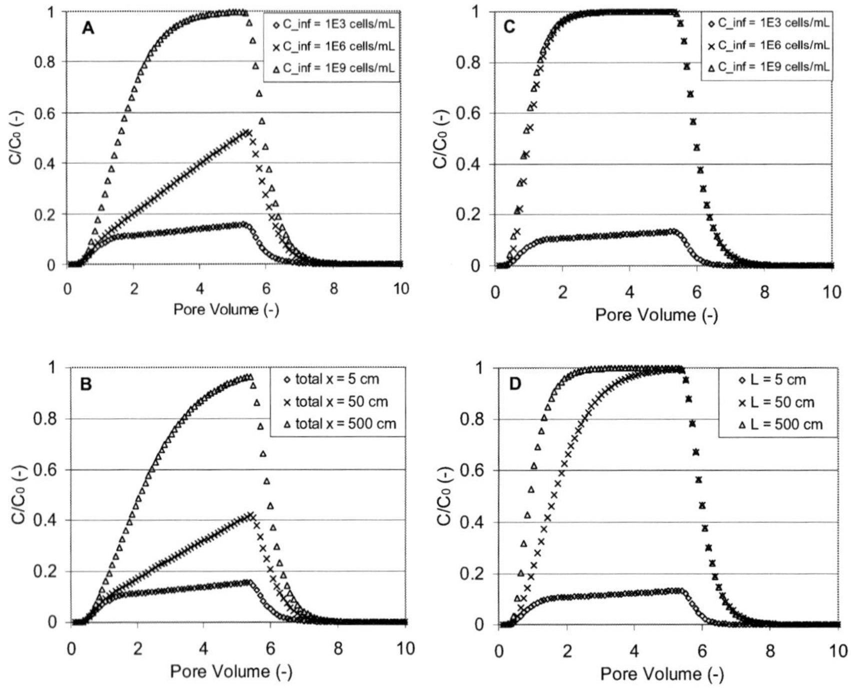

Figure 14. Breakthrough curves (a and b) due to straining defined with the modified Langmuir mass balance differential equation (eq. (9)) and due to straining defined with the Langmuir mass balance differential equation (eq. (10); c and d). Values used are given in Table 3

The parameters used in eq. (10) (Fig. 14c and d) are similar to those in Table 3, except for the value of the fitting parameter, a. We assigned a the value of 2.0, instead of 0.717, because for a good comparison, the breakthrough curves obtained with eq. (10) had to closely resemble breakthrough curves obtained with eq. (9). Comparing Fig. 14a with Fig. 14c did indeed reveal very similar breakthrough curves for the case of 10^3 cells/mL. The major difference was that the normalized concentrations of the 10^6 and 10^9 cells/mL cases of Fig. 14c almost increased to 1 within 2 pore volumes, while this took at least 4 pore volumes for the same cases determined with eq. (9) (Fig. 14a). A similar pattern was observed when the transport distance was increased to 500 cm (Fig. 14b and 14d). Straining site saturation according to the Langmuirian model (eq. (10)) occurred much faster than site saturation according to the modified Langmuir model (eq. (9)), and, as a result, the normalized concentrations in Fig. 14 reached unity much faster.

Table 3. Parameters used for the mass balance calculations in the spreadsheet model. The breakthrough curves are given in Fig. 14

GENERAL FLUID CONDITIONS			**BACTERIA PARAMETERS**		
Temperature	283	K	Density of bact.	1.05	g/cm^3
Density of water	1	g/cm^3	Diameter	1	micron
TRANSPORT PARAMETERS			**STRAINING PARAMETERS**		
Total distance	5	cm	a	0.717	-
Total time	1200	s	b	1.2	-
Velocity (pore water)	5.00E-02	cm/s	Straining rate coeff.	4.50E-02	1/cm
Dispersivity	0.5	cm			
Diffusion coeff.	1.00E-05	cm2/s			
SEDIMENT PARAMETERS					
Porosity	0.4	-			
Bulk density	1.4	g/cm^3			
Grain diameter	40	micron			

4.5 Discussion

The model we used, consisting of two rate coefficients, one of which (the straining rate coefficient) reduced in a modified Langmuirian way due to site saturation or pore space filling, determined both attachment and straining accurately. Firstly, the coefficients of determination between measured and modelled breakthrough data were very high (usually > 0.95), as were the T-values. Secondly, both strained and attached bacteria masses modelled agreed well with the measured strained and attached bacteria masses determined in various ways. Thirdly, neither detachment nor decay were important. We concluded that the mass balance equations for retained bacteria we proposed (eq. 3 and 9) could account for all our observations. Another finding was that in our columns, regardless of grain size, the sticking efficiency decreased with increasing column height. This decrease confirmed the results of *Simoni et al.* [1998], *Baygents et al.* [1998], *Redman et al.* [2001], *Li et al.* [2004] and *Tufenkji and Elimelech* [2004a, 2005, 2005a], who found that the attachment rate coefficient decreased with distance from the column inlet. The indirect evidence for an attachment rate decrease was the standard deviation of the zeta potential (21.7 mV) of *E. coli*, which was a measure of the variation in the surface charge density. It is these variations in charge that might be responsible for the variations seen in experimentally determined sticking efficiencies (*Dong* [2002]). The measured decrease in sticking efficiency also justified our decision to allow the attachment rate coefficient to vary with distance from the column inlet. We defined the cells recovered from the flow reversal experiments and from the extrusion experiments as being strained. However, although these cells were retained in the column without being attached, such retention is not *per se* limited solely to straining alone, i.e. to instances where cells cannot enter pores because the pore throats or pore entrances are too small. Apart from straining, the retention of non-attached cells in the absence of adhesive forces might have included retention in stagnation zones behind sand grains, or retention of cells not in direct contact with the collector surface in localized pockets near the grain surface, as demonstrated by the collision images of *Keller and Auset* [2006]. We did not consider this to be a major problem, since classical colloid filtration theory does not include retention processes that do not lead to attachment to the collector surface. Our results indicate that non-adhesive retention was considerable, and should be included in mass-balance considerations.

Although we were able to develop a fairly simple and straightforward mass balance for the straining process (eq. 9) that can be used for many applications involving the transport of *E. coli* in angular, saturated sand, the physical implications of the fitting parameters (with values 0.045, 0.717, and 1.20) is still unresolved and needs attention in future research. The first fitting parameter (0.045) is probably a measure of the amount of colloids retained per unit of time. If the value increases, then more straining takes place; we would expect this to happen if the colloids are larger than the ones we used. The second and third parameters (0.717 and 1.20) are probably a measure of the pore space available for straining in the sediment.
With regard to straining, our study contrasts with some of the findings of *Bradford et al.* [2002, 2003, 2005] and *Bradford and Bettahar* [2005, 2006]: we found straining was not confined to the column inlet, but took place over the entire column length. However, like Bradford, we also found there was more strained material near the column inlet than further away. Whereas in Bradford's studies, straining diminished exponentially with distance, in our case the diminution was almost linear. It seems likely that colloids near the column inlet have better access to small pore spaces, which is why more straining takes place.
Our study is an extension of earlier work [*Chapter 5*]. From those experiments, all carried out at very high bacteria concentrations (10^8 and 10^9 cells/mL) with another type of quartz sand (60-200 μm) than the one used in the current study, it was concluded that the pore space available for straining was constant. In the present study we found that for concentrations between 10^3 and $5x10^7$ cells/mL, the pore space available for straining depended heavily on the concentration of bacteria in the influent. However, for the high concentration experiments, models fitted with HYDRUS were barely sensitive to fitted parameter values describing straining (k_{str} and B). A conclusion drawn from this is that the sediments we used in the current study also had a maximum pore space available for straining, which filled rapidly at very high bacteria concentrations, resulting in breakthrough curves that could be simply fitted with one rate coefficient. The maximum S_{str}, or S_{max}, as a function of transport distance, could be determined from eq. (9), assuming $C = 5x10^8$ cells/mL, $\frac{\partial S_{str}}{\partial t} = 0$, and fitting parameters $a = 0.717$, and $b = 1.20$ (Fig. 15). If the average S_{max} is taken as $5x10^7$ cells/g, then the maximum pore space available for straining was in the order of 0.01%, assuming $\rho_{bulk} = 1.4$ g/mL and the height and diameter of an organism is 2 by 1 μm. This pore space was in the same range as determined by *Foppen et al.* [2005; Chapter 3].

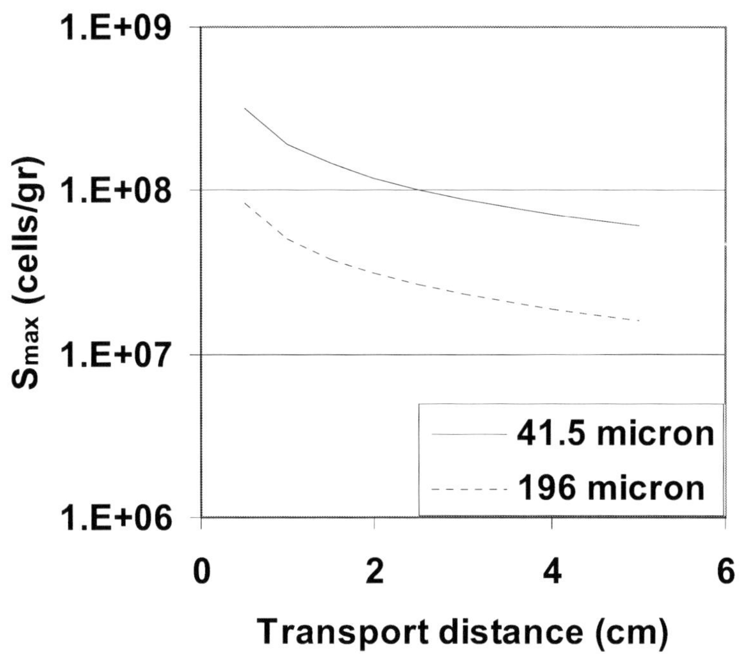

Figure 15. The maximum number of cells/gram sediment, determined from the mass balance of strained bacteria with modified Langmuirian site saturation (eq. (9)

Based on a comparison of the available literature, *Foppen and Schijven* [2006; Chapter 2] proposed a straining rate coefficient that depended on the ratio of colloid to collector diameter, pore water flow velocity, and a dimensionless straining correction factor. The current study has clearly indicated the need to include a reduction factor that depends not only on the collector diameter, but also on the bacteria concentration of the suspension, and the transport distance.

4.6 Conclusions

The main implications of our results are two-fold. Firstly, the occurrence of straining in column experiments can be easily checked and also quantified by simply reversing the flow and collecting the reversed flow effluent. This could be of valuable assistance in interpreting breakthrough curves and profiles of retained particles. Secondly, since straining is an important process in fine-grained sediments, straining is likely to occur and to be of significance during the transport of bacteria in saturated porous media in the natural environment. Our model could be of assistance in both evaluating the significance of straining and in incorporating the straining process in transport modeling simulations.

CHAPTER 5:

TRANSPORT OF *E. COLI* IN COLUMNS OF GEOCHEMICALLY HETEROGENEOUS SEDIMENT

Foppen, J.W.A., J.F. Schijven, 2005. Transport of *E. coli* in columns of geochemically heterogeneous sediment. Wat. Res. 39 (2005) 3082-3088.

Abstract

To elucidate the parameters determining the transport of *E. coli* in aquifers, the attachment of *E. coli* in low concentrations to column sediments was investigated. The sediments comprised 0.18-0.50 mm quartz sand, grains coated with goethite, calcite grains or grains of activated carbon (AC), in varying fractions ($\lambda = 0, 0.05, 0.1, 0.2, 0.4, 0.7, 1.0$) and all of similar diameter to the quartz sand. The weighted sum of favourable and unfavourable sticking efficiencies (α_{total}) showed that upon increasing the fraction of favourable mineral grains (λ) there was an initial rapid increase, which then slowed down. This was most pronounced in the AC experiments, followed by the calcite experiments and then the goethite experiments. We ascribe this non-linear relation to surface charge and hydrophobic heterogeneity of the *E. coli* population.

5.1 Introduction

The main source of drinking water in some developing countries in arid regions is groundwater abstracted from dug or drilled wells. Households in these countries commonly dispose of their solid and liquid waste via soakaways (or pit latrines) but if, as is often the case, the distance from pit latrine to abstraction well is less than 200 m, there is then a real risk of abstracting pathogens (e.g. Lewis et al., 1980; Schijven, 2001). In this regard, *E. coli* is a valuable indicator of faecal contamination, especially because of its ease of detection. There are many factors that are important in the subsurface transport of bacteria and one such factor is grain surface charge heterogeneity. In recent years, the role of grain charge heterogeneity in colloid attachment and detachment processes during transport through saturated porous media has been assessed in greater detail in a series of studies using artificial colloids at high concentrations (10^9 to 10^{14} cells/mL). This work resulted in the development of a theory for predicting transport of colloids in heterogeneously charged porous media, based on the single collector removal efficiency (e.g. Song et al., 1994; Johnson et al., 1996; Ryan and Elimelech, 1996; Elimelech et al., 2000; Bhattacharjee et al., 2002). The study we describe here set out to test this theory by using *Escherichia coli* at lower concentrations (10^3 to 10^4 cells/mL), which are more appropriate to natural conditions, in experiments with columns containing various sediment mixtures of quartz sand with a) goethite-coated sand, b) grains of calcite and c) grains of activated carbon.

5.2 Theory

The attachment rate coefficient, k_a (d^{-1}), which is used in transient mass balance calculations for bacteria retained on the solid matrix, is defined as (e.g. Tufenkji and Elimelech, 2004)

$$k_a = \frac{3(1-\theta)}{2a_c} v\eta \quad (1)$$

where θ is the effective porosity (-), a_c is the grain size diameter (m), v is the pore water flow velocity (m/d) and η is the dimensionless single collector removal efficiency (SCRE). The SCRE is a parameter representing the ratio of the rate of particles striking a collector to the rate of particles approaching the collector. Johnson et al. (1996) and Elimelech et al. (2000) have shown that when the geochemical composition of the sediment is known, grain surface charge heterogeneity can be included in the formulation of the SCRE by

$$\eta = \eta_0 [\lambda \alpha_f + (1-\lambda)\alpha_u] = \eta_0 \alpha_{total} \qquad (2)$$

where η_0 is the single collector contact efficiency (SCCE), determined from physical considerations (e.g. Tufenkji and Elimelech, 2004), α_f and α_u are the sticking efficiencies to favourable and unfavourable attachment sites respectively, and λ is a dimensionless heterogeneity parameter describing the fraction of aquifer grains composed of minerals favourable for attachment or grains coated with favourable attachment patches. For *E. coli*, which at typical groundwater pH values (6-8) is mostly negatively charged, favourable attachment sites are positively charged (e.g goethite and calcite), while unfavourable sites are negatively charged (e.g. quartz).

Values for the sticking efficiency, α_{total}, can be obtained from column experiments for given physicochemical conditions (i.e. suspended particles, porous medium, and solution chemistry; Tufenkji and Elimelech, 2004)

$$\alpha_{total} = -\frac{2}{3}\frac{a_c}{(1-\theta)L\eta_0}\ln(\frac{C}{C_0}) \qquad (3)$$

where C/C_0 is the column outlet normalised bacteria concentration at the initial stage of the particle breakthrough curve and L is the filter medium packed length (m). We determined the SCCE, η_0, with the Tufenkji-Elimelech correlation equation (Tufenkji and Elimelech, 2004)

$$\eta_0 = 2.44 A_S^{1/3} N_R^{-0.081} N_{Pe}^{-0.715} N_{vdW}^{-0.052} + 0.55 A_S N_R^{1.675} N_A^{0.125} + 0.22 N_R^{-0.24} N_G^{1.11} N_{vdW}^{-0.053} \qquad (4)$$

where A_S is a porosity-dependent parameter of Happel's model, N_R is an aspect ratio, N_{Pe} is the Peclet number, N_{vdW} is the van der Waals number, N_A is the attraction number and N_G is the gravity number.

The evaluation of the colloid transport theory in heterogeneously charged porous media for low concentrations of biocolloids concentrates on the question of whether eq. (2) is indeed valid. To elucidate this, we performed a series of laboratory column experiments in which goethite coated sand, calcite grains and activated carbon grains were added to quartz sand in increasing fractions (λ = 0, 0.05, 0.1, 0.2,

0.4, 0.7, 1.0). Goethite and calcite were chosen because they are abundant in aquifers. The activated carbon not only represented the organic material present in aquifers but also addressed hydrophobic parts of the cell surface (the hydrophilic parts of the cell are affected by goethite and calcite).

5.3 Materials

5.3.1 Bacterial suspensions

E. coli strain ATCC25922 was grown in 50 mL of Nutrient Broth (Oxoid CM001) for 24 hours at 37 °C. Bacteria were washed and centrifuged (3000 rpm) three times in a 6.5 mM $NaHCO_3$ solution to which a few drops of 1 M HCl were added to create a HCO_3^-/H_2CO_3 buffer with a constant pH. EC values of the suspension ranged between 575-600 µS/cm, while the pH value ranged between 8.0 and 8.2. One experiment was carried out with a solution consisting of 112 mg/L Na-humic acid, 8.5 mg/L KH_2PO_4, 21.75 mg/L K_2HPO_4, 17.7 mg/L Na_2HPO_4, 27.5 mg/L $CaCl_2$, 11 mg/L $MgSO_4$ and 15 mg/L NaCl (Powelson and Mills, 2001) to assess the occurrence of straining of bacteria.

5.3.2 Column sediment mixtures

A commercially available 99.1% quartz sand (Eurogrid B.V.) with a grain size of 180-500 µm was used. The median of the grain size number distribution was 235 µm. Metal and/or metal oxides were removed from the grain surface by washing the sand in a 0.2 M citrate buffer solution at 80 °C to which sodium dithionite was added (Chu et al., 2001), followed by rinsing with deionised water until the electric conductivity of the water was close to zero and the sand samples were odourless.

Goethite coated sand was prepared via slow oxidation (48 hr) of $FeSO_4 \cdot 7H_2O$ dissolved in initially oxygen free distilled water (Schwertmann and Cornell, 1991). After coating, the goethite concentration was determined (in triplicate) with an ascorbate extraction step (Kostka and Luther, 1994) to be 5.24 mg Fe/g sand. Granules of marble (Merck, Germany) were ground with a mortar and sieved in size fractions. The same procedure was applied to granules of activated carbon (AC) for which Chemviron Carbon Type F400 was used. This type of AC is produced by steam activation of bituminous coal. The pH_{PZC} of this AC is 7.2 (Al-Degs et al., 2000), while the average pore diameter is 27 Å (Lee et al., 2004).

In order to maintain a similar amount of grains of the column sediment, mixtures were prepared by replacing a known amount of quartz sand from a known size fraction with an amount of mineral with similar size fraction according to

$$W_{min} = \frac{\rho_{min}}{\rho_q} * W_q \qquad (5)$$

where W is the weight (g) and ρ is the specific weight (g/mL). The subscripts "min"and "q" denote added mineral and removed quartz sand. Table 1 shows all applied sediment mixtures.

Table 1: *Experimental conditions and values of calculated parameters. (Subscripts "eff"denote effluent; values used for determining the single collector contact efficiency, η_0, are: a_c = 0.235 mm, a_p = 1.14 µm, particle density = 1055 kg/m^3, fluid density = 1000 kg/m^3, fluid viscosity = 1.005*10^3 kg/m s, temperature = 278 K, Hamaker constant = 6.5*10^{-21} J (Walker et al., 2004))*

Experiment	C_0 (cells/mL)	Q (mL/min)	θ (-)	θ_t (-)	EC_{eff} (µS/cm)	pH_{eff}	C/C_0 (-)	η_0 (-)	α_{tot} (-)
Clean Sand									
Humic acid solution	2300	0.46	0.38	0.38	179	7.4	0.890	0.056	0.026
$NaHCO_3$ solution	2300	0.45	0.38	0.38	590	8.2	0.410	0.057	0.196
Goethite									
5%	2100	0.44					0.349	0.058	0.231
10%	2100	0.44					0.348	0.058	0.232
20%	2100	0.43	0.38	0.38	589-601	8.1-8.2	0.248	0.058	0.301
40%	2100	0.46					0.097	0.056	0.528
70%	2100	0.43					0.069	0.058	0.577
100%	7850	0.49					0.074	0.053	0.618
Calcite									
5%	3000	0.45					0.155	0.057	0.416
10%	3000	0.45					0.153	0.057	0.420
20%	3000	0.45	0.38	0.38	635-645	8.2-8.3	0.124	0.057	0.467
40%	3000	0.45					0.066	0.052	0.606
70%	3000	0.51					0.059	0.057	0.693
100%	7850	0.48					0.018	0.054	0.937
Act. Carbon									
5%	4600	0.46	0.38	0.40			0.087	0.057	0.542
10%	4600	0.45	0.39	0.42			0.058	0.059	0.612
20%	4600	0.44	0.39	0.46	590-594	8.1-8.2	0.039	0.063	0.668
40%	4600	0.43	0.40	0.54			0.011	0.068	0.860
70%	4600	0.45	0.42	0.65			0.012	0.072	0.835
100%	7850	0.40	0.44	0.77			0.006	0.083	0.862

5.3.3 Porosity of the column sediments

Total porosity of the quartz sand (clean or coated) and of the 100% calcite was determined gravimetrically to be 0.38. For these types of sediment, total and effective porosity were assumed to be equal. The effective porosities of mixtures of these minerals were also assumed to be 0.38 (Table 1). For AC, total porosity (θ_t) is defined as

$$\theta_t = \theta_b + (1-\theta_b)\theta_p \qquad (6)$$

where θ_b is the bed porosity (-) and θ_p the internal or particle porosity (-). The total porosity of AC was determined gravimetrically to be 0.77, while the particle porosity for Chemviron Carbon Type F400 is 0.59 (Lee et al., 2004). Using eq. (6), the bed porosity for the 100% AC was calculated to be 0.44. In aqueous solutions, the effective porosity of the AC bed is equal to the total porosity. However, *E. coli* cannot enter the internal pores of the AC (with an average diameter of 27 Å) and therefore effective porosity for *E. coli* was equal to the bed porosity. For mixtures of AC and clean sand, the effective porosity for *E. coli* was calculated as the weighted mean of the clean sand porosity and 100% AC bed porosity (Table 1). Since the aqueous solution used the total porosity, as the AC fraction increased the pore water flow velocity decreased. Therefore the pore water flow velocity, v, required to calculate the fluid approach velocity from which the SCCE can be determined, was reduced by a factor $\frac{\theta_b}{\theta_t}$.

5.3.4 Zeta potential

A zeta-meter similar to the one made by Neihof (1969) was used. Movement of bacteria was visible on a video screen attached to a camera mounted on top of a light microscope (Olympus EHT) in phase contrast mode. Particle mobility values were obtained from measurements on at least 50 particles. Velocity measurements were used to calculate the zeta potential with the Smoluchowski-Helmholtz equation.

5.3.5 Bacterial cell size determination

A camera (Olympus DP2) on top of a light microscope (Olympus BX51) in phase contrast mode connected to a computer was used to take images of *E. coli* cells suspended in the $NaHCO_3$ solution ($10^8 - 10^9$ cells/mL). The images were imported into an image-processing program (DP-Soft 2) and analysed using the built-in routines. The average *E. coli* length was 1.95 μm and the average diameter was 0.67 μm. This corresponded to an equivalent spherical diameter of 1.14 μm (Rijnaarts et al., 1993).

5.3.6 Column experiments

Three to five perspex columns (with an inside ∅ of 5 cm and a height of 5 cm) were incrementally packed under saturated conditions with one of the sediment mixtures to a column length of 2 cm. The columns were connected to a pump (Watson-Marlow 101U/R) operating at a rate of 0.40-0.50 mL/min. Prior to each experiment, the columns were flushed for at least 12 hours in order to remove fines. At the beginning of each experiment, the flushing solution in the column above the sediment was removed by suction until the water level was exactly level with the top of the sediment. Then, on top of each column a 3 cm layer of $NaHCO_3$ - *E. coli* suspension with a concentration of 10^3 to 10^4 cells/mL was added, by very gentle sprinkling. This layer was maintained throughout the experiment by manually adding *E. coli* suspension. At the end of the "loading" phase, the same procedure was followed in order to replace the bacterial suspension by a $NaHCO_3$ solution without *E. coli*. Every 2-3 pore volumes the effluent EC, pH and Ca-concentration (for the calcite experiments) and influent *E. coli* concentration was determined. EC was measured with a Tetracon 325 probe, pH with a SenTix 21 probe and Ca-concentration with a Perklin-Elmer AAS 3110. The *E. coli* concentration of both influent and effluent was determined with plates of Chromocult™ agar (Merck). Electrophoretic mobilities were determined every 3 pore volumes in a $NaHCO_3$ - *E. coli* suspension with a concentration of around 10^8 cells/mL. All experiments were carried out at 4° C, in order to discourage the growth and decay of *E. coli.*

5.4 Results

5.4.1 Conditions during the experiments

The pH of the effluent was 8.1-8.3 with effluent EC values of 595 μS/cm (Table 1). The EC values of the calcite experiments were higher (640 μS/cm), due to dissolution of 4-6 mg/L Ca^{2+}. In the $NaHCO_3$ solution, the zeta potential of *E. coli* remained constant throughout each experiment, at a mean value of –62 mV. The standard deviation was 14 mV (Fig. 1). The zeta potential values of *E. coli* in humic acid were less negative (-51 mV) and the range was smaller (standard deviation = 7 mV).

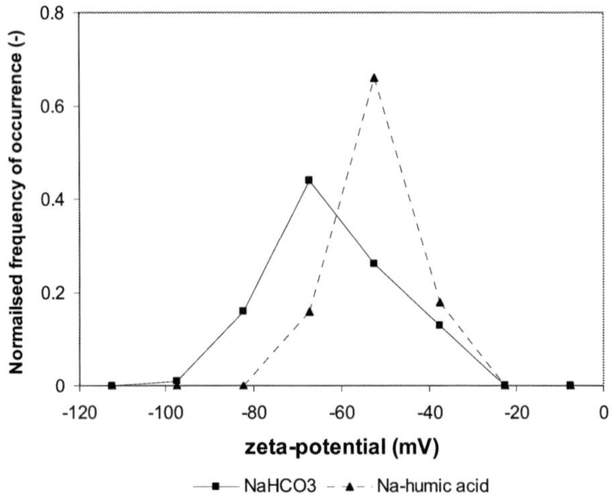

Fig. 1: *Distribution of zeta potential values of E. coli ATCC25922 in a $6.5*10^{-3}$ M $NaHCO_3$ solution and in a 112 mg Na-humic acid solution*

5.4.2 Breakthrough curves

In all cases the decay of influent concentrations was negligible and C_0 was constant (Table 1). From the breakthrough curves (see Figs. 2-5), three phases could be distinguished in general. The first was the clean-bed phase during the first 3 pore volumes (PV), corresponding with the rising limb of the breakthrough curves. The increase in cell recovery during this phase was rapid, without retardation, and tracer-like. Secondly, there was the "post clean-bed phase", which lasted from 3 to 10 PV. For the clean sand - $NaHCO_3$ experiment, a plateau was visible at C/C_0 values of around 0.4 at 3 PV rising to 0.5 at 10 PV (Fig. 2). For the humic acid experiment, breakthrough after 3-4 PV was close to 1, indicating that straining could be neglected and that the clean sand – $NaHCO_3$ plateaus were due to unfavourable attachment of bacteria to the quartz sand. Upon addition of mineral (from $\lambda = 0.05$ to $\lambda = 1.0$; Fig. 3-5 and Table 1), cell recovery during both the clean-bed and post clean-bed phases decreased; the decrease was least pronounced in the goethite experiments and most pronounced in the AC experiments. Thirdly, during the deloading phase, cell recoveries fell sharply to low values ($< 10^{-2}$; see Fig. 6) in all experiments within 5 PV after deloading started.

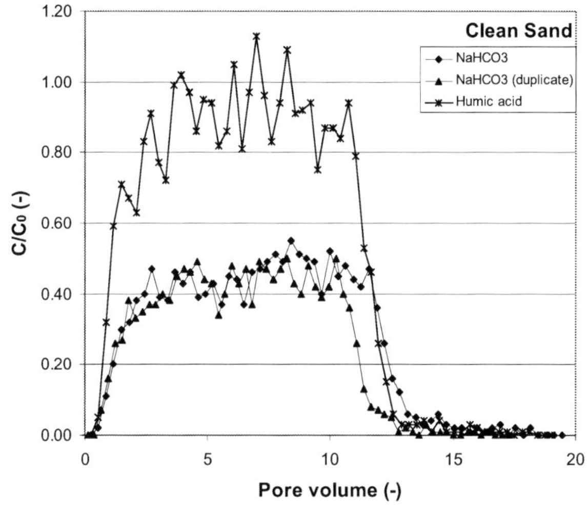

Fig. 2: *E. coli breakthrough curves for the humic acid experiment and the clean sand $NaHCO_3$ experiment*

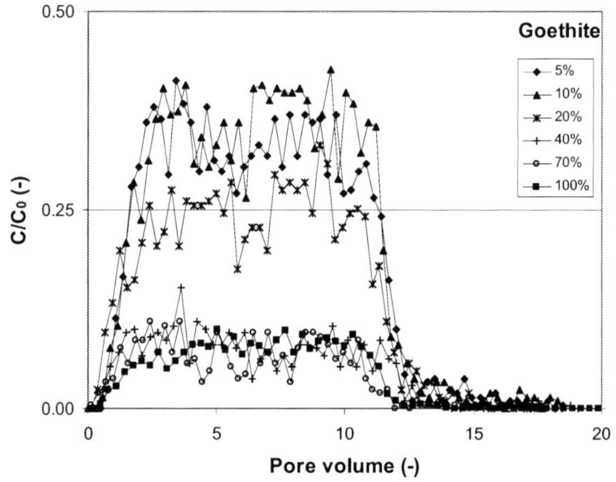

Fig. 3: *E. coli breakthrough curves for various quartz sand and goethite-coated quartz sand sediment mixtures*

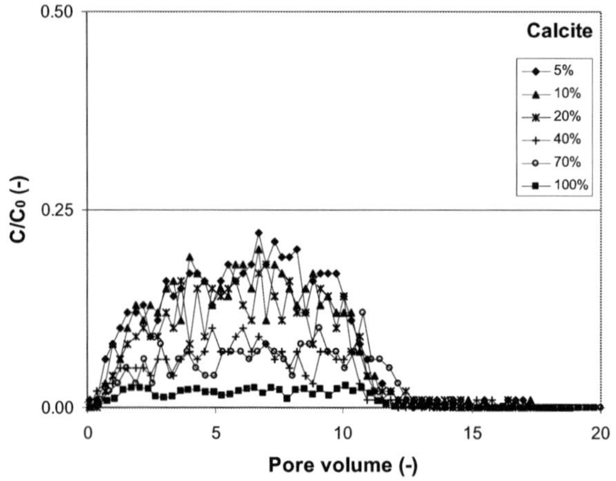

Fig. 4: E. coli breakthrough curves for various quartz and calcite sediment mixtures

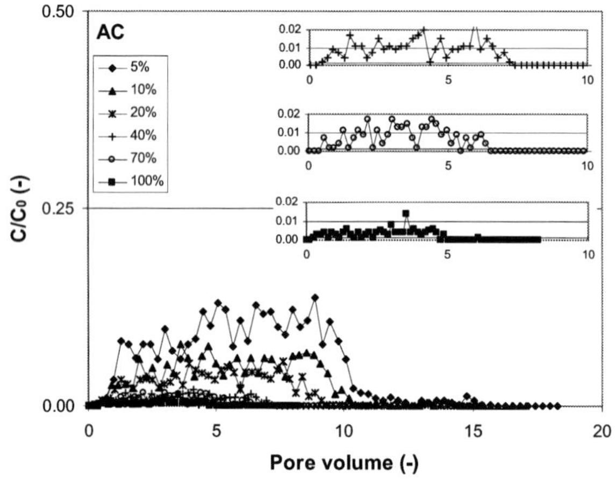

Fig. 5: E. coli breakthrough curves for various quartz and AC sediment mixtures. Inset: separate breakthrough curves for 40%, 70% and 100% AC (from top to bottom)

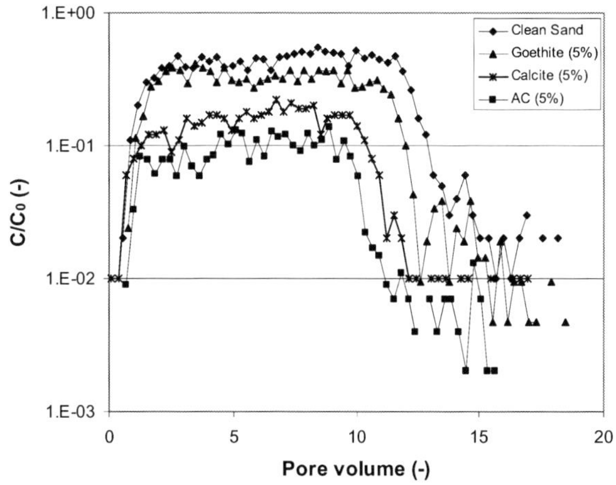

Fig. 6: *Comparison of E. coli breakthrough curves (on log scale) for the clean sand, 5% goethite, 5% calcite and 5% AC experiments*

5.4.3 Sticking efficiency

Table 1 shows C/C_0 (averaged over the interval 3-4 PV) , SCCE and total sticking efficiency (α_{total}) values. Figure 7 shows the relation between α_{total} and λ. The value of α_{total} was 0.196 for the clean sand - $NaHCO_3$ experiment (unfavourable attachment) and 0.937 for the 100% calcite experiment. From Fig. 7, two important observations were made.

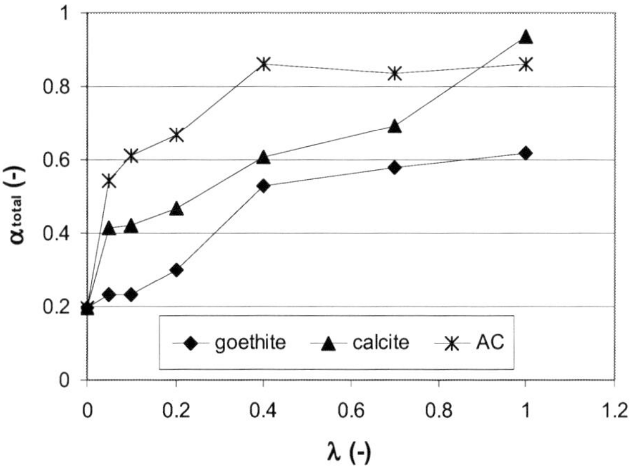

Fig. 7: *Relation between calculated total sticking efficiency (α_{total}) and geochemical heterogeneity parameter (λ) for all experiments*

Firstly, in the goethite experiments, α_{total} rose gradually from 0.191 to 0.528 with increasing λ and levelled out at $\lambda \geq 0.4$. Secondly, for both the calcite and the AC experiments, an initial rapid increase of α_{total} at low λ values (< 0.05 for calcite and < 0.1 for AC) was followed by a gradual increase at higher λ values (0.1-1.0). In addition, like goethite, α_{total} for AC became constant at $\lambda \geq 0.4$. Also, the rapid increase at low λ values was more pronounced in the AC experiments than in the calcite experiments.

5.5 Discussion

The non-linear dependence of the sticking efficiency on λ, shown in Fig. 7, could be explained by two factors:

1. AC offered the most favourable conditions for attachment, followed by calcite and then goethite;
2. The *E. coli* population might have been heterogeneous.

Attachment of *E. coli* to goethite and calcite was most probably determined by charge differences between *E. coli* and the mineral. The pH_{PZC} of goethite (α-FeOOH), is known to be 9.0 -9.1 (Gaboriaud and Ehrhardt, 2003), while the pH_{PZC} of calcite ranges between 9.5 and 10.8 (Somasundaran and Agar; 1967). Because these pH_{PZC} values are higher that the solution pH (8.1-8.3), the minerals were positively charged, and since the pH_{PZC} of calcite is higher than goethite, this could account for the higher sticking efficiency values for calcite determined in the experiments. In the calcite experiments, 4-6 mg/L Ca^{2+} dissolved, and therefore more *E. coli* attached to both quartz and calcite, as predicted by the Schultz-Hardy rule (e.g. Sawyer et al., 1994). However, [Ca] was low ($\approx$ 0.15 mmol/L) and it is doubtful that such a low concentration could have caused the rapid increase of α_{total} at low λ.

The nature of the attachment to AC may be different. It is unlikely that charge differences were important, since the pH_{PZC} ($\approx$ 7.2) of the AC was below the solution pH (= 8.2). In contrast to the relatively simple surfaces of goethite and calcite, where it is primarily the hydroxyl groups that determine the charge at the surface, the carbon surface can contain not one but at least five markedly different types of surface groups, such as carboxylic, lactonic, phenolic, carbonyl and etheric types (Al-Degs et al., 2000), which results in a high adsorption or attachment capacity, especially for organic components. Therefore, we assume that the attachment mechanism was non-polar. Organic parts on the outer membrane of *E. coli* resulting in a certain degree of hydrophobicity are not uncommon for *E. coli*, although the strain is known to be more hydrophilic (Noda and Kanemasa, 1986). The hydrophobicity is attributable partly to the lipid-A part of endotoxins (also known as lipopolysaccharides) and partly to non-ionic parts of the polysaccharides extending a few nm from the outer membrane into the environment outside the cell.

The variation in sticking efficiency between the three mineral types could also be due to grain surface irregularities. Gimbel and Sontheimer (1980; in Matthess et al., 1991) found that grain surface roughness influenced the sticking efficiency. Tufenkji et al. (2004) have even reported that the shape of the individual grains can contribute significantly to the porous medium straining potential, as it may give rise to a very wide pore size distribution. Although we believe that in our case straining did not play a role, the grain surfaces of the goethite, calcite and AC were certainly different, and so were the irregularities on these surfaces. Activated carbon has a particularly irregular grain surface, consisting of a network of pores; this probably contributed to the observed high total sticking efficiency.

As noted above, in addition to these favourable conditions for attachment caused by charge differences, hydrophobic interactions and/or grain surface irregularities, the *E. coli* population might have been heterogeneous. Clear proof of heterogeneity with respect to surface charge was the large *E. coli* zeta potential range (from –20 to –110 mV; Fig. 1). At low λ values, geochemical heterogeneity seemed to determine the removal of *E. coli*, as long as there was an apparent abundance of sufficiently charged *E. coli* that could attach. At higher λ values, the characteristics of the *E. coli* population increasingly determined the attachment: the most easily attachable *E. coli* with sufficient charge had already attached and the less attachable *E. coli* with little or no charge remained in solution. It seems possible that colloid population heterogeneity might also play an important role in the hydrophobic attachment of *E. coli* to AC, resulting in a non-linear dependence of the sticking efficiency on λ. Also Dong (2002) found that for strain *Comamonas* sp. DA001 the variation of surface charge density in a monoclonal bacterial population is a cause for the orders of magnitude variation of experimentally determined collision efficiencies.

We observed no linear relation between α_{total} and λ, as would be predicted by eq. (2). We therefore conclude that this equation is not valid within the boundary conditions of our set of experiments. Clearly, more research is required to elucidate the mechanism behind the relationship between sticking efficiency and geochemical heterogeneity.

CHAPTER 6:

THE EFFECT OF HUMIC ACID ON THE ATTACHMENT OF *ESCHERICHIA COLI* IN COLUMNS OF GOETHITE COATED SAND

Foppen, J.W.A., Y. Liem, J.F. Schijven, 2007b. The effect of humic acid on the attachment of *Escherichia coli* in columns of goethite coated sand. Wat. Res. (subm.).

Abstract

Though coliform bacteria are used worldwide to indicate faecal pollution of groundwater, the parameters determining the transport of *Escherichia coli* in aquifers are relatively unknown. To investigate the effect of dissolved organic carbon (DOC) on the attachment of *E. coli* to saturated goethite-coated sand, we carried out column experiments with *Escherichia coli* with and without humic acid (HA) in monovalent and divalent salt solutions. To characterize sorption of DOC and attachment of *E. coli*, we measured the pH of the influent and effluent, the cation concentrations and the zeta potential of particles. Depending on the chemistry of the *E. coli* suspension, the normalized breakthrough concentrations were over 80 times higher in columns treated with HA compared with columns not treated with HA. However, this difference was not constant: there were time-dependent variations in attachment of *E. coli* to the collector surface, and in the chemical composition of the bacterial suspension. Reduction in removal occurred because HA altered the surface charge of the collector and also sterically hindered *E. coli*. In addition, reduction of removal in a $CaCl_2$ bacterial suspension was probably caused by site competition mechanisms between HA and Ca^{2+} ions. Our results indicate that in the presence of DOC, the concept of geochemical heterogeneity in explaining attachment of biocolloids has limited relevance.

6.1 Introduction

Total Coliform (TC) and Faecal Coliform (FC) concentrations are commonly used as indicators of pathogens related to faecal contamination of groundwater, as they are simple and relatively cheap to determine and give a good indication of the microbiological contamination. The main bacterial strain in TC and FC is *Escherichia coli*. When assessing the risk of microbiological pollution or predicting how far a wastewater plume and its microbiological load can travel in aquifers, it seems appropriate first to assess the behaviour of *E. coli* in terms of its potential to travel in aquifers. Earlier, two of us (Chapter 5) noted that one of the factors influencing *E. coli* retention in aquifers is the mineralogy of the aquifer sediment and the grain charge heterogeneity. We found that in sediments comprised of 0.18–0.50 mm quartz sand coated with goethite in varying fractions (0, 5%, 10%, 20%, 40%, 70%, and 100%), the sticking efficiencies of *E. coli* increased concomitantly with the fraction of quartz sand coated with goethite. (The sticking efficiency is a parameter representing the ratio of the rate of particles sticking to a collector to the rate of particles striking a collector.)

The importance of dissolved organic matter (DOM) in groundwater systems is widely recognized, because the sorption of hydrophobic organic compounds and trace metals onto DOM may significantly increase their aqueous solubility (Yamamoto et al., 2002). DOM in groundwater is a complex mixture of acidic, neutral and basic organic poly-molecules (Thurman, 1985). The concentrations of DOM vary widely, from as high as 100–200 mg/L near the water table to less than 0.5 mg/L in deeper groundwater (Thurman, 1985). The DOM in wastewater is composed of humic materials, polysaccharides, polyphenols, proteins, lipids and heterogeneous organic molecules (Fujita et al., 1996; Ma et al., 2001; Imai et al., 2002) with concentrations ranging from 1–100 mg/L, depending on the treatment undergone. As a result of the importance of DOM for the transport of certain types of contaminants, the sorption of DOM to iron oxides in aquifers has received considerable attention (e.g. Chi and Amy, 2004; Gu *et al.*, 1994; Amirbahman and Olson, 1993). Various authors (Parfitt *et al.*, 1977; Tipping, 1981; Davis, 1982; Gu et al., 1994, Chi and Amy, 2004) have noted that the dominant sorption mechanism is ligand exchange between OH groups at the surface of the iron oxide and carboxyl groups of the DOM, accompanied by an increase in pH.

Equilibrium batch experiments have revealed that DOM tends to compete with viruses for attachment sites on the surface of soil particles and by so doing reduces virus attachment (Gerba, 1984). In column experiments, Foppen et al. (2006; Chapter 7) also found site competition mechanisms between DOM and a bacteriophage (PRD1): introducing DOM to columns of goethite-coated sand grains reduced the removal of PRD1 by 5 log units.

As the effect of DOM on the attachment of *E. coli* to iron oxides had not been studied (Chapter 2), we set out to study the effect of humic acid on the attachment of *E. coli* in columns of goethite-coated sand. We assessed colloid transport and fate by measuring temporal changes in colloid effluent concentrations. The status of the collector surface was investigated by measuring temporal changes of pH and cation concentrations in column influent and effluent solutions. To do so, we measured the zeta potential of the collector surface and bacteria and assessed the characteristics of *E. coli* detachment.

6.2 Materials and methods

6.2.1 Bacterial suspensions

Bacterial suspensions of *Escherichia coli* ATCC 25922 were prepared in a manner similar to that described in Chapter 5. Bacteria were suspended and diluted in one of two solutions (0.015 M $NaNO_3$ or 0.015 M $CaCl_2$) to concentrations of 10^8 cells/mL.

6.2.2 Humic Acid (HA) solution

Humic acid (5 gL^{-1}; ACROS Organic, USA) in the form of sodium salt was dissolved in 0.015 M $NaNO_3$ and filtered through a 0.2 μm glass fibre membrane filter to obtain a dissolved organic carbon (DOC) concentration of 997.44 mgL^{-1}, determined with a TOC analyser (O-I Corporation, USA). Prior to filtration, 1 L milliQ water was flushed through each membrane filter to minimize DOC contributions from the membrane filter. The solution was diluted with 0.015 M $NaNO_3$ to achieve a DOC concentration of 150 mgL^{-1}.

6.2.3 Porous media

A commercially available 99.1% quartz sand (Eurogrid B.V.) with a grain size of 180–500 μm was used. The median of the grain size number distribution was 235 μm. Total porosity of the quartz sand (clean or coated) was determined gravimetrically to be 0.38. Prior to the experiments, the sand was washed in two steps with a 0.2 M citrate buffer and then with sodium dithionite ($Na_2S_2O_4$) to remove metal and/or metal oxides from the grain surface (Chapter 7). After washing, a portion of the sand was coated with goethite, using $FeSO_4{\cdot}7H_2O$ and $NaHCO_3$ (Schwertmann and Cornell, 1991; Chapter 7).

6.2.4 Zeta-potential measurements

A zeta meter similar to the one made by Neihof (1969) was used. The movement of particles (*E. coli*, HA, and goethite colloids) was observed on a video screen attached to a camera mounted on a light microscope (Olympus EHT) in phase contrast mode. Particle mobility values were obtained from measurements on at least 50 particles. Velocity measurements were used to calculate the zeta potential with the Smoluchowski–Helmholtz equation.

6.2.5 Equilibrium sorption experiments

Recirculating column experiments (Chi and Amy, 2004) were performed to determine sorption and desorption kinetics and an equilibrium isotherm for HA. The borosilicate glass columns had an inner diameter of 2.5 cm (Omnifit, Cambridge, UK) with polyethylene frits (25 μm pore diameter) and one adjustable end-piece. Using vibration to minimize any layering or air entrapment, the columns were packed wet with (iron-coated) sand, . The experiments used 15 g of goethite-coated sand, and a flow rate of 0.5 mL/min; the effluent from the column was fed back into the column via a closed loop. A recirculating column set-up mimics the groundwater environment more closely, because the solid material remains motionless during the experiments, and no artefacts (such as disaggregation of particles or dissolution of minerals) are induced by shaking. The HA sorption isotherm was obtained by equilibrating the same amount of sediment with different concentrations of HA solutions (the concentrations ranged from 2 to 200 mg/L DOC) in 0.015 M $NaNO_3$ for 1 day in each column. A desorption isotherm was obtained by displacing the HA solution for 4 days with a 0.015 M $NaNO_3$ solution containing no HA.

6.2.6 Column experiments.

Triplicate column experiments were conducted in columns similar to those as described above. Two types of experiments were carried out (Table 1):

1. to assess the attachment characteristics of *E. coli* to:
   - goethite-coated sand and a pore fluid consisting of 0.015 M $NaNO_3$ (exp #1);
   - goethite-coated sand treated with 150 mgL^{-1} HA prior to application of *E. coli* in 0.015 M $NaNO_3$ (exp #2);
   - goethite-coated sand and a pore fluid consisting of 0.015 M $CaCl_2$ (exp #3);
   - goethite-coated sand treated with 150 mgL^{-1} HA prior to application of *E. coli* in 0.015 M $CaCl_2$ (exp #4) and
   - goethite-coated sand treated with 0.015 M $NaH_2PO_4 \cdot H_2O$ prior to *E. coli* application (exp #5);
2. to assess *E. coli* detachment characteristics (exp. #6). To do so, after equilibrating a column for 100 PV with 0.015 M $NaNO_3$, the column was flushed for 10 pore volumes with *E. coli* in 0.015 M $NaNO_3$. Then, the column was flushed for 5 PV with 0.015 M $NaNO_3$ containing no *E. coli*, followed by 5 PV demineralized water, and, finally, 5 PV 0.005 M KOH. The flushing with demineralized water was to reduce the ionic strength of the solution, thereby increasing the thickness of the electrical double layer

surrounding the iron-coated sand grains and *E. coli*, possibly causing *E. coli* captured in the secondary energy well to become detached. The KOH step (at the high pH of 10) was to detach *E. coli* by making the surface charge of goethite negative (the pH at the point of zero charge of goethite ranges between 8–9; Cornell and Schwertmann, 1992). This provided an indication of the *E. coli* reversibly attached to positively charged sites of the goethite surface.

In all experiments, the column sediment was 0.5 cm thick, and the pump discharge was 0.5 $mLmin^{-1}$. During the experiments, the pH of the influent and effluent was measured with a SenTix 21 probe. The effluent pH was measured in-line directly after leaving the experimental column. To account for possible variations in pH caused by release of iron, the total iron concentrations were determined on a Perkin–Elmer AAS 3110. In exps. #3 and #4, [Ca^{2+}] was also measured with the AAS. The *E. coli* concentrations were determined with a UV spectrophotometer (CECIL-CE 1021) at 254 nm. For the experiments with HA, breakthrough was determined with the same UV spectrophotometer at 254 nm. Since the presence of HA could influence the optical density of *E. coli*, samples were taken every several pore volumes, while *E. coli* was being flushed through the columns. The samples were filtered through 0.2 μm glass fibre membrane filters to remove *E. coli*. The DOC concentration of the filtrate was determined, to ascertain the release of DOC attributable to attachment of *E. coli*. Prior to this filtration, 1 L milliQ water was flushed through the membrane filters to minimize DOC contributions from the membrane filter itself. For exp. #6 involving KOH and demineralized water, in order to account for possible absorbance losses due to lysis or osmotic shock of the bacteria cells, we prepared bacterial suspensions of various concentrations and determined the absorbance over time.
Bacterial suspensions of *Escherichia coli* ATCC 25922 were prepared in a manner similar to that described in Chapter 5. Bacteria were suspended and diluted in a 0.015 M $NaNO_3$ or in a 0.015 M $CaCl_2$ solution to concentrations of 10^8 cells/mL.

Table 1. Chemical compositions of the solutions used and number of pore volumes (PV) flushed during the preparation phase, E. coli application phase, and deloading phase of the column experiments (in triplicate)

Exp. #	Preparation phase	Application phase	Deloading phase
1	100 PV 0.015 M $NaNO_3$	75 PV 0.015 M $NaNO_3$ + 10^8 cells/mL *E. coli*	50 PV 0.015 M $NaNO_3$
2	75 PV 0.015 M $NaNO_3$ + HA (as 150 mgL^{-1} DOC), then 100 PV 0.015 M $NaNO_3$	75 PV 0.015 M $NaNO_3$ + 10^8 cells/mL *E. coli*	50 PV 0.015 M $NaNO_3$
3	75 PV 0.015 M $CaCl_2$	75 PV 0.015 M $CaCl_2$ + 10^8 cells/mL *E. coli*	50 PV 0.015 M $CaCl_2$
4	75 PV 0.015 M $NaNO_3$ + HA (as 150 mgL^{-1} DOC), then 75 PV 0.015 M $CaCl_2$	75 PV 0.015 M $CaCl_2$ + 10^8 cells/mL *E. coli*	50 PV 0.015 M $CaCl_2$
5	75 PV 0.015 M NaH_2PO_4, then 75 PV 0.015 M $NaNO_3$	75 PV 0.015 M $NaNO_3$ + 10^8 cells/mL *E. coli*	50 PV 0.015 M $NaNO_3$

6.3 Results and discussion

6.3.1 Characterization of HA sorption to the collector surface

The equilibrium sorption of HA to goethite was according to a Langmuir type of sorption isotherm, with high sorption affinity at low HA concentrations, and a decreasing sorption affinity at higher HA concentrations (Fig. 1). This type of isotherm is typical for monolayer sorption with a finite collector surface or a finite number of sites for HA to adsorb on. In addition, the desorption experiments showed that after 4 days there was little release of HA, i.e. most of the HA had been irreversibly adsorbed and did not desorb.

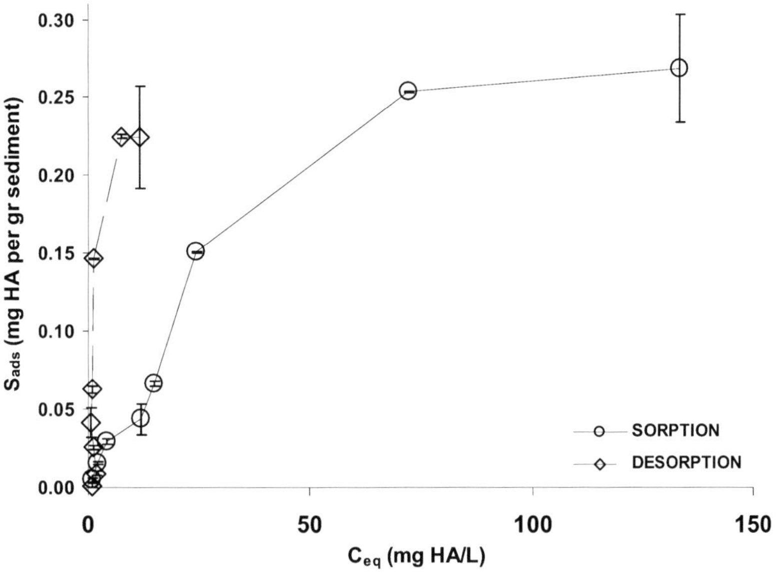

Figure 1. Sorption of humic acid (HA) to goethite at various concentrations (circles) and humic acid still attached to goethite (diamonds) after 4 days of desorption

When HA was flushed through a column, its breakthrough (Fig. 2) was immediate and without significant retardation, but the rising and falling limbs of the breakthrough curves were not similar in shape, indicating that sorption to the collector surface was not an equilibrium process, but kinetically determined. In addition, as is clear from Fig. 2, flushing of HA was continued until breakthrough was close to 100% ($C/C_0 = 1$). During the HA flushing, the pH of the effluent was higher than the pH of the influent (Fig. 3, lower graph, from 0 to 75 PV). Iron concentrations remained below the detection limit and had no influence on the measured effluent pH. This suggests that the collector surface was able to adsorb protons or to release hydroxyls, which increased the pH as pore water passed through the column. Chi and Amy (2004) have reported that hydroxyl release and a rise in pH are characteristic of the adsorption of natural organic matter (NOM) in columns of iron-coated sand, due to ligand exchange in which hydroxyl groups on the surface of iron are exchanged for NOM. However, in our case, a rise in pH also occurred when the columns were flushed with $NaNO_3$ without HA (Fig. 3, upper graph, from 0 to 100 PV).

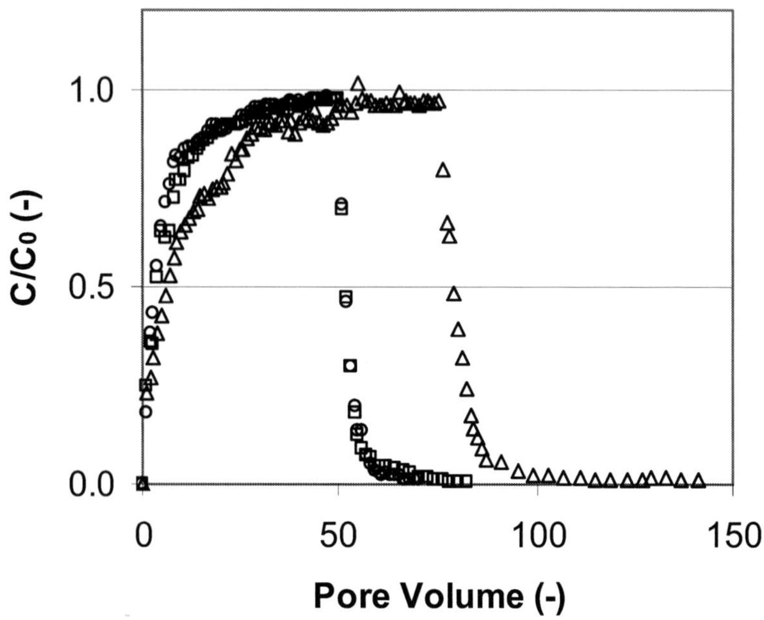

Figure 2. Examples of humic acid breakthrough curves

We attribute the pH increase to the ability of goethite to adsorb protons (e.g. Meeussen et al., 1996). Since the disparity in influent and effluent pH was greater for the HA case than for the case without HA, we conclude that the surface of the collector was altered by HA in such a way that its affinity to adsorb protons was appreciably increased.

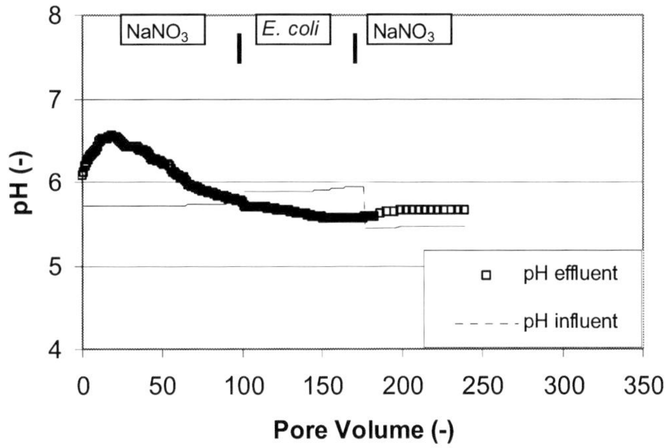

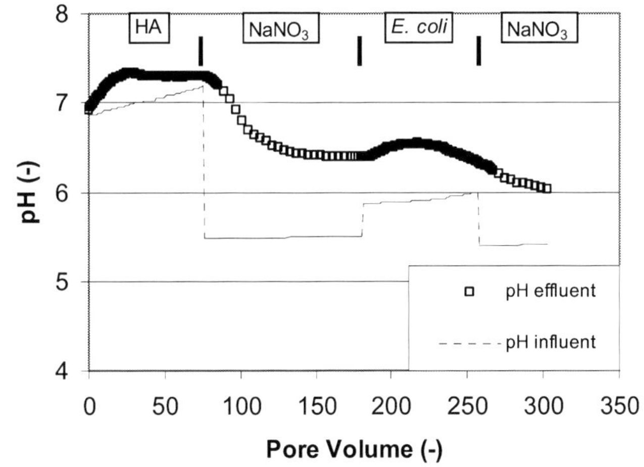

Figure 3. Curves of effluent and influent pH values. Upper graph: exp #1 ($NaNO_3$, no HA); lower graph: exp #2 (first HA, then $NaNO_3$)

6.3.2 Attachment of *E. coli*

Initially, breakthrough in exp #1 ($NaNO_3$, no HA) was rapid and without retardation (Fig. 4). Then, C/C_0 values continued to rise gradually: from 0.4 at PV 10 to 0.6 at PV 75. During the deloading phase (PV > 75), *E. coli* concentrations fell sharply to values below $C/C_0 = 0.05$. As no significant "tailing" was observed, we infer that detachment was negligible. To examine the nature of the retention of *E. coli* in this system, we performed a separate experiment in which a column was flushed with demineralized water after *E. coli* application (Fig. 5). Usually, because the ionic strength of the solution decreases, the repulsive double-layer energy barrier increases, thereby releasing bacteria residing in the secondary energy minimum. In our experiment, however, only a few bacteria detached; clearly, the secondary energy minimum did not play an important role in our experiments.

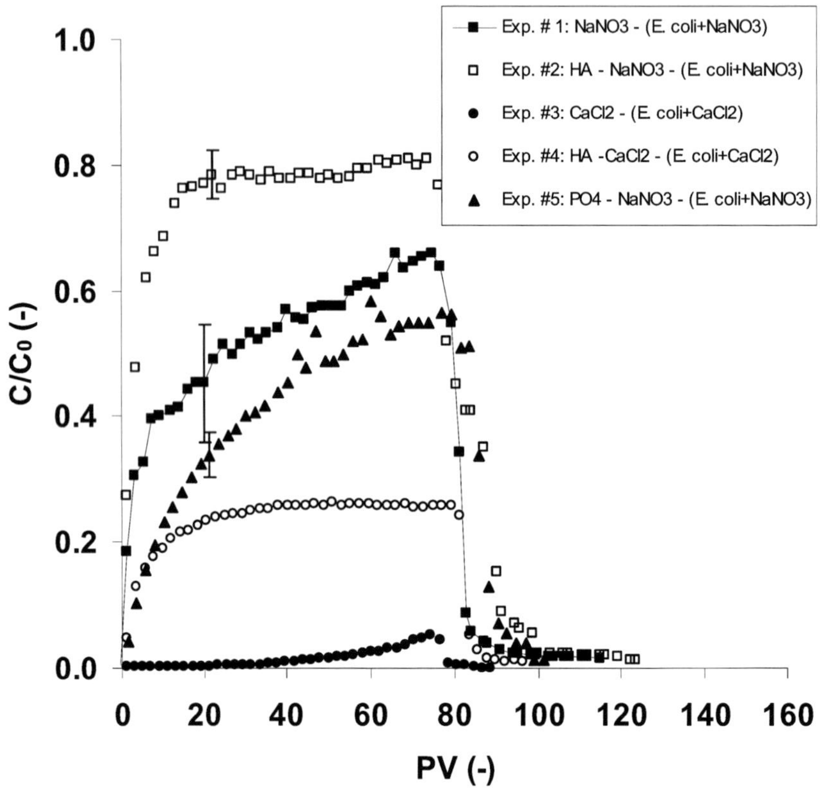

Figure 4. Breakthrough curves of Escherichia coli for exp. #1-5

Subsequently flushing the column with a dilute KOH solution raised the pH to 10, and we expected that the pH-dependent positive surface charge of goethite would become negative. Given the tiny fraction of bacteria released (< 2% of the total bacterial mass applied), we infer that the energy well was so deep that only a few bacteria were able to escape from it, even when its depth was reduced by flushing with demineralized water and dilute KOH. The majority of the bacteria were irreversibly retained in the primary energy minimum, most probably as a result of attractive Van der Waals and electrical double-layer forces. Straining is unlikely to have been an important bacterial retention mechanism, because in similarly shaped and sized quartz sand, not coated with iron, breakthrough in 2 cm columns for comparable conditions was nearly complete ($C/C_0 \approx 0.98$; Chapter 5).

In exp #2 (first HA, then $NaNO_3$), the C/C_0 values increased rapidly to much higher values (0.7–0.8) than in exp #1, and then from PV 20 onwards remained almost constant at around 0.8. The DOC values in the filtrate of the occasionally filtered effluent samples were below the detection limit, and confirmed that HA release during *E. coli* flushing was negligible. We conclude that the HA pretreatment affected the attachment of *E. coli*, causing some 20%–40% of the total bacterial mass applied to the column not to attach, but to elute from the columns.

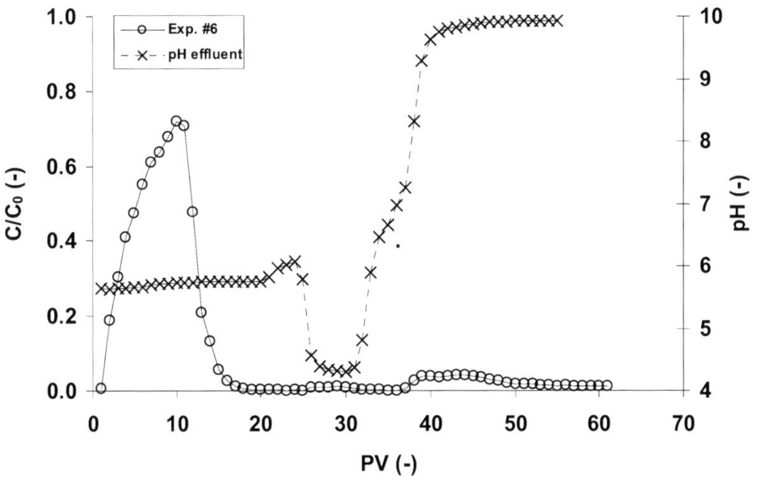

Figure 5. Relative E. coli concentration and measured pH of the effluent of exp. #6 during flushing of demineralized water ($pH_{influent}$ ~4.2) and KOH ($pH_{influent}$ ~10) to assess the type of attachment of E. coli

In exp #3 ($CaCl_2$, no HA), the C/C_0 values ranged between 0.01–0.04, and increased steadily with pore volume. Although *E. coli* was much less negatively charged in $CaCl_2$ (-18.5 mV) than in $NaNO_3$ (-48.3 mV; Table 2), there was more attachment of *E. coli* to the positively charged goethite than in exp #1 ($NaNO_3$, no HA). We were therefore unable to use the zeta potential, which is a measure of the surface charge of *E. coli*, to predict attachment processes. It seems that the presence of Ca was important: the mean Ca concentration in the effluent was found to be 595–600 mg/L (Fig. 6), which was less than the influent concentration (C_0 = 605 mg/L). This indicated that 1–2% of the influent Ca concentration sorbed to the goethite surface, and could have affected attachment of *E. coli*. In exp #4 (first HA, then $CaCl_2$), the C/C_0 values rapidly increased to 0.25 and then remained constant. Thus, pretreating the collector surface with HA caused breakthrough in our relatively small columns of goethite-coated sand, with a total height of 0.5 cm to be over 80 times more than in the column that had not been pretreated with HA. When the reduction in *E. coli* removal was defined as the ratio of the averaged *E. coli* breakthrough concentrations from the HA pretreated columns to the averaged concentrations from the untreated columns at a certain moment in time (Fig. 7), then the reduction in removal attributable to the HA pretreatment of the goethite-coated columns was not constant in time. Initially, the reduction in removal increased, peaking between PV 10–30, but then it decreased to ~5 in the $CaCl_2$ experiment and to less than 1 in the $NaNO_3$ experiment. This variation of the reduction of *E. coli* removal demonstrates the time-dependent variations in attachment of *E. coli* to the collector, which are probably due to attachment site saturation processes. Had we continued the $CaCl_2$

experiments, the reduction in removal might have fallen to less than 1, indicating less breakthrough for the HA pretreated column than for the non-pretreated column. These results demonstrate the complex influence of HA on attachment of *E. coli*, and also the time-dependent nature of the influence of HA adsorbed to the collector surface on attachment of *E. coli* to the complex HA–goethite collector surface.

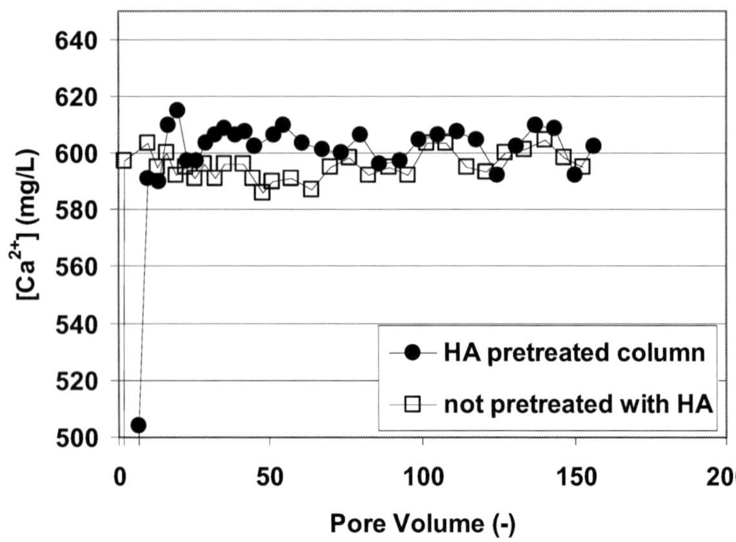

Figure 6. [Ca^{2+}] breakthrough concentrations for exp #3 and exp #4. The [Ca^{2+}] of the influent was 605 mg/L

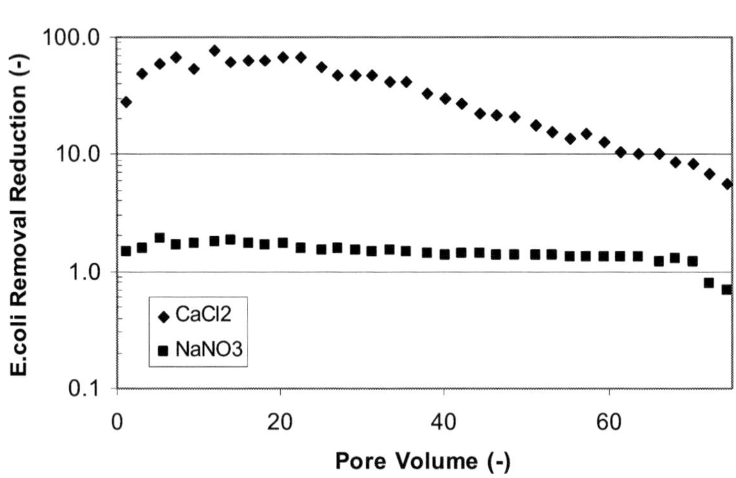

Figure 7. E. coli removal reduction defined as the ratio of the averaged E. coli breakthrough concentrations from the HA pretreated columns and the averaged concentrations from the untreated columns at any moment in time

6.3.3 Explaining the influence of HA on *E. coli* attachment

Three processes might be responsible for the reduced attachment of *E. coli* when HA is present on the collector surface:

1. An alteration in the surface charge of the collector as a result of HA sorption, which discourages attachment;
2. Competition between HA and *E. coli* for the same attachment sites on the goethite surface;
3. A reduction or even elimination of the attractive energy well near the surface of the collector, caused by sorbed HA layer blanketing the goethite-coated grain. As the *E. coli* cannot reach the well, there is less attachment. This process is also known as steric hindrance (Franchi and O'Melia, 2003).

These three processes will now be discussed in detail.

1. Alteration of surface charge. Though the goethite surface was positively charged (+49.4 mV in $NaNO_3$ and +52.7 mV in $CaCl_2$), the zeta potential of the HA was negative (-66.4 mV; Table 2), indicating that upon adsorption of HA, the collector surface charge altered and the conditions became unfavourable for attachment. Further evidence of a change in the surface charge is provided by the pH values. In the columns pretreated with HA, the differences in the pH of the column influent and effluent were greater than in the columns not pretreated with HA (Figs. 3 and 8), indicating that the pretreated columns were better able to capture protons, i.e. were more negatively charged.

Table 2. Zeta-potential values of E. coli, goethite, and humic acid in $NaNO_3$ and in $CaCl_2$

Suspension	Zeta potential of particle (mV)
E. coli in 0.015 M $NaNO_3$	-48.3 ± 7.2
E. coli in 0.015 M $CaCl_2$	-18.5 ± 3.9
Humic acid in 0.015 M $NaNO_3$	-66.4 ± 19.1
Goethite in 0.015 M $NaNO_3$	+49.4 ± 5.7
Goethite in 0.015 M $CaCl_2$	+52.7 ± 5.8

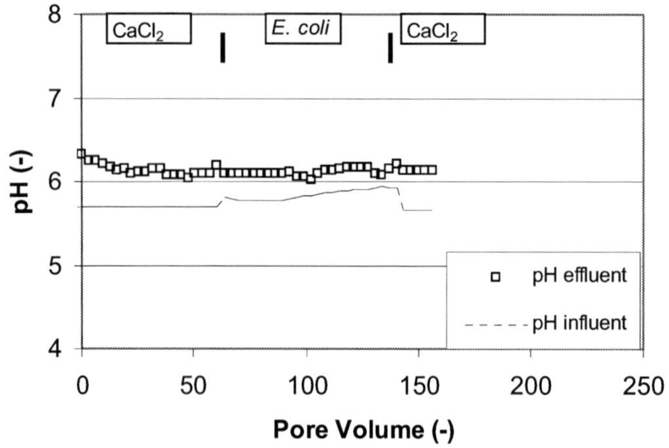

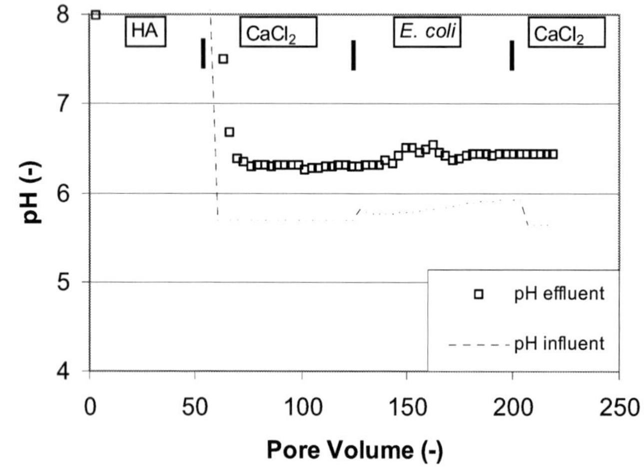

Figure 8. *Curves of effluent and influent pH values. Upper graph: exp #3, lower graph: exp. #4*

2. Competition between HA and *E. coli*. Organic material, such as HA, is a well-known ligand exchanger (Parfitt et al., 1977; Tipping, 1981; Davis, 1982; Gu et al., 1994, Chi and Amy, 2004). The main mechanism of ligand exchange is the replacement of hydroxyl (OH^-) groups at the goethite surface by an adsorbing organic molecule. The mechanism is also known as chemisorption or inner sphere adsorption (Cornell and Schwertmann, 1992). The molecule is usually tightly bound and not easily replaceable. Since *E. coli* has many carboxyl (COOH) groups that form part of the chain of lipopolysaccharides protruding into the environment from the outer membrane of the *E. coli* (Kastowsky et al., 1991, 1993), it is possible that a similar ligand exchange mechanism could have taken place during attachment of *E. coli*, but since HA had already occupied these sites, there was less *E. coli* attachment. In the literature, phosphate sorption is also often referred to as a ligand exchange process (Cornell and Schwertmann, 1992). Infrared studies have shown that each PO_4 molecule replaces two singly coordinated OH groups to form a bond, while desorption of PO_4 from goethite is extremely slow and partly irreversible. The breakthrough curve of *E. coli* in exp #5 (first PO_4, then $NaNO_3$; Fig. 4) was similar in shape to that of exp # (no prior PO_4), but the C/C_0 values were less. Assuming that phosphate sorption was a ligand exchange process in our case too, we conclude that ligand exchange was probably not occurring upon *E. coli* attachment. The *E. coli* and PO_4 did not compete for ligand exchange sites, and therefore, the reduced *E. coli* removal attributable to HA is unlikely to have been caused by competition for this type of site. The situation was different for the $CaCl_2$ experiments. In exp #4 (first HA, then $CaCl_2$), the sorption of Ca^{2+} ions was negligible, as evidenced by the influent and effluent concentrations being identical (Fig. 6); thus, the Ca^{2+} ions were unable to adsorb to the goethite surface after HA pretreatment, neither did Ca adsorb to the HA surface. This lack of Ca sorption might have been why there was less *E. coli* removal in exp #4, and therefore site competition between HA and Ca was an important process, eventually leading to less attachment of *E. coli* in the column pretreated with HA.

3. Blanketing of the energy well. Fig. 9 shows a DLVO energy profile (Fig. 9) for the interaction between goethite surface and *E. coli* interaction in $NaNO_3$, constructed using the same equations used by Loveland et al. (1996). It can be seen that there was a very large primary energy minimum less than a nanometer from the surface of the collector, extending 10 nm away from that collector surface. Although we could not measure the thickness of the HA mono-layer adsorbed to the goethite surface, we suppose that some of these organic molecules were within the nm size range, making the HA mono-layer as thick as or thicker than the diffuse layer containing the attractive energy well. Attachment would then be less, either because the depth of the attractive energy well was reduced by the size and thickness of the layer of adsorbed HA around the grain, or because the energy well had been completely blanketed by HA and could therefore not be reached by the *E. coli*.

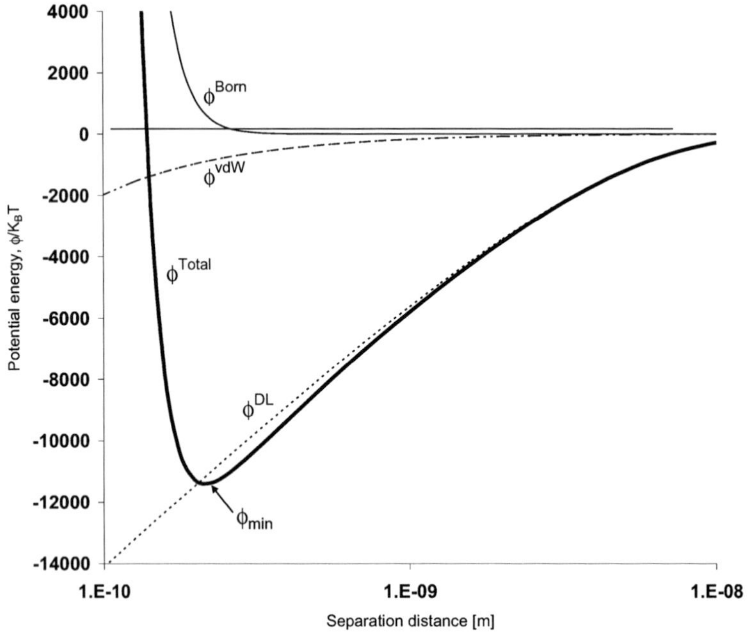

Figure 9. *DLVO energy as a function of separation distance between E. coli and the goethite collector surface. The total potential energy (ϕ^{Total}) is the sum of the double layer potential energy (ϕ^{DL}), the van der Waals potential energy (ϕ^{vdW}) and the Born potential energy (ϕ^{Born}). The total potential energy curve is characterized by an attractive primary well (ϕ_{min1}) at close distance to the collector surface. Values used were: colloid diameter = 1.5 μm, colloid surface potential = -48 mV, collector surface potential = +49 mV, ionic strength = 0.015 M, temperature = 293 K, Hamaker constant = $6.5x10^{-21}$ J, characteristic wavelength = 10^{-7} m, Born collision parameter = $5x10^{-10}$ m. The potential energy is normalized by $k_B T$ (Ryan and Elimelech, 1996)*

6.4 Conclusions

One of the implications of our study is that HA altered the surface of goethite and reduced the attachment of *E. coli*. Depending on the chemistry of the *E. coli* suspension, breakthrough in our relatively small columns consisting of goethite-coated sand with a total height of 0.5 cm was over 80 times more than in columns that had not been pretreated with HA. In addition, the reduction in removal due to the HA pretreatment of the goethite-coated columns was not constant in time: it began by increasing to a maximum between PV 10–30 but then decreased to ~5 for the $CaCl_2$ experiment and less than 1 for the $NaNO_3$ experiment. This time-dependent variation in the reduction of *E. coli* removal is probably the result of attachment site saturation processes. The processes responsible for reducing the removal were surface charge alteration of the collector by HA, and steric hindrance of *E. coli* by HA. In addition, in the case of a $CaCl_2$ bacterial suspension, the reduction in removal was probably caused by site competition mechanisms between HA and Ca^{2+} ions. In our experiments there was evidence for the occurrence of all three mechanisms, but we were unable to distinguish the relative importance of these three processes for reducing *E. coli* attachment.

Since it is likely that *E. coli* travels underground in plumes of wastewater containing dissolved organic material, it is probable that DOM sorption to the collector occurs, and therefore the concept of geochemical heterogeneity (e.g. Song *et al.*, 1994; Johnson *et al.*, 1996) is probably of limited relevance for explaining the attachment of mostly negatively charged *E. coli* – especially in relation to the presence of patches of positively charged iron oxyhydroxides on aquifer grains. Of course, DOM in wastewater is an extremely complex mixture of humic materials, polysaccharides, polyphenols, proteins, lipids and heterogeneous organic molecules (Fujita et al., 1996; Ma et al., 2001; Imai et al., 2002), and therefore our results cannot simply be extrapolated to wastewater infiltration cases. More experimental research is required to elucidate the effect of organic material sorption on biocolloid attachment and detachment processes in aquifers.

CHAPTER 7:

EFFECT OF GOETHITE COATING AND HUMIC ACID ON THE TRANSPORT OF BACTERIOPHAGE PRD1 IN COLUMNS OF SATURATED SAND

Foppen, J.W.A., S. Okletey, and J.F. Schijven, 2006. Effect of goethite coating and humic acid on the transport of bacteriophage PRD1 in columns of saturated sand. J. Contam. Hydrol. 85 (2006) 287-301.

Abstract

The transport of bacteriophage PRD1, a model virus, was studied in columns containing sediment mixtures of quartz sand with goethite-coated sand and using various solutions consisting of monovalent and divalent salts and humic acid (HA). Without HA and in the absence of sand, the inactivation rate of PRD1 was found to be as low as 0.014 day^{-1} (at 5 ± 3 °C), but in the presence of HA it was much lower (0.0009 day^{-1}), indicating that HA helps PRD1 to survive. When the fraction of goethite in the sediment was increased, the removal of PRD1 also increased. However, in the presence of HA, C/C_0 values of PRD1 increased by as much as 5 log units, thereby almost completely eliminating the effect of addition of goethite. The sticking efficiency was not linearly dependent on the amount of goethite added to the quartz sand; this is apparently due to surface charge heterogeneity of PRD1. Our results imply that in the presence of dissolved organic matter (DOM), viruses can be transported for long distances thanks to two effects: attachment is poor because DOM has occupied favourable sites for attachment, and inactivation of virus may have decreased. This conclusion justifies making conservative assumptions about the attachment of viruses when calculating protection zones for groundwater wells.

7.1 Introduction

Schijven *et al.* (2006) have calculated the virus protection zones needed around shallow unconfined aquifers in the Netherlands to ensure a 95% certainty that an infection risk of 10^{-4} per person per year will not be exceeded. They concluded that protection zones of 1 to 2 years' travel time are needed, which is 6 to 12 times the currently applied travel time of 60 days. In their calculations, they assumed that virus attachment would be minor if no sites are available for attachment, as had been the case in the anoxic part of an aquifer in an earlier study of deep well injection (Schijven *et al.*, 2000). Attachment might also be minor if the attachment sites are blocked by organic matter (e.g. Pieper *et al.*, 1997). It was, however, also recognised that the most important parameters determining the size of the protection zone are the inactivation and attachment rate coefficients. If favourable sites for attachment are present (in the form of ferric oxyhydroxides, for example), transport of virus may be significantly limited.

One of the many factors important in the underground transport of bacteria and viruses is grain surface charge heterogeneity. The latter may result when diverse functional groups are exposed on adjacent facets of mineral surfaces or when the minerals contain inherent or superficial chemical impurities. The most common sources of surface charge heterogeneity in underground aqueous environments are oxides of iron, aluminium, and manganese, because these compounds carry a positive surface charge at near-neutral pH and are generally present as coatings on silicate mineral grains (Johnson *et al.*, 1996).

The detailed studies done in recent years on the role of grain charge heterogeneity in colloid attachment and detachment processes during transport through saturated porous media have resulted in a theory for predicting the transport of colloids in heterogeneously charged porous media, based on the single collector removal efficiency (e.g. Song *et al.*, 1994; Johnson *et al.*, 1996; Ryan and Elimelech, 1996; Elimelech *et al.*, 2000; Bhattacharjee *et al.*, 2002). Most of these studies were carried out using artificial colloids with exactly similar surface charge at high concentrations (10^9 to 10^{14} cells/mL) in solutions consisting of simple monovalent or divalent electrolytes. However, Simoni *et al.* (1998) clearly demonstrated that when a biocolloid (in their case, *Pseudomonas sp.*) was used instead of an artificial colloid, population heterogeneity greatly influenced the transport through sand columns; they attributed this to differences in the lipopolysaccharide composition in the outer membrane of the bacteria. Dong (2002) and Foppen and Schijven (2005; Chapter 5) also found that population heterogeneity was an important factor in the transport of bacteria in saturated porous media.

The importance of dissolved organic matter (DOM) is widely recognised, because the sorption of hydrophobic organic compounds and trace metals onto DOM may significantly increase their aqueous solubility (Yamamoto *et al.*, 2002). DOM in groundwater is a complex mixture of acidic, neutral, and basic organic poly-molecules (Thurman, 1985). Concentrations vary widely and may be as high as 100-200 mg/L near the water table to less than 0.5 mg/L in deeper groundwater (Thurman, 1985). Often pathogens travel through soils and aquifers in a plume of wastewater. DOM in wastewater is composed of humic materials, polysaccharides, polyphenols, proteins, lipids, and heterogeneous organic molecules (Fujita *et al.*, 1996; Ma *et al.*, 2001; Imai *et al.*, 2002; Ilani *et al.*, 2005) with concentrations ranging from 1-100 mg/L, dependent on types of treatment. Because of the importance of DOM for the transport of certain types of contaminants, the sorption of DOM to various kinds of metal oxides in aquifers has received considerable attention in literature (e.g. Chi and Amy, 2004; Kaiser, 2003; Gu *et al.*, 1994, 1996; Amirbahman and Olson, 1993). The ligand exchange mechanisms between OH groups at the surface of the metal oxide and carboxyl groups of the DOM, accompanied by an increase in pH has been mentioned several times as the dominant sorption mechanism (Parfitt *et al.*, 1977; Murphy *et al.*, 1990; Tipping, 1981; Davis, 1982; Gu, 1994).

From equilibrium batch experiments, it has become clear that DOM tends to compete with viruses for attachment sites on the surface of soil particles and by so doing reduces virus attachment (Gerba, 1984). DOM can also be used to elute previously attached viruses (Sobsey *et al.*, 1980). Furthermore, Dizer *et al.* (1984) showed that in sand columns at pH 7, the attachment rates of poliovirus 1, coxsackieviruses A9 and B1, echovirus 7 and rotavirus SA11 were lower in percolating secondary effluent than in percolating groundwater, tertiary effluent and distilled water.

To date, little research has been done on the combined effect of favourable attachment sites (e.g. from ferric oxyhydroxides) and DOM on the attachment of viruses. Therefore, the study we describe here set out to test the theory for predicting the transport of colloids in heterogeneously charged porous media, as mentioned above. We used PRD1, a model virus, at low concentrations (10^5 virus/mL) in various solutions consisting of monovalent and divalent salts plus DOM, in experiments with columns containing various sediment mixtures of quartz sand and goethite-coated sand.

7.2 Theory

The attachment rate coefficient, k_a (d^{-1}), which is used in transient mass balance calculations for bacteria retained on the solid matrix, is defined as (e.g. Tufenkji and Elimelech, 2004)

$$k_a = \frac{3(1-\theta)}{2a_c} v\eta \qquad (1)$$

where θ is the effective porosity (-), a_c is the grain size diameter (m), v is the pore water flow velocity (m/d) and η is the dimensionless single collector removal efficiency (SCRE). The SCRE is a parameter representing the ratio of the rate of particles striking and sticking to a collector to the rate of particles approaching the collector. Johnson *et al.* (1996) and Elimelech *et al.* (2000) have shown that when the geochemical composition of the sediment is known, grain surface charge heterogeneity can be included in the formulation of the SCRE by

$$\eta = \eta_0 [\lambda \alpha_f + (1-\lambda)\alpha_u] = \eta_0 \alpha_{total} \qquad (2)$$

where η_0 is the dimensionless single collector contact efficiency (SCCE), determined from physical considerations (e.g. Tufenkji and Elimelech, 2004), α_f and α_u are the dimensionless sticking efficiencies to favourable and unfavourable attachment sites respectively, and λ is a dimensionless heterogeneity parameter describing the fraction of surface area favourable to attachment. For PRD1, which at typical groundwater pH values (6-8) is mostly negatively charged (Harvey and Ryan, 2004; Loveland *et al.*, 1996; Schijven, 2001), favourable attachment sites are positively charged (e.g goethite), while unfavourable sites are negatively charged (e.g. quartz).

Values for the sticking efficiency, α_{total}, can be obtained from column experiments for given physicochemical conditions (i.e. suspended particles, porous medium, and solution chemistry; Tufenkji and Elimelech, 2004)

$$\alpha_{total} = -\frac{2}{3}\frac{a_c}{(1-\theta)L\eta_0}\ln\left(\frac{C}{C_0}\right) \qquad (3)$$

where C/C_0 is the column outlet normalised bacteria concentration at the initial stage of the particle breakthrough curve and L is the packed length of the filter medium (m). We determined the SCCE, η_0, with a short form of the Tufenkji-Elimelech correlation equation (Tufenkji and Elimelech, 2004)

$$\eta_0 = 2.44 A_S^{1/3} N_R^{-0.081} N_{Pe}^{-0.715} N_{vdW}^{-0.052} \qquad (4)$$

where A_S is a porosity-dependent parameter of Happel's model, N_R is an aspect ratio, N_{Pe} is the Peclet number and N_{vdW} is the van der Waals number. Eq. (4) is to account for virus transport caused by diffusion. The Tufenkji-Elimelech correlation equation also accounts for collisions due to interception and sedimentation. However, because pore water flow velocities were low and the virus is small (< 100 nm), the diffusion mechanism is by far the most dominant and therefore we ignored interception and sedimentation.

The evaluation of the colloid transport theory in heterogeneously charged porous media concentrates on the question of whether eq. (2) is indeed valid. To elucidate this, we performed a series of laboratory column experiments in which 5 chemically different PRD1 suspensions (Table 1) were flushed through sediment mixtures of quartz sand with varying fractions of goethite-coated sand ($\lambda = 0, 0.025, 0.05, 0.075, 0.1, 0.5$). The choice of the 5 solutions was to simulate various uncontaminated and contaminated groundwater environments. Goethite was chosen because it is the commonest iron mineral in aquifers.

7.3 Materials and Methods

7.3.1 PRD1 suspensions

Highly concentrated suspensions of PRD1 were prepared as described by ISO 10705-1 (1995) using host *Salmonella typhimurium* LT2 and pre-diluted approximately 10^4 times to a stock suspension containing 10^9 plaque-forming particles per mL (pfp/mL) that was stored in the dark at 4 °C. Five chemically different solutions were prepared (Table 1). The NaHCO3 and NaHCO3-LHA solutions were assumed to be representative of PRD1 in uncontaminated groundwater, while the AS-LHA, AS-HHA and AS-P-HHA were assumed to

Table 1: Chemical compositions of solutions used (LHA: low Na-humic acid concentration; AS: artificial sewage; HHA: high Na-humic acid concentration; P: phosphate buffer)

Exp. #	Solution	Composition
1	NaHCO3	6.5 mM $NaHCO_3$
2	NaHCO3-LHA	6.5 mM $NaHCO_3$, 11 mg/L Na-Humic Acid
3	AS-LHA	0.25 mM $CaCl_2$, 0.92 mM $MgSO_4$, 10.56 mM NaCl, 11 mg/L Na-Humic Acid
4	AS-HHA	0.25 mM $CaCl_2$, 0.92 mM $MgSO_4$, 10.56 mM NaCl, 110 mg/L Na-Humic Acid
5	AS-P-HHA	0.25 mM $CaCl_2$, 0.92 mM $MgSO_4$, 10.56 mM NaCl, 0.06 mM KH_2PO_4, 0.13 mM K_2HPO_4, 0.13 mM Na_2HPO_4, 110 mg/L Na-Humic Acid

represent PRD1 travelling in a plume of wastewater. To simulate DOM, lignite-based Na-humic acid (ACROS Organics, Belgium) was used. The total organic carbon (TOC) concentration resulting from dissolving Na-humic acid was determined with a 700 Model Analyser (O-I Corporation, USA). All the chemicals used to prepare the various suspensions were obtained from J.T. Baker (Mallinckrodt Baker Inc.). Seeding suspensions of PRD1 were prepared by diluting each of the five chemical solutions with 0.1 mL of the stock suspension.

7.3.2 Column sediment mixtures

A commercially available 99.1% quartz sand (Eurogrid B.V.) with a grain size of 180-500 μm was used. The median of the grain size number distribution was 235 μm. Total porosity of the quartz sand (clean or coated) was determined gravimetrically to be 0.38. Prior to the experiments, the sand was washed to remove metal and/or metal oxides from the grain surface (Chu *et al.*, 2001). Five hundred mL 0.2 M citrate buffer containing 44.1 g/L sodium citrate ($Na_2C_6H_5O_7 \cdot 2H_2O$) and 10.5 g/L citric acid ($H_2C_6H_5O_7$) was added to 2 L centrifuge bottles containing 300 g of water-washed sand. The bottles were put in a shaking water bath at 80° C. After the solution reached 80° C, 15 g of sodium dithionite ($Na_2S_2O_4$) was added to each bottle. The bottles were removed from the shaker and hand-shaken vigorously for several seconds. They were then returned to the shaker and shaken rapidly for 1 min, after which the speed was reduced and the samples were digested for 15 min with occasional rapid shaking. The solution was decanted from the sand and the procedure repeated twice more, followed by rinsing the sand with deionised water until the electric conductivity of the water was close to zero and the sand samples were odourless.

Iron-coated sand: After washing, a portion of the sand was coated with goethite (Schwertmann and Cornell, 1991). To do so, 13.9 g of unoxidised crystals of $FeSO_4 \cdot 7H_2O$ was dissolved in 1 L distilled water through which N_2 gas had been bubbled for 30 minutes prior to the coating process, in order to remove oxygen. To the solution 300 g of citric acid/dithionite washed sand and 110 mL of 1 M $NaHCO_3$ solution were added. Then the N_2 gas was replaced by air with a flow rate of 30-40 mL/min, while stirring continuously. Oxidation of Fe(II) was complete after 48 hours, during which time the suspension changed colour from green-blue to ochre. After coating, the amorphous iron (III) concentration of the sand was determined (in triplicate) with an ascorbate extraction step developed by Kostka and Luther (1994) to be 5.24 mg Fe/g sand (determined with an ICP (Perkin-Elmer Plasma 1000)).

7.3.3 Column experiments

Five perspex columns (with an inside ∅ of 5 cm and a height of 8 cm) were incrementally packed under saturated conditions with one of the sediment mixtures to a column length of 7 cm. The columns were connected to a pump (Watson-Marlow 101U/R) operating at a rate of 0.45-0.60 mL/min. Prior to each experiment the columns were flushed for at least 12 hours in order to remove fines. At the beginning of each experiment, the flushing solution in the column above the sediment was removed by suction until the water level was exactly level with the top of the sediment. Then, a 1 cm layer of PRD1 suspension was gently sprinkled on top of each column. This layer was maintained throughout the experiment by adding

PRD1 suspension manually. At the end of the "loading" phase, which took 3 pore volumes, the same procedure was followed in order to replace the viral suspension by a $NaHCO_3$ solution without PRD1. Deloading also took 3 pore volumes. Every 1-2 pore volumes the effluent EC and pH were determined. EC was measured with a Tetracon 325 probe and pH with a SenTix 21 probe. All experiments were carried out at 4° C, in order to discourage the possible decay of PRD1.

7.3.4 PRD1 quantification

The PRD1 in the effluent was assayed according to ISO 10705-I (1995) using *Salmonella typhimurium* LT2 as the host, omitting nalidixic acid in the top agar layer. Virus concentrations from effluent samples were determined by step dilutions in peptone physiological saline (PFS) and assayed on a basal medium made up of Tryptone Yeast Glucose Agar (TYGA) in 9 cm petri dishes. Plates were incubated overnight (about 12 hours) at 37 °C . PRD1 was quantified by counting the plaques formed in the TYGA.

7.3.5 Inactivation experiments

To determine the inactivation rate coefficient of PRD1, 1 L NaHCO3 and AS-P-HHA solutions were dosed with a 10^5 virus/mL PRD1 concentration. The suspensions were kept in glass bottles wrapped with black plastic and kept at 5°C. Samples were taken frequently within an 80-days period and the PRD1 concentration was determined according to procedures outlined above.

7.4 Results

7.4.1 Conditions during the experiments

The EC values of the effluent ranged from 580 μS/cm for the NaHCO3 and NaHCO3-LHA experiments to 1490 μS/cm for the AS-P-HHA experiments, while the pH varied between 6.9 and 8.3 (Table 2). The EC and pH were constant during each column experiment. Inactivation rates of PRD1 (Fig. 1) varied from a very low 0.0009 day^{-1} for the AS-P-HHA suspension to 0.014 day^{-1} for the NaHCO3 suspension. From this we concluded that inactivation during each experiment (which lasted 8-12 hours) was negligible.

Table 2: Experimental conditions and values of calculated parameters (values used for determining the single collector contact efficiency, η_0, are: $a_c = 0.235$ mm, $a_p = 67$ nm, fluid viscosity = $1.005*10^3$ kg/m s, temperature = 278 K, Hamaker constant = $6.0*10^{-21}$ J (Loveland et al., 1996))

Exp. #	Solution	Porous Media	Pump Velocity (ml/min)	EC (effluent) (μS/cm)	pH (effluent)	Average C/Co (2-3.5 PV)	λ	η_0	α_{total}
1	**NaHCO3**	Clean Sand	0.50		8.3	5.52E-01	0.000	0.433	0.0546
		2.5% Fe	0.52		8.2	2.81E-04	0.025	0.421	0.0682
		5% Fe	0.60	590 - 620	8.2	1.04E-04	0.050	0.380	0.0787
		7.5% Fe	0.50		8.2	3.31E-05	0.075	0.433	0.0980
		10% Fe	0.56		8.2	1.75E-05	0.100	0.399	0.0914
		50% Fe	0.60		8.1	1.60E-04	0.500	0.380	0.0791
2	**NaHCO3-LHA**	Clean Sand	0.52		8.1	9.13E-01	0.000	0.421	0.0009
		10% Fe	0.48	580 - 610	8.1	7.99E-01	0.100	0.446	0.0019
		50% Fe	0.47		8.1	7.94E-01	0.500	0.452	0.0019
3	**AS-LHA**	Clean Sand	0.48		6.9	9.14E-01	0.000	0.446	0.0007
		2.5% Fe	0.50		6.9	8.42E-01	0.025	0.433	0.0014
		5.0% Fe	0.52	1290 -1320	6.9	5.22E-01	0.050	0.421	0.0054
		7.5% Fe	0.49		6.9	1.82E-01	0.075	0.439	0.0146
		10% Fe	0.50		6.9	2.07E-02	0.010	0.433	0.0319
		50% Fe	0.46		6.9	2.54E-03	0.050	0.459	0.0498
4	**AS-HHA**	Clean Sand	0.48			8.74E-01	0.000	0.446	0.0011
		10% Fe	0.47		7.1	8.79E-01	0.100	0.452	0.0010
		50% Fe	0.48	1290 -1320		8.62E-01	0.500	0.446	0.0012
5	**AS-P-HHA**	Clean Sand	0.53		7.3	8.31E-01	0.000	0.415	0.0015
		2.5% Fe	0.53		7.3	8.89E-01	0.025	0.415	0.0010
		5.0% Fe	0.49		7.3	8.52E-01	0.050	0.439	0.0014
		7.5% Fe	0.49	1300-1490	7.3	7.39E-01	0.075	0.439	0.0025
		10% Fe	0.48		7.3	8.34E-01	0.010	0.446	0.0015
		50% Fe	0.46		7.2	5.03E-01	0.050	0.459	0.0056

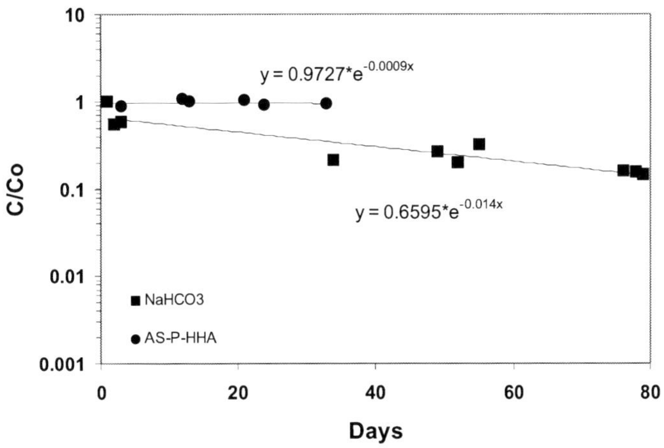

Fig. 1: *Inactivation of PRD1 in NaHCO3 and in AS-P-HHA solutions. Inactivation rates were 0.0061 d^{-1} for PRD1 in the NaHCO3 solution and 0.0004 d^{-1} for PRD1 in the AS-P-HHA solution*

7.4.2 Breakthrough curves

In general, three phases could be distinguished from the breakthrough curves (see Figs. 2-6). During the first 2 pore volumes (PV), corresponding with the rising limb of the breakthrough curves. The increase in phage recovery during this phase was rapid, unretarded, and tracer-like. The second phase lasted from 2-3.5 PV.

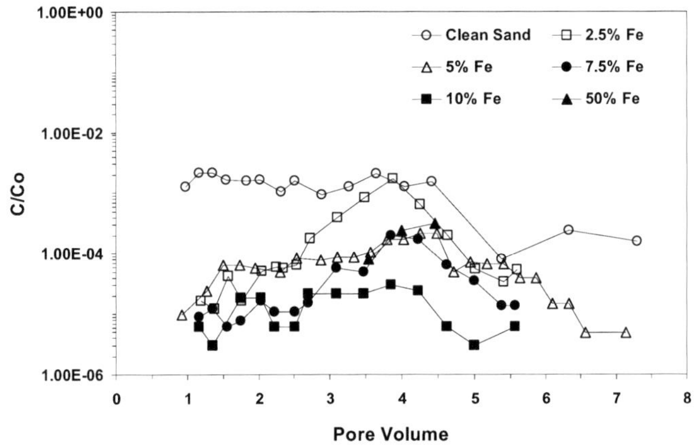

Fig. 2: *PRD1 breakthrough curves for the NaHCO3 solution for various quartz sand and goethite-coated sediment mixtures (λ = 0, 0.025, 0.05, 0.075, 0.1, 0.5)*

In all breakthrough curves a plateau was visible with more or less constant C/C_0 values. The height of the plateau varied, depending on suspension used and on the fraction of goethite added to the sediment. Maximum C/C_0 values were lowest in the

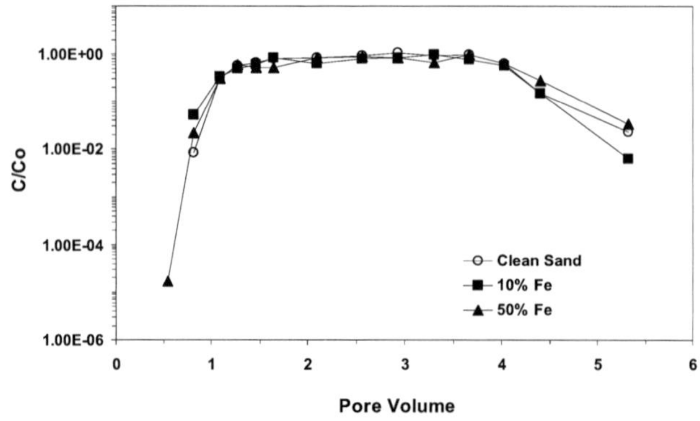

Fig. 3: *PRD1 breakthrough curves for the NaHCO3-LHA solution for various quartz sand and goethite-coated sediment mixtures (λ = 0, 0.1, 0.5)*

NaHCO3 experiment with goethite-coated sand (as low as ~ 10^{-5}; Fig. 2). The NaHCO3 clean quartz sand experiment also resulted in low C/C_0 values (~ 10^{-3}), which we found surprising because the retention of PRD1 in clean quartz sand experiments is usually extremely low (Schijven, 2001): in our small (7 cm) columns, the removal of PRD1 in quartz sand was 3 log units. Apparently, this type of "clean" quartz sand still contained impurities that caused this removal. Intriguingly, when only a small amount of humic acid was added to this solution (NaHCO3-LHA with 11 mg/L Na-Humic Acid, equivalent to 3.3 mg/L TOC; Fig. 3) in all cases breakthrough was complete within 2 pore volumes, regardless how much goethite-coated sand had been added to the sediment. Even the 50% goethite fraction resulted in complete breakthrough of PRD1 without significant retention. When the ionic strength was increased from $6.5*10^{-3}$ M (NaHCO3-LHA) to $14.99*10^{-3}$ M (AS-LHA; Fig. 4), also adding Ca and Mg to the suspension led the maximum C/C_0

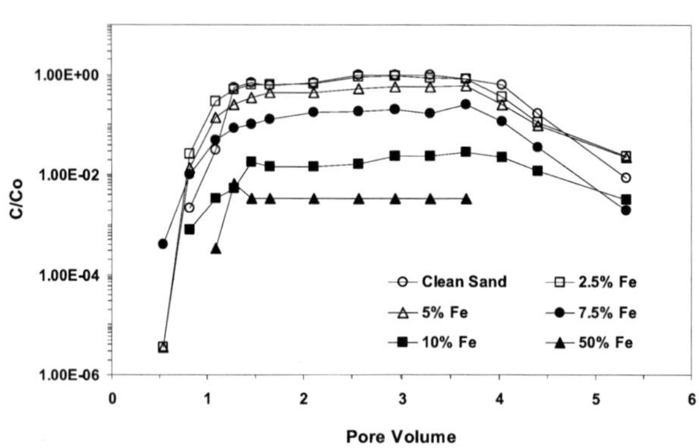

Fig. 4: *PRD1 breakthrough curves for the AS-LHA solution for various quartz sand and goethite-coated sediment mixtures (λ = 0, 0.025, 0.05, 0.075, 0.1, 0.5)*

values to fall to ~10^{-3} for the 50% goethite coated sand case. When the humic acid concentration was increased to 33 mg/L TOC (AS-HHA; Fig. 5), breakthrough was

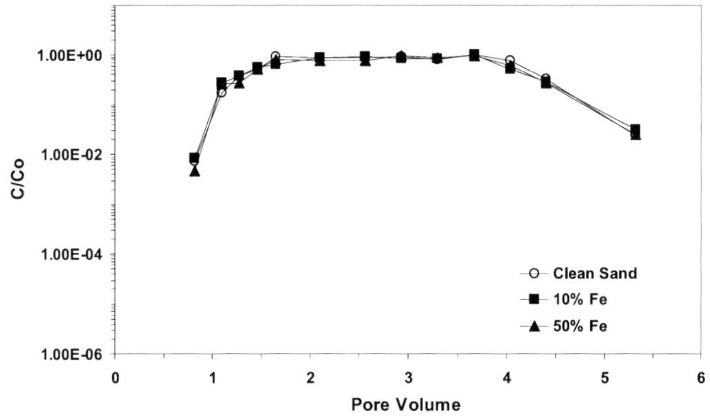

Fig. 5: *PRD1 breakthrough curves for the AS-HHA solution for various quartz sand and goethite-coated sediment mixtures (λ = 0, 0.1, 0.5)*

again complete without significant retention, and upon addition of phosphate (AS-P-HHA; Fig. 6), the maximum C/C_0 values fell to ~10^{-1} for the 50% goethite-coated sand case. During the third phase (the deloading phase), cell recoveries fell by around 2 log units within 3 pore volumes after deloading started. Due to time constraints, the sampling of the column effluents was not extended, but our experience (Schijven *et al.*, 2002) leads us to expect that following the three pore volumes after deloading, the tails of the breakthrough curves would become flatter. Commonly, detachment is slower than attachment. So, in our case, where inactivation may also be ignored, the level of the maximum breakthrough concentration was primarily determined by the rate of attachment to the sand grains.

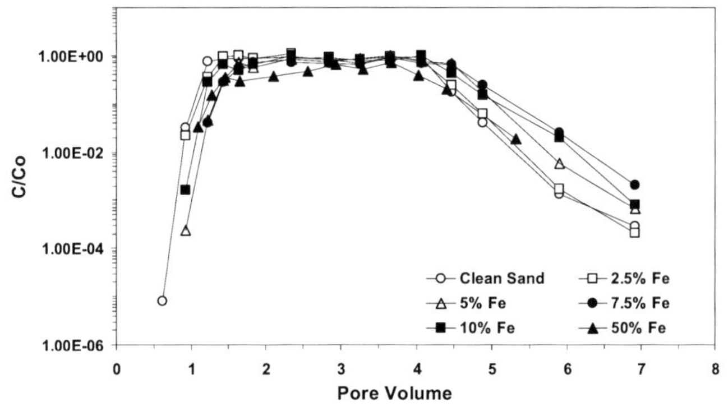

Fig. 6: *PRD1 breakthrough curves for the AS-P-HHA solution for various quartz sand and goethite-coated sediment mixtures (λ = 0, 0.025, 0.05, 0.075, 0.1, 0.5)*

7.4.3 Sticking efficiency

Table 2 shows C/C_0 (averaged over the interval 2-3.5 PV), SCCE and total sticking efficiency (α_{total}) values. Figure 7 shows the relation between α_{total} and λ. Total sticking efficiency varied between a minimum of 0.0007 for the clean sand NaHCO3-LHA case and as much as 0.0682 - 0.0980 for the NaHCO3 suspension goethite-coated sand cases. From Fig. 7, two important observations were made. Firstly, in all experiments, α_{total} varied enormously (over a 2 log unit range), but regardless of the chemical composition of the suspension, when the humic acid concentration was high (or high enough), there seemed to be a minimum sticking efficiency of 0.001 ± 0.0005. Secondly, in the NaHCO3 experiments, the NaHCO3-LHA and the AS-LHA experiments, α_{total} rose gradually with increasing λ and levelled out at $\lambda \geq 0.1$. For the AS-HHA experiments α_{total} did not increase with increasing λ, while for the AS-P-HHA experiments, α_{total} continued to increase gradually with increasing λ,without levelling off.

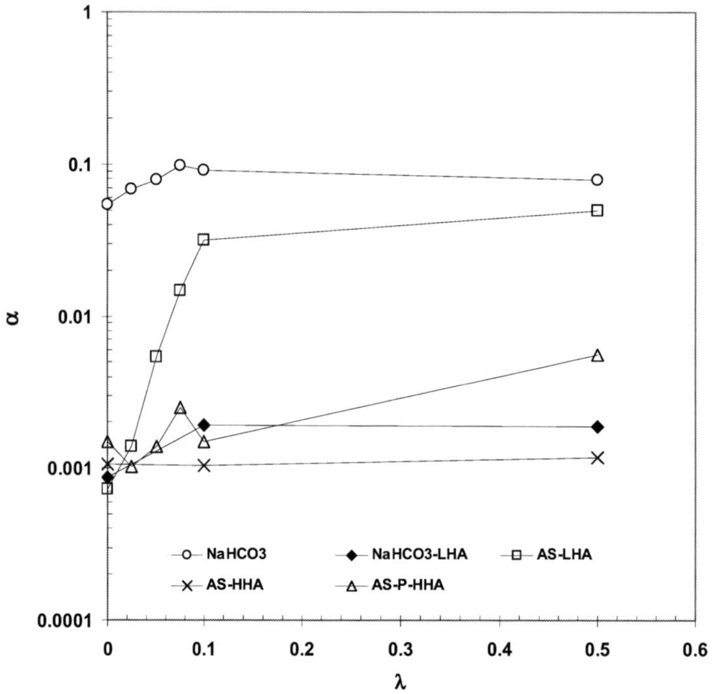

Fig. 7: Relation between calculated total sticking efficiency (α_{total}) and geochemical heterogeneity parameter (λ) for all experiments

7.5 Discussion

7.5.1 The effect of solution chemistry and humic acid on the total sticking efficiency

Although the sand was almost 100% quartz and had been thoroughly washed prior to the column experiments, removal of PRD1 was still considerable, resulting in a high sticking efficiency of 0.0546 (Table 1). We attribute this to microscopic chemical heterogeneities on the surface (and thus the surface charge) of the grains. Others (Tufenkji and Elimelech, 2004a, 2005, 2005a; Li *et al.*, 2004; Walker *et al.*, 2004; Dong *et al.*, 2002; Elimelech *et al.*, 2000) have also found that commonly used collectors (such as soda lime glass beads and borosilicate microscope cover slips) possess microscopic surface charge heterogeneities, which may have a profound influence on colloid adhesion behaviour. Adding goethite to the quartz sediment resulted in more attachment. The pH_{PZC} of goethite (α-FeOOH), is known to be 9.0 -9.1 (Gaboriaud and Ehrhardt, 2003) and since this pH_{PZC} value was higher than the solution pH (6.9-8.3), the mineral was positively charged. At the solution pH values we used, PRD1 is negatively charged ($pH_{PZC}<3$; Harvey and Ryan, 2004) and therefore the addition of goethite led to more attachment. Adding only a few mg/L TOC (NaHCO3-LHA suspension; Fig. 3) almost stopped removal. We think this was caused by sorption of DOM to the goethite surface, which can be explained by ligand exchange (e.g. Parfitt *et al.*, 1977; Murphy *et al.*, 1990; Tipping, 1981; Davis, 1982; Gu, 1994, 1996; Chi and Amy, 2004;)

$$[Mineral]\!\!-\!OH + {}^{-}OOC\!-\![DOM] \rightarrow [Mineral]\!\!-\!OOC\!-\![DOM] + OH^{-} \qquad \text{(R1)}$$

DOM and PRD1 either competed for the same attachment sites or, due to sorption of DOM, attachment of PRD1 was sterically hindered. Upon increasing the ion strength and adding a small amount of Ca and Mg (AS-LHA; Fig. 4), attachment increased. It is difficult to explain this by DLVO theory, because the increase in ion strength reduced the double layer repulsion only slightly (in the case of similarly charged PRD1 and quartz mineral surface) and did not deepen the primary energy well (in the case of a negatively charged PRD1 surface and positively charged goethite surface). Neither can this increased attachment be attributed to the shielding effect that Zhuang and Jin (2003) described. In fact, those authors found the opposite: in their case, due to an increase in ionic strength the sorption decreased when particles and the surface were oppositely charged, because the increasing salt concentration in aqueous solutions shields the electronic charges of both the particles and the sorbing surface and thereby reduces electrostatic attraction. We think that increased attachment was caused by a combination of cation bridging and ligand exchange of PRD1 according to:

$$[Mineral]\!\!-\!OH + {}^{-}OOC\!-\![PRD1] + M^{2+} \rightarrow [Mineral]\!\!-\!O\!-\!M\!-\!OOC\!-\![PRD1] + H^{+} \qquad \text{(R2)}$$

in which M^{2+} is a divalent cation, like Ca^{2+} or Mg^{2+}. When the DOM concentration increased (AS-HHA; Fig. 5), cation-bridging with DOM became more important (e.g. Thurman, 1985; Chi and Amy, 2004):

$$\begin{aligned}&[Mineral]\!-\!OH + {}^{-}OOC\!-\![DOM] + M^{2+} \rightarrow \\ &\qquad [Mineral]\!-\!O\!-\!M\!-\!OOC\!-\![DOM] + H^{+}\end{aligned} \quad \text{(R3)}$$

Due to R3, attachment of PRD1 reduced, because DOM was out-competing the PRD1 for the same attachment sites or because of increased sterical hindrance of the DOM.

It is not easy to explain why attachment increased after phosphate had been added to the suspension (AS-P-HHA; Fig. 6), and, because the attachment increase was limited, the question arises whether the results were statistically different from the AS-HHA case. It is known that phosphate is a strong complexing ligand (Chi and Amy; 2004) competing with DOM for the same binding sites on the surface of goethite. Others, such as Griffioen (1994) and Foppen and Griffioen (1995) have reported the uptake of phosphate by iron hydroxides. We therefore expected that PO_4^{3-} ions would in some way compete with both PRD1 and DOM for the same sites, but did not expect that this would result in more attachment of PRD1. A possible explanation could be that in addition to the competition mechanism, there was cation bridging between the phosphate ion and PRD1, similar to reactions R2 and R3. Another explanation could be that due to the sorption of phosphate to the iron surface, less DOM adsorbed, causing less steric hindrance for PRD1, possibly resulting in more PRD1 attachment. Whatever the possible attachment mechanisms, the most important outcome of our experiments was that the effect of geochemical heterogeneity on the transport of PRD1 was clearly shown. However, the presence of DOM, or, more generally, the chemical composition of the solution was the most important factor determining the transport of PRD1, which caused the C/C_0 values in our experiments to increase as much as 5 log units, thereby almost completely eliminating the effect of geochemical heterogeneity.

7.5.2 Non-linear relationship between sticking efficiency and heterogeneity parameter

An important observation was that the sticking efficiency was not linearly dependent on λ (Fig. 7). We believe that the cause may be population heterogeneity of the PRD1 particles, *i.e.* a small fraction of the PRD1 particles did not readily attach to goethite. At low λ values (< 0.1), λ seemed to determine the removal of PRD1, as long as there was an apparent abundance of appropriately charged PRD1 that could attach. At higher λ values, the characteristics of the PRD1 population increasingly determined the attachment: the most easily attachable PRD1 with sufficient negative charge had already attached, and the less attachable PRD1 with little or no charge remained in solution, despite the fact that attractive van der Waals forces were still operative. This non-linear dependence of the sticking efficiency on λ that we found is not included in eq. (2) and therefore that equation is not valid within the boundary conditions of our set of experiments. Clearly, more research is required to elucidate the mechanism behind this non-linear dependence, so that a more adequate relationship between sticking efficiency and geochemical heterogeneity can be formulated.

Reports of observed variation in attachment characteristics due to colloid population heterogeneity in the literature are rare, perhaps because of the limited variation in the zeta potential of the artificial colloids used (e.g. Johnson and Elimelech, 1995; Johnson *et al.*, 1996; Elimelech *et al.*, 2000). Probably one of the best examples of population heterogeneity affecting the transport of colloids in saturated porous media is that described by Simoni *et al.* (1998). They found that only a fraction of the cells of *Pseudomonas* strain B13 was efficiently deposited in sand columns, while the remainder passed through an identical second sand column without hindrance. In an unconfined aquifer at Cape Cod (USA), Blanford *et al.* (2005) found that after an initial decrease, the aqueous PRD1 concentrations remained essentially constant, irrespective of variations in geochemical properties within the aquifer. Therefore, Blanford *et al.* (2005) concluded that apparently a small fraction of viable PRD1 particles persist in the aqueous phase and may travel significant distances in the underground environment. Dong (2002) also found that the variation of surface charge density in a monoclonal bacterial population of strain *Comamonas* sp. DA001 helped explain the large variation (several orders of magnitude) of the experimentally determined collision efficiencies.

7.5.3 Implications for pathogen transport in aquifers

Although the inactivation rate of PRD1 (at 5 ± 3 °C) was low in all cases, it was significantly lower in the presence of DOM. It appears that DOM has a protective effect on the survival of PRD1. Both Ryan *et al.* (2002) and Blanford *et al.* (2005) also found that inactivation rates of PRD1 suspended in water can be reduced in the presence of anionic surfactants, which are a major component of DOM. If the same applies to pathogenic viruses too, then this implies that in the presence of DOM such viruses can be transported longer distances underground.

Sticking efficiencies of 0.001 under field conditions are high enough to reduce virus concentrations effectively (Schijven *et al.*, 1999, 2000, 2002). If attachment sites are absent or inaccessible, the sticking efficiencies may be as low as 10^{-4} to 10^{-6} (Schijven and Hassanizadeh, 2002). In our column experiments, sticking efficiencies were at least 10^{-3} in those cases where reduction of concentrations was negligible. Lower estimates of α could not be made, possibly due to a scaling problem, i.e. in such short columns (7 cm) the accuracy of estimating α is limited. Analysis of the retained PRD1 concentration profiles would have been helpful to determine the distribution of attachment rate coefficients and associated sticking efficiencies (Li *et al.*, 2004; Tufenkji and Elimelech, 2005 and 2005a).

Our general conclusion is that in the presence of DOM, viruses can be transported for long distances thanks to two effects: attachment is poor because of hypothesized DOM occupation of sites favourable for attachment, and inactivation of virus may have decreased. This vindicates the conservative assumption for attachment (α ranges from 10^{-4} to 10^{-6}) of virus that Schijven *et al.* (2006) used when calculating protection zones for groundwater wells.

CHAPTER 8:

TRANSPORT OF *ESCHERICHIA COLI* IN SATURATED POROUS MEDIA: DUAL MODE DEPOSITION AND INTRA-POPULATION HETEROGENEITY

Foppen, J.W.A., M. van Herwerden, and J.F. Schijven, 2007. Transport of *Escherichia coli* in saturated porous media: dual mode deposition and intra-population heterogeneity. Wat. Res. DOI: 10.1016/j.watres.2006.12.041

Abstract

Because of heterogeneity among members of a bacteria population, deposition rates of bacteria may decrease upon the distance bacteria are transported in an aquifer. Such deposition rate decreases may result in retained bacteria concentrations, which decrease hyper-exponentially as a function of transport distance, and may therefore significantly affect the transport of colloids in aquifers. We investigated the occurrence of hyper-exponential deposition of *Escherichia coli*, an important indicator for fecal contamination, and the causes for such behavior. In a series of column experiments with glass beads of various sizes, we found that attachment of *E. coli* decreased hyper-exponentially, or, on logarithmic scale in a bimodal way, as a function of the transported distance from the column inlet. From data fitting of the retained bacteria concentration profiles, the sticking efficiency of 40% of the *E. coli* population was high ($\alpha = 1$), while the sticking efficiency of 60% was low ($\alpha = 0.01$). From the *E. coli* total population, an *E. coli* subpopulation consisting of slow attachers could be isolated by means of column passage. In subsequent column experiments this subpopulation attached less than the *E. coli* total population, consisting of both slow and fast attachers. We concluded that the main driver for the observed dual mode deposition was heterogeneity among members of the bacteria population. Intra-population may result in some microbes traveling surprisingly high distances in the subsurface. Extending the colloid filtration theory with intra-population variability may provide a valuable framework for assessing the transport of bacteria in aquifers.

8.1 Introduction

Colloid filtration theory (CFT) is a commonly used approach for describing attachment of *E. coli*, one of the most important indicator organisms for fecal pollution, during transport through porous media (Matthess et al., 1988; Powelson and Mills, 2001; Pang et al., 2003; chapter 5 of this study). The most important characteristic of the CFT is that the removal of particles is described by first-order kinetics with a spatially and temporally constant rate of particle deposition, resulting in concentrations of suspended and retained particles that decrease log-linear with distance. However, recent studies (Simoni et al., 1998; Baygents et al., 1998; Redman et al., 2001; Li et al., 2004; Tufenkji and Elimelech, 2004a, 2005b; Tong and Johnson, 2006) have found that colloid retention decreased hyper-exponentially with distance, suggesting that the attachment rate coefficient is not constant and that the CFT is not applicable to those cases. In the absence of straining, which can be defined as the trapping of *E. coli* in pores that are too small to pass through (e.g. Bradford et al., 2003), Tufenkji and Elimelech (2004a) attributed this hyper-exponential deviation from CFT to the concurrent existence of both favorable and unfavorable colloidal interactions with collector surfaces. As supporting evidence, the contribution of fast deposition sites was demonstrated to be equal to the fraction of colloids depositing in the secondary energy well (unfavorable attachment) and on charged metal oxide impurities (favorable attachment) on the glass bead collector surface. In a response to the work of Tufenkji and Elimelech (2004a), Johnson and Li (2005) argued that collector heterogeneity cannot by itself generate the observed deviation, but it can amplify the distribution induced by heterogeneity among the colloidal population.

The occurrence of hyper-exponential retained profiles due to dual mode deposition of specifically *E. coli* during transport in saturated porous media has not yet been studied (Chapter 2). Since bimodal deposition can have a profound effect on the distances this important fecal indicator organism can travel in the subsurface, we investigated the occurrence of dual mode deposition of *E. coli*. In addition, we examined whether the fast depositing bacteria fraction was equal to the bacteria fraction deposited in the secondary minimum and on bulk surface impurities. Also, we examined the presence of attachment variations within our monoclonal bacteria population, and the effect on bacteria breakthrough concentrations.
Column experiments with six well defined narrow glass bead sizes ranging from 38 to 425 micrometer were conducted. We used this range of glass bead sizes, because from theoretical considerations, the attachment rate coefficient increases with decreasing bead size (Fig. 1).

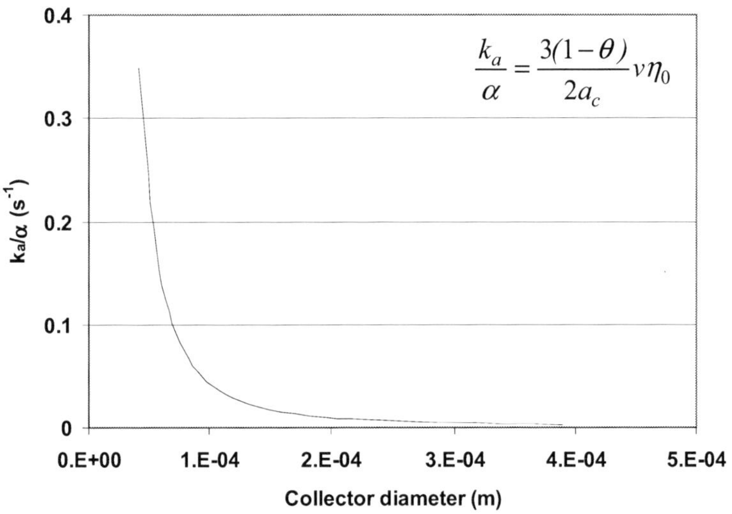

*Figure 1. The ratio of the attachment rate coefficient, k_a, to the sticking efficiency, α, as a function of the collector diameter. Values used in eq. (1) are θ = 0.38 and v = 2.9x10^{-4} ms^{-1}. The value of η_0 for each grain size was calculated with the Tufenkji-Elimelech correlation equation. Values used were: colloid diameter = 1.0 μm, particle density = 1055 kg/m^3, fluid density = 1000 kg/m^3, fluid viscosity = 1.005*10^3 kg/m s, temperature = 278 K, and the Hamaker constant = 6.5x10^{-21} J*

Therefore, we hypothesized that, if dual mode deposition would take place, retained profiles would become increasingly hyper-exponential as a function of transport distance on decreasing bead size. The relation between the attachment rate coefficient, k_a (in s^{-1}), and glass bead diameter, a_c (in m), was defined as (Harvey and Garabedian, 1991; Ryan and Elimelech, 1996):

$$\frac{k_a}{\alpha} = \frac{3(1-\theta)}{2a_c} v\eta_0 \tag{1}$$

where θ is the effective porosity, v is the pore water flow velocity (ms^{-1}), η_0 is the dimensionless single collector contact efficiency (SCCE), and α is the dimensionless sticking efficiency of the colloid. The single collector contact efficiency, η_0, is defined as (Yao et al., 1971; Tufenkji and Elimelech, 2004)

$$\eta_0 = \eta_D + \eta_I + \eta_G \tag{2}$$

where η_D, η_I and η_G, represent theoretical values for the single-collector contact efficiency when the sole transport mechanisms are diffusion, interception, and sedimentation, respectively. The sticking efficiency can be considered constant for various collector diameters in the case where bacteria, collector type and experimental conditions remain constant. So, when a_c decreases, k_a and η_0 increase. The latter increases because the interception and diffusion terms increase when a_c decreases (e.g. Tufenkji and Elimelech, 2004). To determine the fraction of bacteria cells deposited in the secondary minimum and on favorable attachment sites, particle elution experiments with low ionic strength and high pH solutions were carried out. Bacteria intra-population heterogeneity was studied by analyzing breakthrough curve (BTC) variations of suspensions of bacteria with and without fast attaching bacteria. Because we used small glass beads, straining could also take place. By varying the ionic strength of the bacteria suspensions, the relative contribution of straining and attachment associated with the retained profiles was assessed. Also, to qualitatively determine the presence or absence of straining, we made use of an experimentally simple technique, derived from filter backwashing.

8.2 Theory

The one-dimensional macroscopic mass balance equation for mobile bacteria suspended in the aqueous phase without the interference of biological factors such as growth and decay can be expressed as (Cameron and Klute, 1977; Corapcioglu and Haridas, 1984, 1985):

$$\frac{\partial C}{\partial t} = D\frac{\partial^2 C}{\partial x^2} - v\frac{\partial C}{\partial x} - \frac{\rho_{bulk}}{\theta}\frac{\partial S}{\partial t} \tag{3}$$

where C is the mass concentration of suspended bacteria in the aqueous phase (g/mL), D is the hydrodynamic dispersion coefficient (m^2/s) and includes the effects of random motility and chemotaxis, ρ_{bulk} is the bulk density of the porous medium (g/mL), S is the total retained bacteria concentration (g/g) and x is the distance traveled (m). The retained bacteria fraction can be expressed by a first-order kinetic term:

$$\frac{\partial S}{\partial t} = \frac{\theta}{\rho_{bulk}} k_a C \tag{4}$$

If there is steady state and if the influence of hydrodynamic dispersion is negligible, for a continuous particle injection at concentration C_0 (at $x = 0$) and time period t_0, the solution to eqs. 1 and 2 for a column initially free of particles can be written as:

$$C(x) = C_0 \, exp\left[-\frac{k}{v} x \right] \tag{5}$$

and

$$S(x) = \frac{t_0 \theta}{\rho_{bulk}} kC(x) = \frac{t\theta C_0}{\rho_{bulk}} k \, exp\left[-\frac{k}{v} x \right] \tag{6}$$

Eq. 5 is commonly referred to as the classical colloid filtration model and has been extensively used in modeling the transport of colloids and micro-organisms in saturated porous media. If there is kinetic attachment and intra-population heterogeneity, it is assumed that a population of cells consists of two fractions, f and $(1-f)$, for fast and slow removal respectively, each with different attachment rate coefficients, k_{fast} and k_{slow}. Then, eqs. (5) and (6) can be written as (Johnson and Li, 2005):

$$C(x) = C_0 \left(f \, exp\left[-\frac{k_{fast}}{v} x \right] + (1-f) exp\left[-\frac{k_{slow}}{v} x \right] \right) \tag{7}$$

and

$$S(x) = \frac{t\theta C_0}{\rho_{bulk}} \left(k_{fast} f \, exp\left[-\frac{k_{fast}}{v} x \right] + (1-f) k_{slow} \, exp\left[-\frac{k_{slow}}{v} x \right] \right) \tag{8}$$

8.3 Materials and methods

8.3.1 Bacterial suspensions

Bacterial suspensions were prepared in a manner similar to that described in Chapter 5. Bacteria were suspended and diluted in a phosphate-buffered saline (PBS) solution (0.25 g/L NaCl, 0.00145 g/L KH_2PO_4, 0.059 g/L K_2HPO_4) to concentrations of 2-4x10^7 cells/mL (Table 1). The EC value of the suspension was 600 μS/cm, while the pH was 7.5.

8.3.2 Porous media

Soda-lime glass microspheres of various sizes (38-45, 53-63, 75-90, 112-125, 180-212, and 355-425 μm) were obtained from Whitehouse Scientific Ltd. (Chester, U.K.). The chemical composition of the beads reported by the manufacturer was: ~74% SiO_2, ~12% Na_2O, ~8% CaO, ~4% MgO, ~1% Al_2O_3, ~0.3% K_2O, ~0.3% SO_3 and <0.3% Fe_2O_3, B_2O_3, Sb_2O_3, BaO, ZnO, and PbO. Prior to the experiments, the spheres were rinsed sequentially with acetone, hexane and concentrated HCl, followed by repeated rinsing with demineralized water until the electrical conductivity was close to zero (Li et al., 2004). The total porosity of the glass beads was determined gravimetrically to be 0.38.

8.3.3 Column experiments (PBS experiments)

Duplicate column experiments were conducted in borosilicate glass columns with an inner diameter of 2.5 cm (Omnifit, Cambridge, U.K.) with polyethylene frits (25 μm pore diameter) and one adjustable end-piece. The columns were packed wet with one of the glass bead sizes with vibration to minimize any layering or air entrapment. Column sediment length was 5 cm. Prior to each experiment, the packed column was equilibrated by pumping (Watson-Marlow 101U/R) 10 pore volumes of the background PBS solution through the column at a constant discharge rate of 3.5 mL/min. A suspension of *E. coli* of the same background PBS solution was pumped for 3 pore volumes at the same constant discharge rate followed by 5 pore volumes of *E. coli*-free PBS solution. The *E. coli* concentration was determined using optical density measurements (at 410 nm) with a UV-visible spectrophotometer (Perkin-Elmer Lambda 20, Wellesley, MA, USA). Absolute cell numbers were deduced after calibrating with plate counts. An overview of the experimental parameters is given in Table 1.

8.3.4 Column dissection and enumeration of retained particles (PBS experiments)

After completing an experiment, the column was dissected into sections for observing the spatial distribution of bacteria in the column. The bottom end-piece was removed without disturbing the packed bed, and the beads were extruded in 0.5 cm sections. The packed bed remained saturated with PBS during the entire extrusion process so as not to shift or cause release of retained bacteria. Each 0.5 cm section of porous media was placed into a weighed polyethylene container filled with 50 mL demineralized water and 0.1 mL 0.1 M KOH with a pH of 10.5. We tested the effect of adding KOH to an *E. coli* suspension on the optical density of the suspension. For [KOH] < 1.0 mL 0.1 M KOH per 50 mL demineralized water, the optical densities were not altered. Occasionally, the containers were gently shaken by hand. After 1-1.5 h, the optical density of the sample was determined. The remaining glass bead-*E. coli* suspension was placed in an oven at 70 °C for 24 h to evaporate the remaining liquid. The mass of glass beads and the volume of solution in each section were determined by mass balance from the weights of the empty containers, liquid- and bacteria-filled containers, and dry bead-filled containers. In each experiment, the mass balance of *E. coli* was determined by comparing the sum of the number of bacteria in the effluent and the number of bacteria deposited in the column with the total amount of bacteria injected.

Table 1. Type of experiments, experimental conditions, and results (fitted fast fraction, f, mass balances and fractions) of all experiments. Calculated parameters k_{fast} and k_{slow}, and fitted parameter f are given for the case of $\alpha_{fast} = 1$ and $\alpha_{slow} = 0.01$

Experiments and experimental parameters | **Results**

Column experiments (Fig. 3, diamonds and Fig. 5)

Bead size (μm)	Col. length (cm)	Fluid	Discharge (mL/min)	C_0 (* 10^7) (cells/mL)	Max C/C_0 (-)	Mass balance (fraction of influent mass) Effluent	Retained	Total	Calculated η_0 (-)	k_{fast} (s^{-1})	k_{slow} (s^{-1})	Fitted f (-)
355-425	5	PBS	3.36	3.19	0.516	0.55	0.35	0.90	3.90E-03	2.79E-03	2.79E-05	0.915
	5	PBS	3.55	2.90	0.549	0.57	0.48	1.05	3.77E-03	3.99E-03	2.85E-05	1.328
180-212	5	PBS	3.50	3.45	0.445	0.45	0.54	0.99	6.98E-03	1.04E-02	1.97E-04	0.716
	5	PBS	3.58	2.45	0.438	0.45	0.48	0.93	6.90E-03	1.05E-02	9.64E-04	0.935
112-125	5	PBS	3.40	2.82	0.516	0.55	0.68	1.23	1.19E-02	2.84E-02	2.90E-03	0.522
	5	PBS	3.60	3.15	0.548	0.59	0.35	0.94	1.16E-02	2.93E-02	1.70E-04	0.349
75-90	5	PBS	3.23	3.25	0.480	0.50	0.38	0.88	1.85E-02	6.00E-02	3.60E-04	0.322
	5	PBS	3.50	2.56	0.368	0.36	0.45	0.81	1.79E-02	6.30E-02	9.01E-04	0.465
53-63	5	PBS	3.46	3.41	0.420	0.42	0.59	1.01	2.79E-02	1.39E-01	1.39E-03	0.524
	5	PBS	3.50	2.82	0.352	0.35	0.68	1.03	2.78E-02	1.40E-01	4.16E-03	0.373
38-45	5	PBS	3.30	3.24	0.258	0.27	0.91	1.18	4.44E-02	2.93E-01	6.45E-03	0.382
	5	PBS	3.30	4.65	0.392	0.42	0.50	0.92	4.36E-02	3.06E-01	4.11E-03	0.588

Low ionic strength experiments (Fig. 3, squares)

Bead size (μm)	Col. length (cm)	Fluid	Discharge (mL/min)	C_0 (* 10^7) (cells/mL)	Max C/C_0 (-)
355-425	5	demi	7.00	3.52	0.847
180-212	5	demi	7.00	3.60	0.840
112-125	5	demi	7.33	3.30	0.796
75-90	5	demi	7.08	3.37	0.857
53-63	5	demi	6.83	3.63	0.877
38-45	5	demi	7.00	3.30	0.801

Flow reversal experiments

Bead size (μm)	Col. length (cm)	Fluid	Discharge (mL/min)	C_0 (* 10^7) (cells/mL)	Mass balance (fraction of influent mass) Effluent	Retained	Reverse flow	Total
355-425	2	PBS	3.50	83.96	0.73	0.19	0.00	0.92
180-212	2	PBS	3.60	85.49	0.63	0.27	0.00	0.90
112-125	2	PBS	3.65	86.68	0.61	0.30	0.00	0.92
75-90	2	PBS	3.58	81.35	0.52	0.40	0.01	0.92
53-63	2	PBS	3.52	79.98	0.51	0.41	0.02	0.94
38-45	2	PBS	3.60	87.66	0.48	0.48	0.00	0.97

Population heterogeneity experiments (Fig. 4)

Bead size (μm)	Col. length (cm)	Fluid	Discharge (mL/min)	C_0 (* 10^7) (cells/mL)	Susp. 1 average of max C/C_0	Susp. 2 average of max C/C_0
75-90	5	PBS	3.80	1.79	0.55	0.40
53-63	5	PBS	3.81	2.28	0.48	0.32
38-45	5	PBS	3.60	2.88	0.39	0.31

Particle elution experiments (Fig. 6)

Bead size (μm)	Col. length (cm)	Fluid	Discharge (mL/min)	C_0 (* 10^7) (cells/mL)	Eluted bacteria (fraction of retained bacteria) demi	demi+KOH	total
355-425	5	PBS, demi, demi+KOH	3.56	3.97	0.02	0.00	0.02
112-125	5	PBS, demi, demi+KOH	3.63	3.15	0.15	0.03	0.18
38-45	5	PBS, demi, demi+KOH	3.60	3.91	0.12	0.04	0.15

8.3.5 Assessing straining

Low ionic strength experiments. Column experiments were carried out with *E. coli* in demineralized water at a discharge rate of 7.0 mL/min (Table 1). Using demineralized water instead of PBS reduced the ionic strength and increased the double layer repulsion, thereby reducing attachment. Increasing the pore water flow velocity reduces the single-collector contact efficiency of the system (η_0 in eq. (1)), thereby reducing attachment.

Flow reversal experiments. Detachment or resuspension of particles during backwashing of sand filters takes place when hydrodynamic forces exerted by the fluid in the column overcome the adhesive chemical and colloidal forces, which prevent removal (Hirtzel and Rajagopalan, 1985; Raveendran and Amirtharajah,

1995). Hydrodynamic forces include two components, lift and drag. The drag is the force on the particle in the direction of the bulk fluid motion. The lift is the force on the particle in the direction normal to the flow field, causing the particle to drift away from the collector surface. Chemical and colloidal forces include Born repulsion, hydration forces, the van der Waals attractive force, and the double layer interactive force. An overview of both the colloidal forces and the hydrodynamic forces is given in Table 2.

Table 2: Forces controlling the resuspension of colloids during flow reversal (from Raveendran and Amirtharajah, 1995). Explanation of symbols: H = Hamaker constant between the collector and colloid (J), z_{pg} = separation distance between colloidal particle and collector grain (m), a_p = colloidal particle radius (m), ε = permittivity of the medium (CV^{-1}m^{-1}), κ = Debye-Huckel length (m^{-1}), k = Boltzmann constant (J/K), T = absolute temperature (K), Z = charge number or valence of the ion (dim. less), e = charge of the electron (C), ψ_1 and ψ_2 are electrical surface potentials (V), σ_c = collision diameter (m), K and h are empirical constants, μ = dynamic viscosity (kgm^{-1}d^{-1}), ν = kinematic viscosity (m^{2}s^{-1}), $\frac{dv}{dx}$ = velocity gradient, and $v_{colloid}$ = velocity of the colloid (ms^{-1})

Van der Waals (attractive):	$F_W = \frac{H}{6z_{pg}}(\frac{a_p}{z_{pg}} - 1)$
Electrical double layer (attractive or repulsive):	$F_{DL} = -64\pi a\varepsilon\kappa\left(\frac{kT}{Ze}\right)^2 \tanh\left(\frac{Ze\psi_1}{4kT}\right)\tanh\left(\frac{Ze\psi_2}{4kT}\right)\exp(-\kappa z_{pg})$
Born (repulsive):	$F_{Born} = -\frac{Ha_p\sigma_c^6}{180z_{pg}^8}$
Hydration (repulsive):	$F_{Hydr} = -2\pi a_p Kh\exp(-\frac{z_{pg}}{h})$
Hydrodynamic lift:	$F_{Lift} = 81.2a_p^2\mu\nu^{-\frac{1}{2}}\left(\frac{dv}{dx}\right)^{\frac{1}{2}}v_{colloid}$
Hydrodynamic drag:	$F_{Drag} = 1.7009(6\pi a_p\mu v_{colloid})$

If, immediately after a column experiment, the flow direction is reversed, then, from Table 2, the magnitude of all forces has not changed. However, the direction of the hydrodynamic drag and lift forces has changed. If these forces exceed adhesive forces, cells may resuspend. We assume that adhesive forces on strained cells are low or even absent, because the cells are not in direct contact with the collector surface. Also Bradford and workers assumed, that strained colloids were not attached (Bradford et al., 2002). Furthermore, recent images of colloid collisions with the solid interface via interception and diffusion in a complex pore micromodel (Keller and Auset, 2006) also show clusters of microspheres not in direct contact with the collector surface. Therefore, during flow reversal, hydrodynamic forces will exceed adhesive forces of strained cells, resulting in remobilization of these cells. Obviously, in many cases, straining and attachment occur simultaneously and they are difficult to distinguish from each other. Also, during flow reversal, mobilized cells may again be retained in the column due to straining and/or attachment. Therefore, the flow reversal method can only be used in a qualitative way, since it does not give accurate information on the amount of cells being strained in the column. Recently, we applied the flow reversal method to confirm the occurrence of straining of *E. coli* in columns of quartz sand (Foppen et al., 2006a).

For each glass bead size, 2 cm columns were flushed with a 10 times higher *E. coli* mass (compared with the PBS experiments) in PBS for exactly 2 pore volumes followed by flushing for 3 pore volumes with background PBS solution in order to remove mobile *E. coli* cells in the column (Table 1). Then, the flow was immediately reversed for 3 pore volumes and the effluent was collected.

8.3.6 Proving intra-population heterogeneity

For the three smallest bead sizes, column experiments in PBS were carried out, whereby the influent consisted of two *E. coli* suspensions. Suspension 1 was prepared by flushing an *E. coli* suspension (see also section Bacteria Suspensions) through a 1 cm column consisting of the smallest glass bead size (Fig. 2a). We expected that this treatment would have removed the fast attaching bacteria. Suspension 2 was prepared in the same way as described in the section Bacteria Suspensions, and consisted of a mixture of fast and slow attaching bacteria. The two suspensions, with identical optical density values, were flushed through one and the same column, immediately after each other, i.e. first suspension 1 with the slow attachers (t1 in Fig. 2b) followed by suspension 2 with the slow and fast attachers (t2 in Fig. 2b). Because the fast attachers were removed from suspension 1, but not from suspension 2, we expected that the breakthrough concentrations of suspension 1 would be higher than suspension 2.

8.3.7 Retention in secondary energy minima and on collector surface impurities

To determine the bacteria mass present in the secondary minimum and on bulk surface impurities (e.g. metal-oxides), after a column experiment, flushing of the column was continued for an additional pulse of 2 pore volumes with demineralized water, followed by a pulse of 2 pore volumes of demineralized water to which KOH was added (50 mL demineralized water and 0.2 mL 0.1 M KOH) (see Table 1; particle elution experiments). The pulse with demineralized water was expected to release bacteria captured in the secondary energy well, while the pulse with demineralized water + KOH would increase the pH to 10-11, thereby releasing bacteria captured by metal oxides present on the surface of the glass beads.

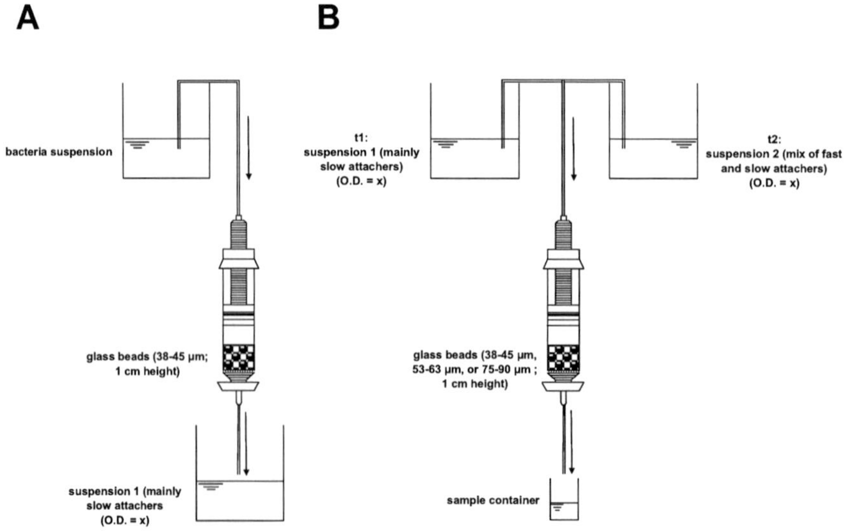

Figure 2. Preparation of the E. coli suspension mainly consisting of slow attachers (a) and principle of a slow/fast experiment, whereby two E. coli suspensions are flushed through one column, immediately after each other (b)

8.3.8 Determining parameters and model fitting

To determine the fraction of fast attaching bacteria, f, eq. (8) was fitted to the measured retained bacteria profiles. For k_{fast} and k_{slow}, eq. (1) was rewritten as:

$$k_{fast} = \frac{3(1-\theta)v}{2a_c}\alpha_{fast}\eta_0 \tag{9}$$

$$k_{slow} = \frac{3(1-\theta)v}{2a_c}\alpha_{slow}\eta_0 \tag{10}$$

The single-collector contact efficiency, η_0, was calculated with the Tufenkji-Elimelech correlation equation (Tufenkji and Elimelech, 2004). We assumed α_{fast} to be equal to maximum (i.e. $\alpha_{fast} = 1$) from which k_{fast} was calculated. Then, from non-linear regression analysis, using the Levenberg-Marquardt method, f and α_{slow} were determined. The convergence threshold of the fitting process was defined in terms of chi-square changes between iterations, which we kept at $< 10^{-7}$. The fitting process was carried out twice. For one set of modeled retained profiles, f and α_{slow} were allowed to vary, while for the second set α_{slow} was fixed to a minimum value, and only f was varied.

8.3.9 Size of *E. coli* cells

To determine η_0, the size of *E. coli* cells was measured in a similar way to that described in Chapter 5.

8.4 Results

8.4.1 Size of *E. coli* cells

The average *E. coli* length was 2.27 μm and the average diameter was 0.90 μm. This corresponded to an equivalent spherical diameter of 1.43 μm (Rijnaarts et al., 1993).

8.4.2 Breakthrough curves of the PBS experiments

Breakthrough of *E. coli* in PBS (curves with diamonds in Fig. 3a-c) for all the glass bead sizes was rapid and without significant retardation. Within 1.5-2 pore volumes a plateau developed with a more or less constant normalized concentration, C/C_0. During flushing with the background solution (the deloading phase), C/C_0 values rapidly fell below the detection limit of the spectrophotometer. C/C_0 values for the plateau phase of the largest glass bead fraction (355-425 μm) were the highest (around 0.55), while the 38-45 μm glass beads had the lowest C/C_0 values for the plateau phase (around 0.35).

8.4.3 Retained profiles of the PBS experiments

In all experiments, the mass balance was found to be within ± 20%, and in most experiments the mass balance was ± 10% (Table 1). Retained concentrations (Fig. 3d-f) were expressed as the number of cells over the total number of cells per gram of dry glass beads. Retained concentrations in the largest glass bead size fraction (Fig. 3d) decreased log-linear with distance from the inlet from around 0.02 at 0.25 cm depth to 0.009 at 4.75 cm depth. When the glass beads were smaller: (1) retained profiles consisted of two log-linear slopes, and (2) the slope nearest to the inlet of the column became steeper. From this, we concluded that retained profiles became increasingly hyper-exponential upon decreasing bead size.

8.4.4 Assessing straining

The *E. coli* breakthrough curves of the experiments in demineralized water (curves with squares in Fig. 3a-c) had the same shape as the PBS breakthrough curves. However, the plateau phase of the breakthrough curves for all the glass bead fractions was between 0.8 and 0.9. So, when bacteria were flushed with demineralized water, only 5-20% of the total bacteria mass applied to the column was retained. The lower plateau phase of the PBS experiments proved that 70-80% of the fraction of bacteria retained in the columns was due to attachment rather than straining, because, for the low ionic strength conditions of the demineralized water experiments, this fraction was able to travel through the column without significant retention and could not therefore be attributed to straining. The physical dimensions of both the pore structure in the column and of the *E. coli* bacteria apparently did not inhibit their transport.

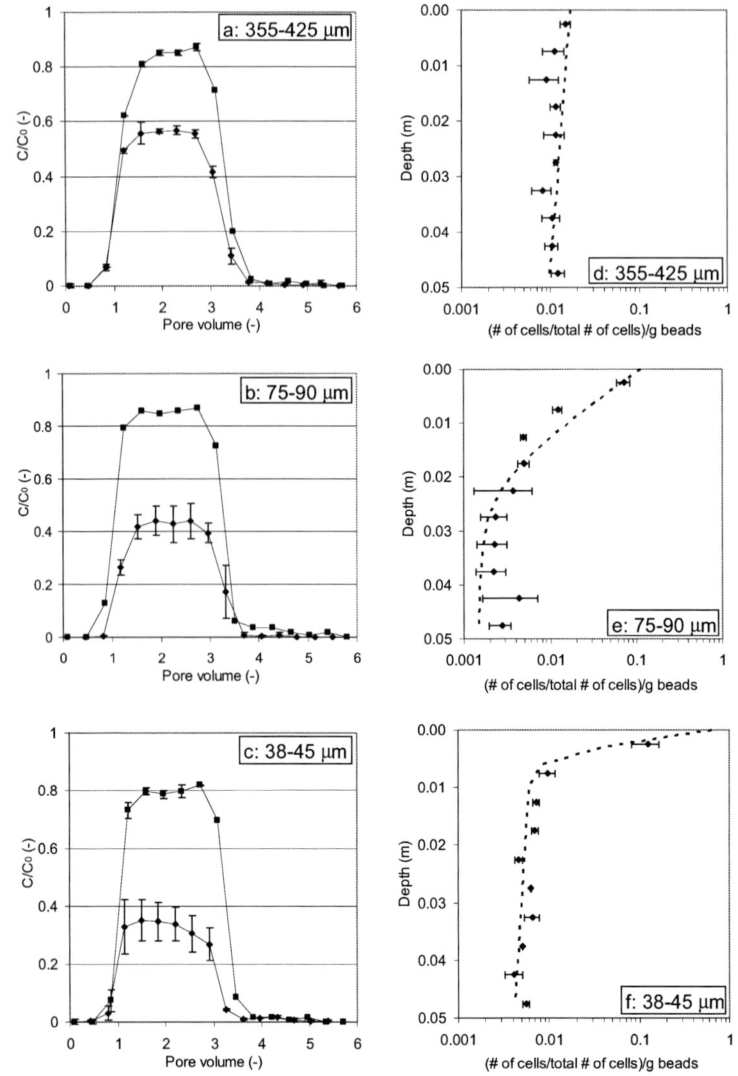

Figure 3a-f. Breakthrough curves (a, b, and c) and spatial distributions of E. coli for experiments in columns packed with a wide range of glass beads (size fractions are given in the upper right of each graph). Spatial distributions (d, e, and f) are given as normalized concentrations per gram of dry glass beads as a function of depth. Key experimental conditions are given in Table 1. Error bars indicate variation between duplicate experiments. Fig. 3a-c: diamonds: E. coli in PBS; squares: E. coli in demineralized water. Fig. 3d-f: diamonds: E. coli in PBS; dashed line: fitted spatial distribution of E. coli

The bacteria mass in the effluent of the reversed flow experiments was invariably low for all glass bead sizes (only 0-2%; see Table 1). So, hardly any bacteria were dislocated or dislodged, which indicated that, during flow reversal, the adhesive forces acting on the bacteria remained higher than the hydrodynamic forces. When inspecting the effluent under a microscope, we never observed sticking of *E. coli* to each other. Therefore, the absence of bacteria in the effluent indicated that bacteria were attached to the glass bead surface.

Our conclusion was that, because only few bacteria were retained in the experiments with demineralized water, and because no bacteria were dislodged during flow reversal, straining did not play a significant role in our experiments. Therefore, the hyper-exponential retained profiles of Fig. 3 were mainly due to attachment of *E. coli*.

8.4.5 Proving intra-population heterogeneity

The previous experiments showed that attachment was the most important process, so we examined whether suspensions of bacteria with and without fast attaching bacteria lead to breakthrough variations. During the first 2.5-4 pore volumes, the column was loaded with a suspension of slow attachers, which resulted in C/C_0 values of the plateau phases ranging from 0.55 for the 75-90 μm glass beads to 0.39 for the 38-45 μm glass beads (Fig. 4 and Table 1). When the flushing suspension was changed for a suspension consisting of a mixture of slow and fast attachers, C/C_0 values decreased significantly to 0.3-0.4. The decreased breakthrough, or increased removal in the same column, proved the existence of heterogeneous attachment characteristics of the bacteria population, in this case resulting from the absence of fast attachers in the first suspension and the presence of those fast attachers in the second suspension.

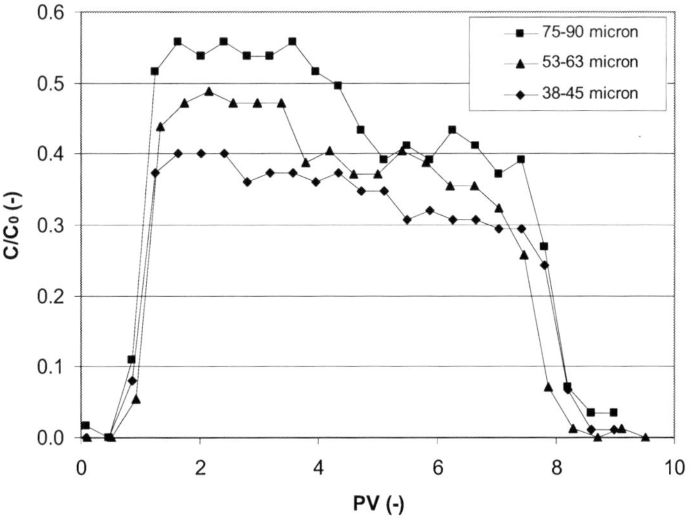

Figure 4. Breakthrough curves for experiments conducted with slow attachers (from 0 to 2.8-4 pore volume) and then with a mixture of slow and fast attachers

8.4.6 Model fitting

When both α_{slow} and f were allowed to vary when fitting the retention profiles of Fig. 3d-f, α_{slow} increased from a minimum of 0.01 for the four smallest glass beads to 1 for the two largest beads. The fast fraction, f, was constant around 0.4 for the four smallest glass beads and increased to 1 for the two largest glass bead fractions (Fig. 5 and Table 1). When α_{slow} was kept constant at the fitted minimum sticking efficiency of 0.01 (Fig. 5), a similar pattern for f was seen. Because f for the four smallest glass bead fractions was constant at around 0.4, we concluded that the fraction of fast attachers in our *E. coli* suspension was 40% and the remaining fraction (slow attachers) was 60%. For the largest two bead fractions, due to the limited heights of the columns (5 cm), the retained profiles were increasingly determined by the fraction of fast attachers only – the slow attachment was hardly taking place and/or completely masked. Extending the columns lengths may have revealed the slow attachment and the hyper-exponential deposition behavior for these two largest glass beads.

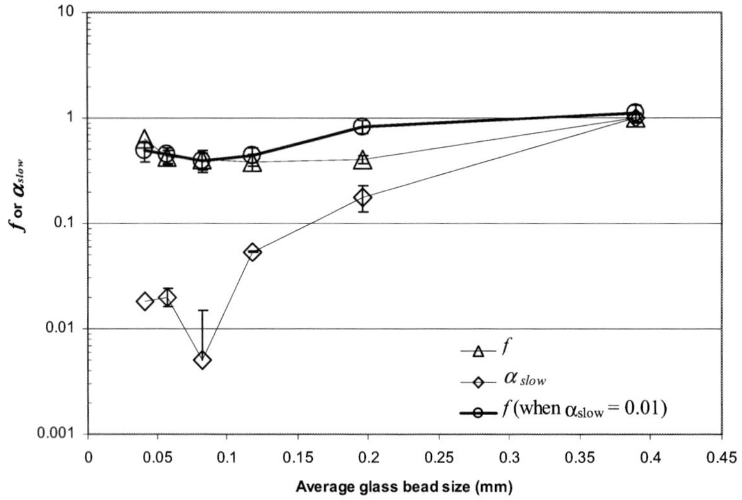

Figure 5. Fitted parameter values for f and α_{slow} as a function of the average bead size. Error bars indicate variation between duplicate experiments

8.4.7 Fraction retained and fast fraction

When a pulse of demineralized water was flushed through the column after a PBS experiment, 2-15% of the *E. coli* retained in the glass beads was released (Fig. 6 and Table 1), probably as a result of release from the secondary energy minimum. When flushing with a pulse of demineralized water with KOH, an additional 0-4 % of the retained *E. coli* was released, probably as a result of release from charged metal oxides. This total of 2-18% was not equal to the fraction of fast depositing bacteria determined from fitting (40%). Especially the very low (2%) bacteria mass obtained from eluting the largest glass bead did not coincide with the fast fraction (~ 100%) determined from fitting. Therefore, in our case, the concurrent existence of both favorable and unfavorable attachment mechanisms, was not able to explain dual mode deposition.

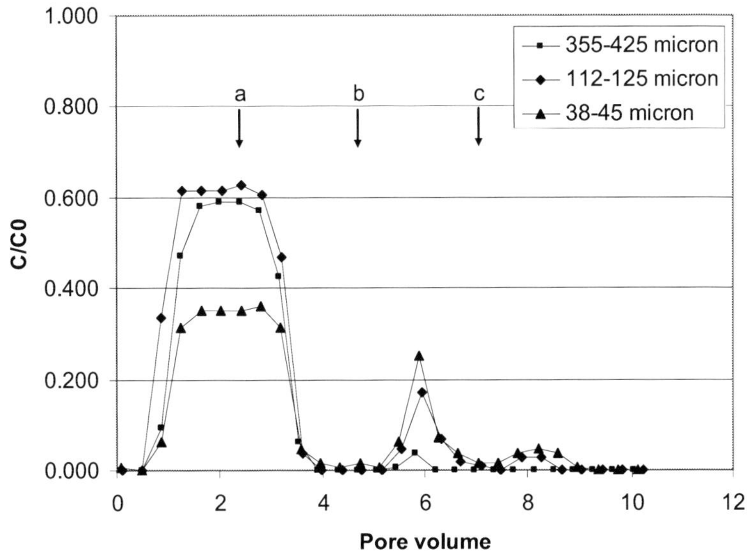

Figure 6. Breakthrough curves for elution experiments conducted with E. coli. Arrows indicate the injection of (a) a particle-free solution at the same ionic strength as the deposition phase; (b) particle-free demineralized water; and (c) particle-free demineralized water and high pH (1 mM KOH, pH 11)

8.5 Discussion

Because the retained profiles for the smaller glass beads consisted of two log-linear slopes, dual mode deposition took place in our experiments. Based on low ionic strength experiments and on flow reversal experiments, we rejected straining as a possible cause for this hyper-exponential retention. Retention in our columns was mainly due to attachment. When collectors were too large (and/or the column length was too small), dual mode deposition was completely masked, because in such cases just one type of (fast) deposition dominated the retention profile, resulting in log-linear retained bacteria masses. Small spherical collectors were required to demonstrate the bimodal attachment characteristics of our monoclonal *E. coli* population, most likely because of the higher interception and diffusion term in the single-collector contact efficiency for small collectors compared to large ones. An important cause for the observed dual mode deposition was the aspect of intra-population heterogeneity. According to the model fitting, the monoclonal *E. coli* population consisted of 40% fast attachers and 60% slow attachers. When removing the fast fraction, breakthrough was higher than of a suspension of a stickier, fast fraction (Fig. 4), which proved the important role that intra-population heterogeneity played during deposition. Baygents et al. (1998) also showed there were intra-populational differences in surface charge densities in the bacteria they used (groundwater isolates strain A1264 and strain CD1), which under the conditions of their experiments lead to bimodal deposition. Simoni et al. (1998) also found sticking efficiency variations for *Pseudomonas sp.* strain B13. They attributed sticking efficiency variations to varying average lipopolysaccharide (LPS) length or variations in the LPS composition leading to heterogeneous physicochemical properties of *Pseudomonas*.

The total fraction of bacteria depositing in the secondary energy well (unfavorable attachment) and on charged metal oxide impurities (favorable attachment) was not equal to the fast fraction (40%) mentioned above. So, the relation between these two fractions, that was so obvious from the results of Tufenkji and Elimelech (2004a), was absent in our case. Similar to Tong and Johnson (2006), we concluded that the fast depositing fraction of the bacteria population neither deposited preferentially within the secondary energy minimum, nor on charged metal impurities. The main driver for dual mode deposition in our case was heterogeneity among members of the bacteria population.

Due to its negative surface and relatively low inactivation rate coefficient (Chapter 2), *E. coli* is an important indicator of fecal contamination of groundwater. We have previously pointed out that sticking efficiencies determined from field experiments are lower than those determined under laboratory conditions (Chapter 2). Here we have shown that this observed difference might well be due to intra-population heterogeneity, thereby enhancing the understanding of the transport of these fecal indicator organisms in aquifers.

8.6 Conclusions

Dual mode deposition of *E. coli* took place in our experiments. An important cause for this dual mode deposition was not due to straining, but due to heterogeneous attachment characteristics of the *E. coli* population we used. Based on particle elution experiments, we ruled out the possibility that the fast attaching, or stickier, bacteria fraction was preferentially deposited in the secondary energy well and on charged metal impurities. Intra-population may result in some microbes traveling surprisingly high distances in the subsurface. Extending the colloid filtration theory with intra-population variability may provide a valuable framework for assessing the transport of bacteria in aquifers.

PART 2

HYDROCHEMICAL AND MICROBIOLOGICAL PROCESSES IN THE AQUIFERS BELOW SANA'A

CHAPTER 9:

MANAGING WATER UNDER STRESS IN SANA'A, YEMEN

Foppen, J.W.A., M. Naaman, and J.F. Schijven, 2005b. Managing urban water under stress in Sana'a, Yemen. Arabian Journal for Science and Engineering, Vol 30, No. 2C, p. 69-83, 2005.

Abstract

Lack of water management in the Sana'a Basin in Yemen has led to mining of groundwater and massive groundwater quality deterioration. In the last four years, from 2001-2005, management of waste water has changed dramatically and entire city neighborhoods have been connected to a conventional sewer system. In this paper, the effects of this measure on long-term groundwater quality development of the aquifers underlying Sana'a and, more specifically, on water quality of the public supply peri-urban wellfields is analyzed. The results, obtained with a transient groundwater model, indicated that by 2020 the construction of the sewerage will have considerably reduced the area polluted by groundwater, but the process is slow. Furthermore, construction of the sewerage will hardly affect the groundwater quality of the wellfields, since flow is not directed towards most of the production wells. The Yemeni authorities should realize that less expensive sanitation alternatives are available, but they need user participation, which, in turn, would raise public awareness that water supplies and sanitation are not to be seen as solely a government responsibility.

9.1 Introduction

The effects of waste water infiltration (primary or secondary, diffuse or as a point source) on groundwater quality have been studied at various locations. One of the major problems with on-site sanitation is nitrate pollution of the groundwater. When Foppen (2002; Chapter 10) used an exploratory one-dimensional transport model of a 200 m column of the aquifer under the urban area of Sana'a, he concluded that upon diffuse infiltration of wastewater the two main hydrochemical processes dominating groundwater quality in this aquifer are nitrification and cation exchange. Water management in Yemen in the last 2 decades has been characterized by ill-defined ministerial mandates and rules and regulations being made, but not implemented or enforced. This lack of water management has led to severe water scarcity. Also, it has been recognized that water scarcity is being accelerated by massive groundwater quality deterioration, especially in the vicinity of major urban areas, mainly due to the lack of wastewater collection systems. However, in the last four years, in response to the unacceptable health risks, odor nuisance and road damage associated with sewage continuously flooding the streets in some areas of Sana'a, the management of waste water has changed dramatically. As a result, entire city neighborhoods have been connected to the sewerage and a new sewage plant has been built just north of the airport (see Fig. 1).

In this paper we analyze the effect of these measures on trends in the long-term groundwater quality of the aquifers under Sana'a. A groundwater model of the greater Sana'a Basin was constructed with MODFLOW and calibrated using PEST. This transient model was used to analyze conservative contaminant transport (of nitrate) resulting from various water management options.

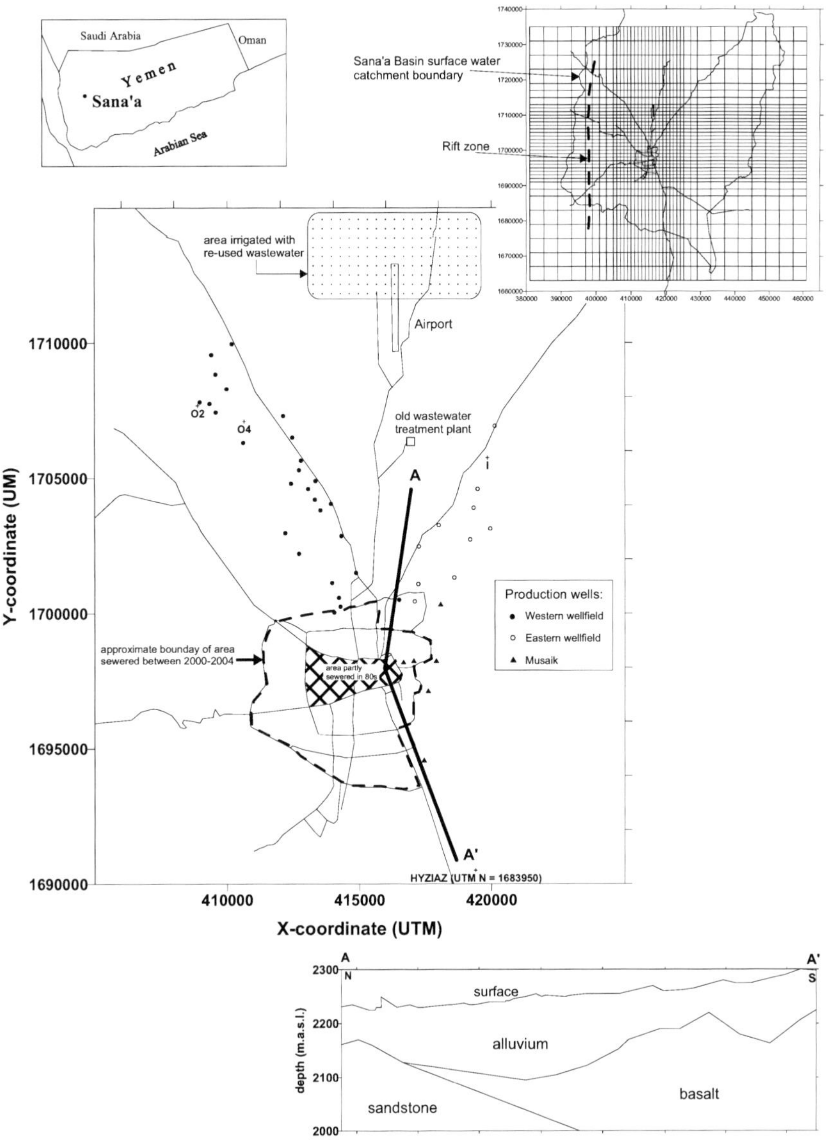

Figure 1: *Location map of Sana'a showing the production wells of the public supply wellfields, the location of the old wastewater treatment plant, the area irrigated with re-used wastewater (originating from the new wastewater treatment plant) and major roads. Wells O2, O4, I and Hyziaz are observation wells (see also Fig. 3). Upper left: Location of Sana'a; upper right: discretization of the area in 38 rows and 37 columns for the MODFLOW model; lower right: cross-section A-A' showing the most important hydrogeologic formations in the Sana'a Basin*

9.2 Area description

9.2.1 Physiography

Sana'a is in the center of the Sana'a Basin (Sana'a Basin catchment boundary is drawn in Fig. 1), an intermontane plain of some 3,200 km^2 in the central Yemen Highlands. The plain is about 2,200 meters above sea level but to the west, south and east is flanked by mountains rising to about 3,000 m.a.s.l. During the last 20 years Sana'a city has grown at an annual rate of around 10%: there were 80,000 inhabitants in 1970, 1,000,000 in 1994 and the 2005 estimate is around 2,000,000 inhabitants, although official figures are not available. As a consequence, the area of the city has grown tremendously in the last 20 to 30 years.

In the 1980s a sewer system was constructed in the old and densely populated center of Sana'a. The wastewater was collected in sewage ponds a few kilometers north of Sana'a. The coverage of the sewage network was not complete and pit latrines were still very common, especially in the western part of the sewered area. In 2000 a new wastewater treatment facility was built just north of the airport. Since its completion, a large part of the city has been connected to the sewerage. During the previous sewerage construction works in the 1980s households had been given the choice of being connected, but this time no such option was available: connection to the sewerage is mandatory in the greater city center (area within the dashed line of Fig. 1).

9.2.2 Water balance of the Sana'a Basin

The main aquifer in the Sana'a Basin is the Cretaceous sandstone (the Tawilah Formation; see cross-section in Fig. 1), which has a relatively high premeability and secondary porosity. The aquifer is heavily exploited throughout the Basin. In rift zones, the sandstone has been hydraulically "disconnected" by basalt that has erupted to the surface (see Fig. 1, upper right).

The average yearly precipitation in the period 1972-1990 was 235 mm (Foppen, 1996). Infiltration of surface flows in ephemeral wadis is believed to be the most important and least predictable component of recharge from precipitation in the Sana'a Basin (Alderwish and Dottridge, 1998). Mean recharge in the Sana'a Basin is estimated at 4-8% of the precipitation measured at Sana'a. The volume of groundwater abstracted from the aquifers in the Sana'a Basin was estimated during a number of well inventories in the Sana'a Basin (in 1973, 1984, 1994-1995 and in 2001). The total number of wells has grown from a few hundred in 1973 to around 6000 in 2001 (see Fig. 2), while the abstracted volume has grown from 60 Mm^3/yr to 370 Mm^3/yr. Of this amount, around 85% is for agricultural use. The main irrigated crops are the cash crops qat and grapes (Hamdi, 2000). Around half of the domestically used water is supplied by wellfields around the city that are controlled and maintained by the public supplier. The most important are the Western wellfield, the Eastern wellfield and the Musaik wellfield (see Fig. 1). The other half is supplied by the private sector, which sells water in bottles, jerry cans or tankers. Alderwish and Dottrigde (1998) estimated that 59% of the water used for domestic supply infiltrated via cesspits into the urban aquifers as wastewater. Hamdi (2000) estimated the total volume of wastewater infiltration in the city in 1995 was 10-20 Mm^3.

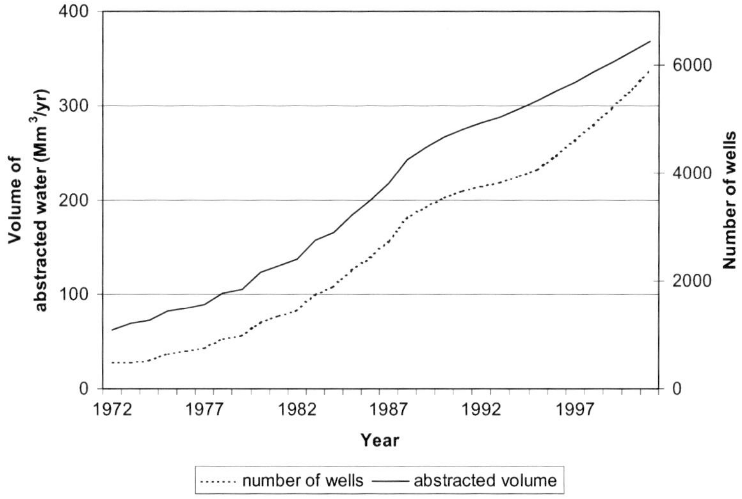

Figure 2: *Estimated number of wells in the Sana'a Basin (based on well inventories in 1972, 1984, 1993-1995 and in 2001) and total abstracted volume of groundwater (in* Mm^3/yr*)*

As a result of the rapid and more or less uncontrolled increase in the number of wells and volume of abstracted water, water tables have fallen drastically (see Fig. 3). In the sandstone aquifer the rate of decline since the early 1970s has varied from place to place and has ranged from 1.5 to 4 m/yr.

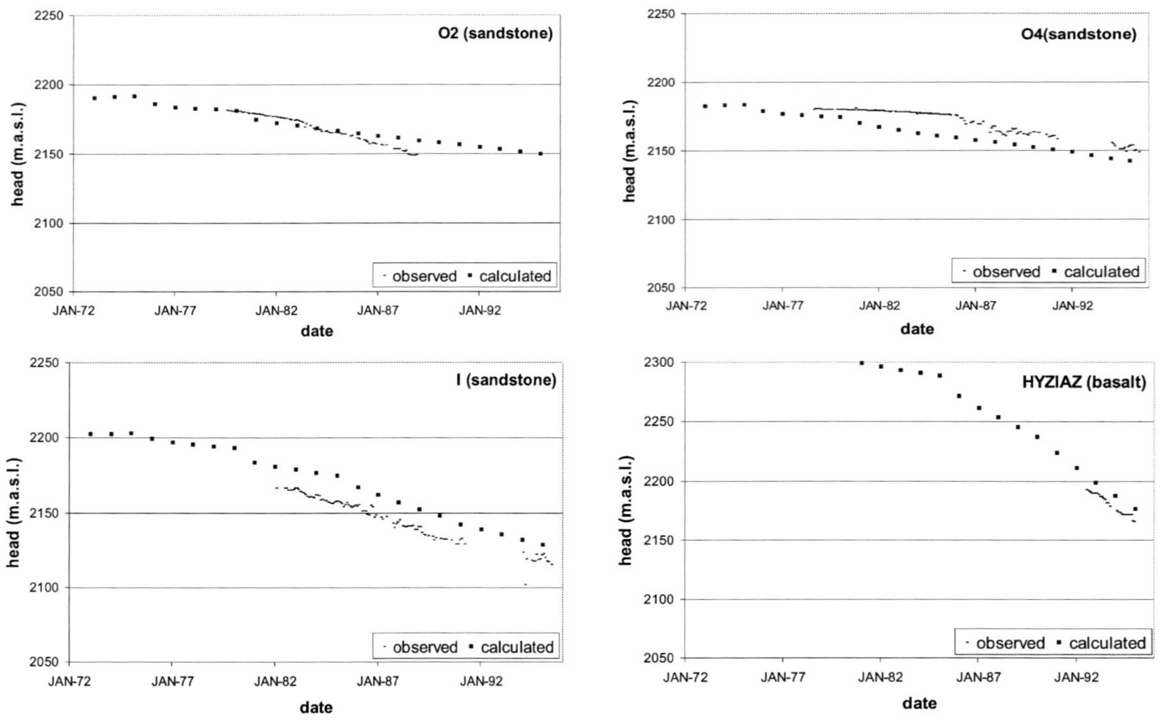

Figure 3: *Hydrographs of observation wells in the Tawilah sandstone (wells O2, O4, and I) and in the Tertiary Volcanics (line: measured head; square: calculated head)*

9.2.3 Hydrochemistry of the urban aquifers

Groundwater in the aquifers below Sana'a typically has high concentrations of almost all major cations and anions (Chapter 10). The [NO_3^-] ranged from 1-3 mmol/l, while NH_4^+ was absent (see Fig. 4). Upon infiltration of wastewater from cesspits to the saturated zone nitrification occurs, whereby NH_4^+ is oxidized first to NO_2^- and then to NO_3^-. However, the [NO_3^-] in groundwater was far less than the average [NH_4^+] in wastewater (10 mmol/L), indicating that there is a sink for N species between the cesspit and the saturated zone. Besides the effects of dispersion and diffusion on [NO_3^-], the most likely sink is cation exchange of NH_4^+.

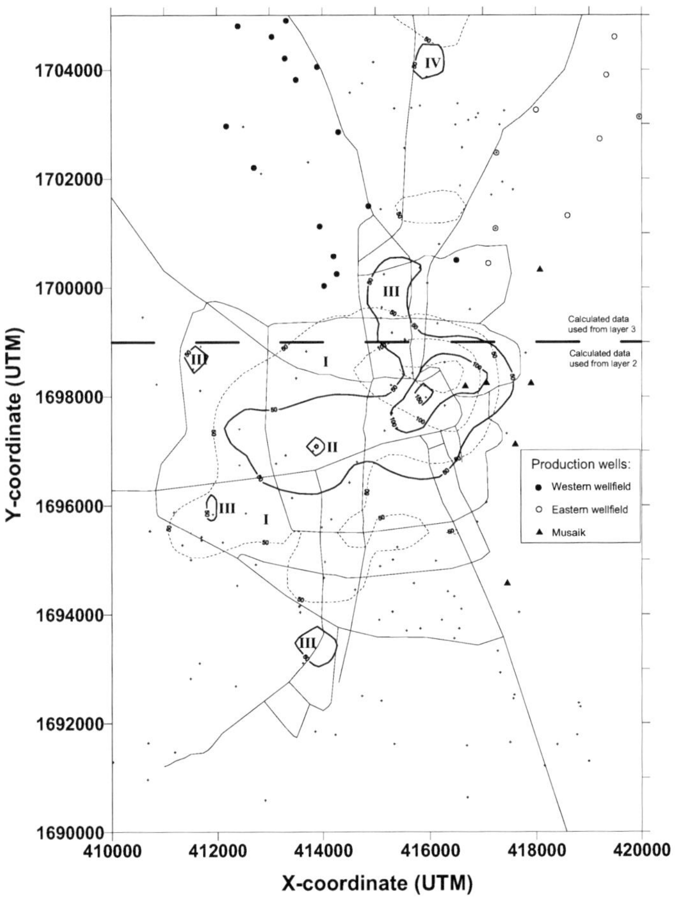

Figure 4: Nitrate concentration in the aquifers in Sana'a in 1995 (in mg/L; solid line: measured; dashed line: calculated)

9.3 Methods

9.3.1 Model set-up

The model area (see Fig. 1) covers some 6,400 km^2, half of which consists of the Sana'a Basin (3,200 km^2). The city is in the center of the model area. The area was discretized into 38 columns and 37 rows, while the cell width ranged from 1 km in the center of the model to 4 km near the model boundaries. The grid is oriented N-S, allowing adjustments to be made for anisotropies due to the generally N-S faults and fissures. The aquifer system was schematized into 3 separate model layers (Alluvium, Volcanics, and Sandstone). Initial horizontal permeability estimates to start the calibration process were obtained from a limited number of pumping tests carried out in the Western and Eastern wellfields (Foppen, 1996). As no data were available on vertical permeability values, at the beginning of the calibration process they were assumed to be similar to the values used by Foppen (1996).

The model boundaries for each layer consisted of head-dependent flow cells, in which a flux across the boundary is calculated given a boundary head value. Boundary head values were assumed to be equal to the elevation (in m.a.s.l.) of the bottom of layer 2 plus 0.75 of its thickness. The steady state model was calibrated for the year 1972 with the head data from Italconsult (1973), while the transient model was calibrated with the well-inventory data from 1984 (Mosgiprovodkhoz, 1986), and from the period 1993-1995 (Foppen, 1996), and with 20 hydrographs of observation wells within the Western and Eastern wellfields covering the period 1972-1995 (monthly observations). The transient modeling period was separated into 10 stress periods, each consisting of 5 time steps of one year (except stress period 1, which was 3 years, from 1973-1975). Per stress period, data on abstractions for layers 1, 2 and 3 were adjusted in order to simulate the growth in the abstracted volume of groundwater in the Sana'a Basin (see Fig. 2). Irrigation return flow was assumed to be 25% of the entire water volume abstracted for agriculture and kept pace with the increase in abstracted volume. Irrigation return flow was simulated by recharge assigned to wells in the uppermost layer in the model. Infiltration of wastewater was simulated by means of a number of recharge wells located in layer 1, the Quaternary Alluvium, and keeping pace with the growth of the city (Fig. 1). Calibration of both steady state model and transient model was carried out with a non-linear parameter estimation code (PEST; Webtech360, 2003). The parameter space was determined by minimum and maximum parameter values that were set at 0.001 and 1000 times the initial parameter value at the start of the calibration process. If final calibrated parameter values at the end of the calibration process became equal to the minimum or maximum value, the parameter space was increased and/or model areas assumed to be hydrogeologically homogeneous were regrouped in order to ensure maximum and unbiased parameter estimation.

With the calibrated flow files, conservative transport of nitrate was simulated with MT3DMS (Webtech360, 2003). For this, longitudinal dispersivity values of 10% of the total distance traveled (approx. 7500 m) of the plume of contamination were assumed, while transversal dispersivity was assumed to be 10% of the longitudinal dispersivity. The recharge wells in layer 1 to simulate wastewater infiltration were assigned an average value for [NO_3^-] of 248 mg/L (= 4 mmol/L). Calculated nitrate values were compared with values measured in 1995 (Chapter 10).

9.3.2 Scenario calculations

To evaluate the effect of the construction of the sewerage on groundwater quality we modeled two scenarios for the period 2005-2020. In the first scenario (called 'laissez faire') uncontrolled infiltration of wastewater in the city was assumed to continue at the same rate as in 2000. In the second scenario (called 'sewering the city center'), the uncontrolled infiltration of wastewater in the city was stopped by eliminating all recharge wells in the city in model layer 1 in the year 2000 and onwards. In this case, sewage originating from the city was allowed to infiltrate north of the airport near the location of the new wastewater treatment plant (Fig. 1). We realized that the area sewered in the model (the entire city) was larger than the area actually sewered by 2004. However, we felt this was justified, as the Yemeni authorities intend to continue expanding the sewered area for at least the next few years.

Near the location of the new wastewater treatment plant, a scheme for re-use of wastewater for agricultural purposes was anticipated with an areal extent of some 21 km^2 (see Fig. 1). However, due to the reluctance of farmers in the Sana'a Basin to use treated wastewater, we assumed re-use would not be very efficient: 75% of treated wastewater leaving the new wastewater treatment plant was allowed to infiltrate into the aquifers again. Furthermore, the model assumed that as a result of wastewater treatment the [NO_3^-] of the 75% of treated wastewater leaving the plant was half (124 mg/L) of the [NO_3^-] of untreated wastewater infiltrating into the aquifers (248 mg/L).

Finally, in both scenarios agricultural abstractions and related irrigation return flow were kept constant at the year 2000 values, anticipating the implementation and enforcement of the new water law in Yemen that has stipulated strict rules and regulations for the drilling of new wells in the Sana'a Basin.

All calculations were carried out with Modflow Pro Version 7.0.17 (Webtech360, 2003). Within Modflow Pro, use was made of MODFLOW96 for the flow calculations, PEST for the parameter optimization, and MT3DMS for the contaminant transport calculations.

9.4 Results

9.4.1 Calibration of aquifer parameters

Layers 1 and layer 3 were calibrated using one value for horizontal permeability (see Table 1, second row), while layer 2 (Volcanics) had 2 values (one for the plain and one for other areas). Due to the N-S faults, fissures and cracks, the horizontal permeability anisotropy values for layers 2 and 3 were 4 and 4.5. These values indicate that the horizontal permeability from N to S was 4-4.5 times higher than the

Table 1: Values of the calibrated aquifer parameters

Parameter	Layer	Calibrated value	Remarks
Horizontal	1	$6.70*10^{-2}$	
permeability (m/d)	2	$3.00*10^{-3}$	Plain
		$3.60*10^{-2}$	Other areas
	3	$1.80*10^{-1}$	
Horizontal	1	n.a.	
anisotropy (-)	2	4.0	
	3	4.5	
Vertical leakance (d^{-1})	1↔2	$4.60*10^{-5}$	
	2↔3	$6.10*10^{-7}$	Thickness < 250 m
		$5.20*10^{-7}$	Thickness = 250-500 m
		$1.40*10^{-7}$	Thickness = 500-750 m
		$4.40*10^{-8}$	Thickness > 750 m
Specific yield (-)	1	$9.70*10^{-3}$	
	2	$4.97*10^{-3}$	
	3	$1.04*10^{-1}$	
Storage coefficient or	1	n.a.	
storativity (-)	2	$3.74*10^{-7}$	
	3	$9.08*10^{-4}$	

values from W to E given in Table 1, second row. The resistance between layers 1 and 2 was calibrated at 62,500 days (equal to a conductance of $4.60*10^{-5}$ d^{-1}; see Table 1), and due to the extremely varying thickness of the Tertiary Basalt (layer 2), the conductance between layers 2 and 3 was subdivided into 4 regions according to its thickness (< 250m, 250-500m, 500-750m, and >750m). The specific yield of both layers 1 and 2 was low (4.97-$9.70*10^{-3}$); layer 3 had a calibrated specific yield of 10.4%. Finally, storativity in the confined parts of layers 2 and 3 was $3.74*10^{-7}$ for layer 2 and $9.08*10^{-4}$ for layer 3. The Root Mean Squared (RMS) error or standard deviation between measured and calculated heads for the steady state calibration was 25 m (179 pairs of values) and for the transient model was 63 m (292 pairs of values). Given the magnitude of the change in heads over the entire modeled area (around 1000 m), these values were considered to be acceptable.

Comparison of the measured and calculated heads (Fig. 5) revealed no significant deviations between calculated and measured heads in the entire head range measured in the Sana'a Basin. However, as already indicated by the RMS, the points in the transient case (Fig. 5b) were more scattered than in the steady state case (Fig. 5a).

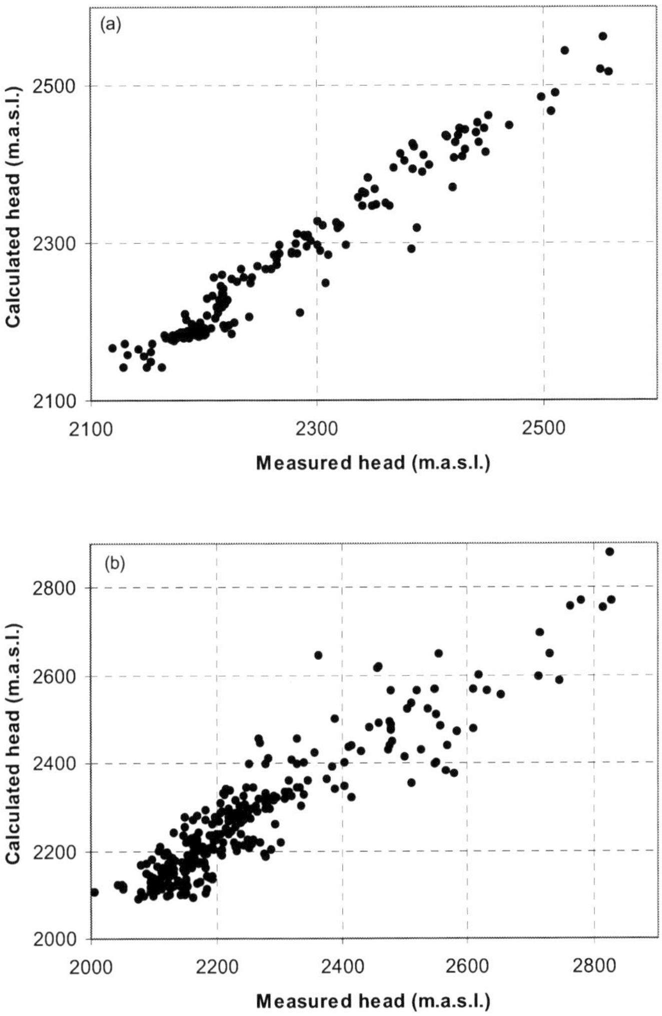

Figure 5: *Calculated versus measured heads resulting from the calibration with PEST (a: steady state calibration; b: transient calibration)*

9.4.2 Basin-wide groundwater depletion

When comparing calculated heads in layer 3 (Tawilah Sandstone) in 1972 with those in 1995, a decline in head in the entire aquifer was apparent. Maximum drawdown was around 100 m 5 km east of the airport; maximum drawdown in the Western wellfield was around 40-70 m, and in the Eastern wellfield around 70-90 m. Calculated drawdowns of a number of selected observation wells in both wellfields are also given in Fig. 3 (squares). Due to the infiltration of domestic sewage, the water table in layer 1 (Quaternary Alluvium) had risen some 20 m in the central parts of Sana'a. Since the Musaik wellfield is close to the city, drawdown in layer 2 was negligible, because the effect of overabstraction outside the city on the one hand was almost counterbalanced by the effect of domestic sewage infiltration.

9.4.3 Sewage infiltration and nitrate concentration

Calculated nitrate concentrations in layer 2 (south of the line Y = 1 699 000) and layer 3 (north of the same line) were combined and compared with measured nitrate concentrations in drilled wells in 1995 (see Fig. 4). Since the wells had mainly been drilled to depths between 200 and 400 m, in order to accurately compare values the two (southward dipping) layers were combined into one map. The calculated nitrate values agreed reasonably with measured values in terms of maximum concentration (maximum isoline in both cases is 150 mg/L) and in terms of extent of the area affected by sewage infiltration. However, there were a number of differences between calculated and measured nitrate values: i) the calculated nitrate values in the northeast and in the south of Sana'a were larger than measured values (see areas marked 'I' in Fig. 4), and ii) measured values in a number of areas throughout the city (marked 'II' and 'III' in Fig. 4) were higher than calculated values. Both differences were mainly caused by variations in sewage infiltration rate and/or variations in infiltrating nitrate concentrations. For instance, in area II in Fig. 4 there were a number of cesspits still in use though in the simulation the infiltration of sewage in that area had already ceased in 1985. Furthermore, areas III in Fig. 4 were associated with the large agricultural areas that still exist within the city; in these areas, infiltration of irrigation return flow has been going on for decades. Finally, area IV in Fig. 4 was associated with the infiltration of sewage at the location of the old wastewater treatment plant (see also Fig. 1).

9.4.4 Managing the urban aquifers: laissez faire and sewering the city center

The results of both scenario calculations are given in Figs. 6 and 7. In the laissez faire scenario, nitrate concentrations in the aquifer exceeded 50 mg/L throughout the entire urban area, with maximum values of around 150-200 mg/L (Fig. 6a). In the sewering the city center scenario, nitrate concentrations fell considerably, to less than 50 mg/L throughout the city and with maximum values around 70 mg/L just north of the city center (Fig. 6b). So, remediation of the aquifer complex occurred, but the process is slow and it will take decades before contaminant concentrations have fallen to acceptable levels.

In Fig. 7, the averaged nitrate concentration of each wellfield has been given, taking into account the calculated nitrate concentration and the yearly volume abstracted from each production well. In the laissez faire scenario, mixing the entire abstracted volume caused the nitrate concentration to increase only slightly from around 15 mg/L in 2005 to 20 mg/L in 2020 ('mixture' in Fig. 7a). This surprisingly slow increase was mainly due to the slow increase in nitrate concentration in the Western and Eastern wellfields, the main suppliers of drinking water in Sana'a. This was because the direction of flow out of the city was due North and therefore most of the production wells were not affected by the contaminant plume. Furthermore, production wells in these wellfields tap the Tawilah sandstone aquifer (layer 3) and

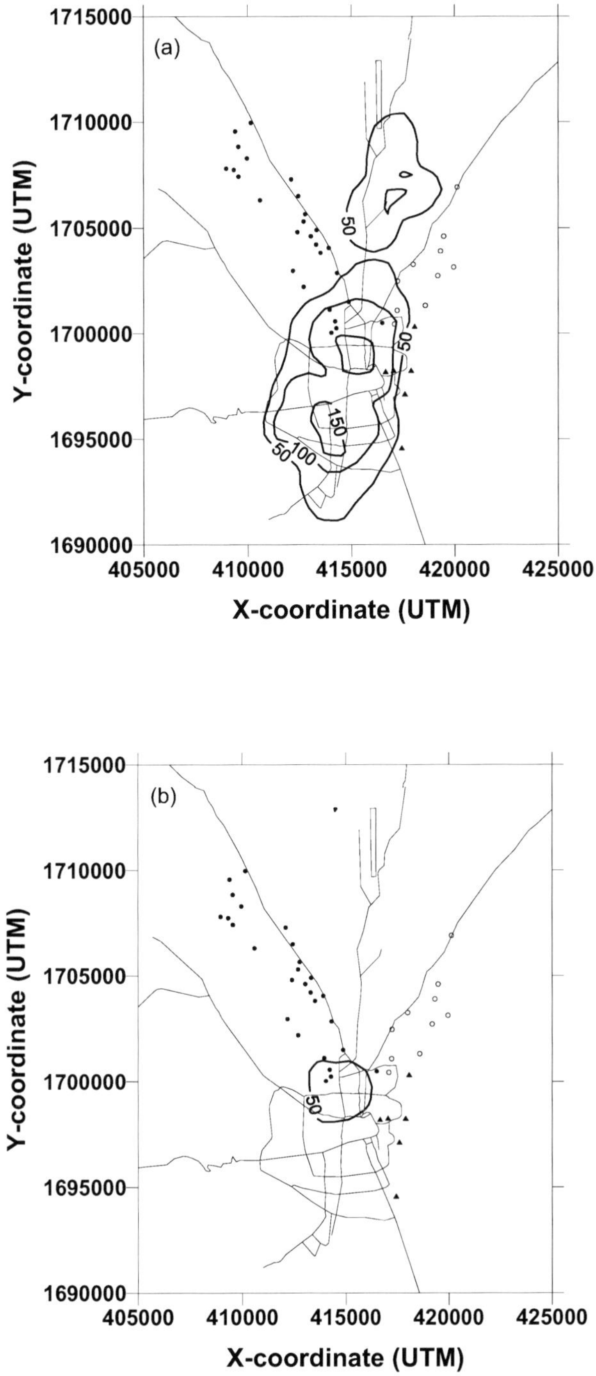

Figure 6: *Calculated nitrate concentration in the aquifers in Sana'a in 2020 (a: laissez faire; b: sewering the city center). Dots indicate the presence of public supply drinking water wells*

due to the relatively high porosity of the sandstone (see Table 1), pore water flow velocities are relatively low. Therefore, nitrate concentrations increased only very slowly. In the Musaik wellfield, which abstracts water from the Volcanics (layer 2),

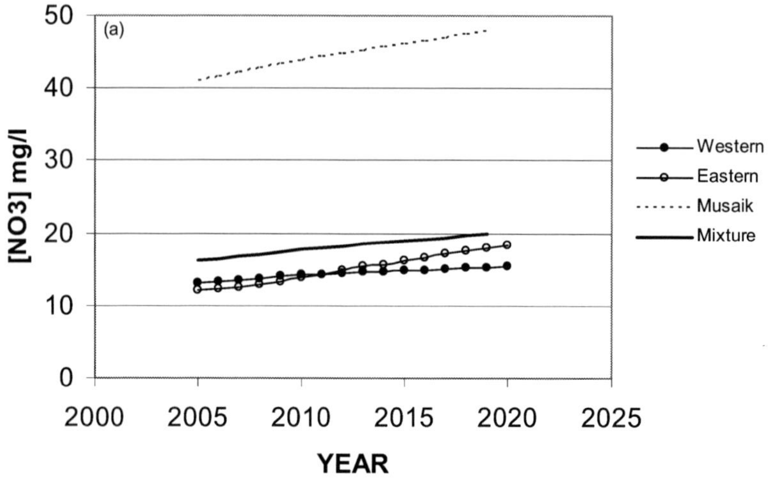

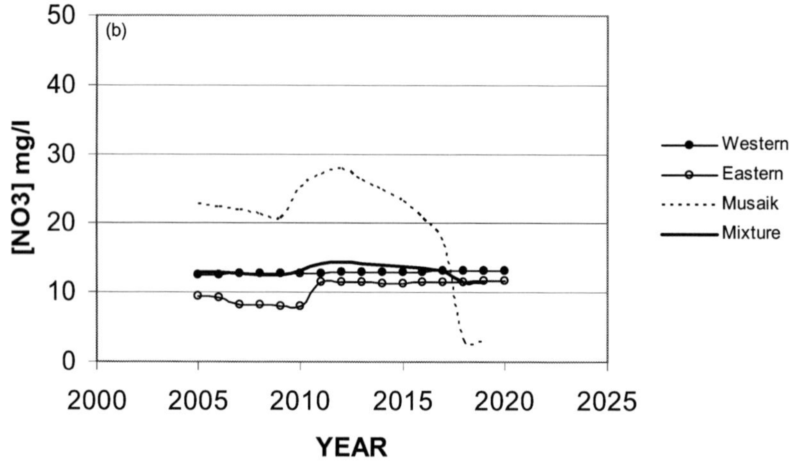

Figure 7: *Development of nitrate concentration in the wellfields, determined as the weighted average of all production wells per wellfield and as the weighted average of the three wellfields ('mixture'; a: laissez faire; b: sewering the city center)*

the averaged nitrate concentration was higher, not only due to the proximity of the production wells to the city, but also due to the layer's relatively low porosity and high pore water flow velocity. In the sewering the city center scenario (Fig. 7b), the rising trend stopped and the nitrate concentration remained more or less constant (see 'mixture' in Fig. 7b) at around 12-13 mg/L in the period 2005-2020. The averaged nitrate concentration in the Musaik wellfield fell from 28 mg/L in 2012 to 3 mg/L at the end of the simulation period.

9.5 Discussion

9.5.1 Model limitations

Despite the reasonable agreement between calculated and measured heads and nitrate concentrations, the model has a number of serious limitations related to the lack of data. Firstly, there are no data on specific storage and specific yield of the aquifers, since all the available pumping tests have been carried out without observation wells. Those data would add to the credibility of the model, since almost all the water in the Sana'a Basin is abstracted from storage. Secondly, there are no data on the vertical permeability of the aquifers, while recharge data are scarce (the only reliable data are from Alderwish and Dottridge, 1998). Another type of limitation is that a number of time-dependent parameters were made constant. More specifically, the total population of Sana'a, and therefore the total water demand were assumed to be constant and so were crop water requirements in the entire Sana'a Basin. These assumptions are certainly not true in the Sana'a Basin and therefore the results of the scenario calculations can only be indicative.

9.5.2 Advantages and disadvantages of the sewerage construction

Prolonging the lifetime of the public supply wellfields (Western, Eastern and Musaik wellfields) does not seem to be an advantage of the construction of the sewerage. The only (and very important) advantage is the reduction of health risk for the population of Sana'a. In the past, in a number of areas throughout the city cesspits became clogged very frequently and sewage flooding the streets with associated health risks, odor nuisance and road damage were facts of life in those areas. Nowadays, sewage flooding the streets does not occur any more.
However, there seem to be three important disadvantages of the sewerage. In the first place a decline of the phreatic water table of some 20-40 m has occurred in the Quaternary Alluvium. As a result, the average daily yield of many dug wells has fallen sharply. Nowadays most of those wells abstract water for only 1-2 hours per day, whereas this used to be 10-18 hours. This could have serious consequences for the agricultural yield of the so-called awqaf (single: waqf), which are abundant in Sana'a. A waqf is a small plot of agricultural land (usually less than one hectare) including a number of houses and a mosque. In Islamic culture the word waqf has the meaning of holding certain property and preserving it for the confined benefit of certain philanthropy and prohibiting any use or disposition of it outside that specific objective. Usually each waqf has a dug well, which is used for ritual ablution, domestic purposes and irrigation. The socio-economic consequences of a drastic decline in agricultural yield of the awqaf could be substantial, since a certain part of the population in the older parts of Sana'a seems to rely on these agricultural plots.

A second disadvantage of the sewage network is that there is hardly any control over the re-use of the enormous supply of treated water north of the airport. According to Hamdi (2000), the uncontrolled re-use of treated sewage will not have any beneficial effect on sustainable (renewable) water resources management. It will merely increase agricultural demand for water, as the treated sewage is commonly used by farmers who cannot afford facilities to abstract groundwater. Furthermore, a lack of control in combination with a lack of knowledge of proper wastewater re-use might lead to environmental degradation, soil salinization, and adverse public health effects (dissemination of waterborne diseases) for agricultural communities re-using the waste water.

A third disadvantage is that construction of a sewage network is a very expensive option that requires maintenance, both financially and technically. From other urban areas in the world it is known that conventional sewerage is by far the most expensive option. Furthermore, conventional sewerage is basically a water-driven system that performs best at high specific water consumption rates (preferably over 70-80 L/cap./d). In view of the critical situation with respect to water resources availability within the Sana'a Basin, it was probably unwise to consider the city-wide application of conventional sewerage.

In our opinion, a wastewater holding tank with an overflow construction to small bore sewerage could be considered as a serious alternative to the conventional wet off-site sanitation, especially in the areas of basalt outcrops. This type of sewerage is often cheaper than conventional sewerage and also easier to construct in areas of basalt (Sinnatamby et al., 1986; Veenstra and Foppen, 1998). However, for small-bore sewage systems to function effectively, there must be user participation, as the system's performance relies on changes in domestic habits and practices, and a fair degree of local care being taken by the beneficiaries. Therefore, in addition to saving money, another benefit would be that this type of sanitation would help raise the awareness of Sana'a citizens that water supplies and sanitation are crucial to the sustainability of their society, and thus are not to be seen as solely the responsibility of government.

CHAPTER 10:

IMPACT OF HIGH-STRENGTH WASTERWATER INFILTRATION ON GROUNDWATER QUALITY AND DRINKING WATER SUPPLY: THE CASE OF SANA'A, YEMEN

Foppen, J.W.A., 2002. Impact of high-strength wastewater infiltration on groundwater quality and drinking water supply: the case of Sana'a, Yemen. J. Hydrol. 263 (2002) 198-216.

Abstract

In Sana'a, the capital of the Yemen Arab Republic, a major well inventory was compiled in 1995 during which samples were analysed for major cations and anions. Five years later the opportunity was taken to repeat the exercise on a sub-set of the original wells. The results showed that groundwater in the urban area was characterised by high concentrations of almost all major cations and anions due to the continuous infiltration of wastewater into the aquifers via cesspits. The dominant watertype appeared to be $CaCl_2$. The Cl^--concentration ranged from 3-10 mmol/l and NO_3^--concentration ranged from 1-3 mmol/l while NH_4^+ was absent in all samples. It is concluded that cation exchange has taken place. Ca^{2+} in groundwater has been enriched, while Na^+, K^+ and NH_4^+ have been depleted. Groundwater affected by wastewater had pH values of 0.5-1 unit lower than groundwater not affected by wastewater, indicating that acidification has taken place. Over the period between the two surveys, concentrations of almost all major anions and cations increased, while pH decreased, both owing to the continuous infiltration of wastewater. An exploratory one-dimensional transport model of a 200 m column of the aquifer underlying Sana'a showed that, over a 15-year period of continuous wastewater infiltration, a quarter of the NH_4^+ present in raw sewage would oxidise to NO_3^- thereby producing acidity and some 60 per cent would be adsorbed. The model indicates that after 50 years of wastewater infiltration, exchange of NH_4^+ has become limited due to the limited Cation Exchange Capacity (CEC) of the soil. Therefore, more NH_4^+ will be oxidised to NO_3^- and $[NO_3^-]$ in groundwater will rise. At the same time, groundwater in the zone of NH_4^+ oxidation will become very acid due to a lack of buffering minerals. The modelling studies, together with the results from the surveys, tend to indicate that up to 12 per cent of the current population of the city could be dependent on contaminated groundwater for their drinking water supply.

10.1 Introduction

The effects of primary and secondary wastewater effluent recharge on groundwater quality have been studied at several locations (Chen et al., 1974; Aulenbach and Tofflemire, 1975; Mann, 1976; Runnells, 1976; Baxter and Clark, 1984). Canter and Knox (1985) give an overview of the effects of septic tanks on groundwater quality and Feigin et al. (1991) describe the processes and problems associated with irrigation with treated wastewater effluent. Hamad (1993) studied the impact of wastewater effluent used for irrigation on soils in some developing African countries and Somasundaram et al. (1993) give an overview of all aspects of groundwater pollution of the Madras Chennai urban aquifer in India. Those authors check whether parameter values exceed certain guideline values, but do not analyse hydrochemical processes. They also give an overview of case histories of urban groundwater pollution (a few of which are in developing countries), but most of these consider only certain aspects of groundwater contamination (microbiology, nitrogen species, organic compounds, etc.). Eiswirth and Hötzl (1997) analysed the impact of leaking sewers on groundwater quality in the vicinity of an old and very damaged sewer pipe section in the city of Rastatt in Germany. They found that precipitation of iron oxides, anaerobic oxidation, fermentation and ammonification

occurred within a thin anaerobic zone immediately below the leaking sewer, while oxidation of organic matter, dissimilatory nitrate reduction and bicarbonate buffering occurred within an aerobic zone above the capillary fringe. Gooddy et al.(1997) described the geochemical processes occurring beneath unsewered Santa Cruz, Bolivia and Hat Yai, Thailand. They found that dominant reactions involved organic carbon and nitrogen, during progressive consumption of electron acceptors. They also found that if the buffering capacity of the aquifers was exceeded by the acidity produced during these oxidation reactions, the pH fell. Klimas (1997) analysed groundwater quality data in 6 cities in Lithuania and concluded that nitrification usually takes place in the upper (aerobic) parts of the aquifers underlying the cities, while denitrification and sometimes ammonification occurs in the lower parts of the aquifers. Yang et al. (1999) use solute balances for three conservative species originating from wastewater from leaking sewers (Cl^-, SO_4^{2-} and total N) to quantify the long-term groundwater recharge (period 1850-1995) for the city of Nottingham. Jacks et al. (1999) estimated tentative nitrogen budgets for pit latrines based upon nitrogen isotope ratios and ammonia volatilisation measurements. Recently, Zilberbrand et al. (2001) studied the effect of seawater encroachment and associated cation exchange in combination with the nitrifying effect of recharge from cesspits in the Tel Aviv urban area in Israel.
The first objective of this research is to analyse the effects of unsewered urbanisation and infiltration of wastewater on (chemical) groundwater quality in Sana'a, the capital of the Arab Republic of Yemen, located in the southwest of the Arabian Peninsula. Like Zilberbrand et al. (2001), cation exchange, nitrification and the presence of a sink for N-species play an important role. An exploratory hydrochemical model will be used to evaluate these processes. The second objective of this research is to evaluate the implications for future drinking water supply with the results of the hydrochemical model.
First, the physiography of the area is described together with the water supply and wastewater conditions in Sana'a. After an overview of the materials and methods, the results of the well inventory are discussed with emphasis on the processes and temporal trends that occur due to continuous wastewater infiltration. Then, the modelling of the pit latrine-groundwater system is described. Model results are compared with measured data from 1995 and the sensitivity of the model is analysed. Then the model is verified with data from 2000. Finally, the use of the abstracted groundwater that is severely contaminated due to wastewater infiltration is briefly discussed.

10.2 Area description

10.2.1 Physiography

Sana'a is located in the centre of the Sana'a Basin, an intermontane plain of some 3,200 km^2, located in the central Yemen Highlands. The plain has an elevation of about 2,200 metres above sea level (see Fig. 1) but to the west, south and east it is flanked by mountains rising to about 3,000 m.a.s.l. The climate in the Sana'a Basin is semi-arid with an average annual rainfall over the period 1972-1989 of 235 mm at Sana'a Airport.
Three geological formations are important in Sana'a (see also Fig. 1):

- Quaternary Alluvium
- Tertiary Basalt
- Cretaceous Sandstone

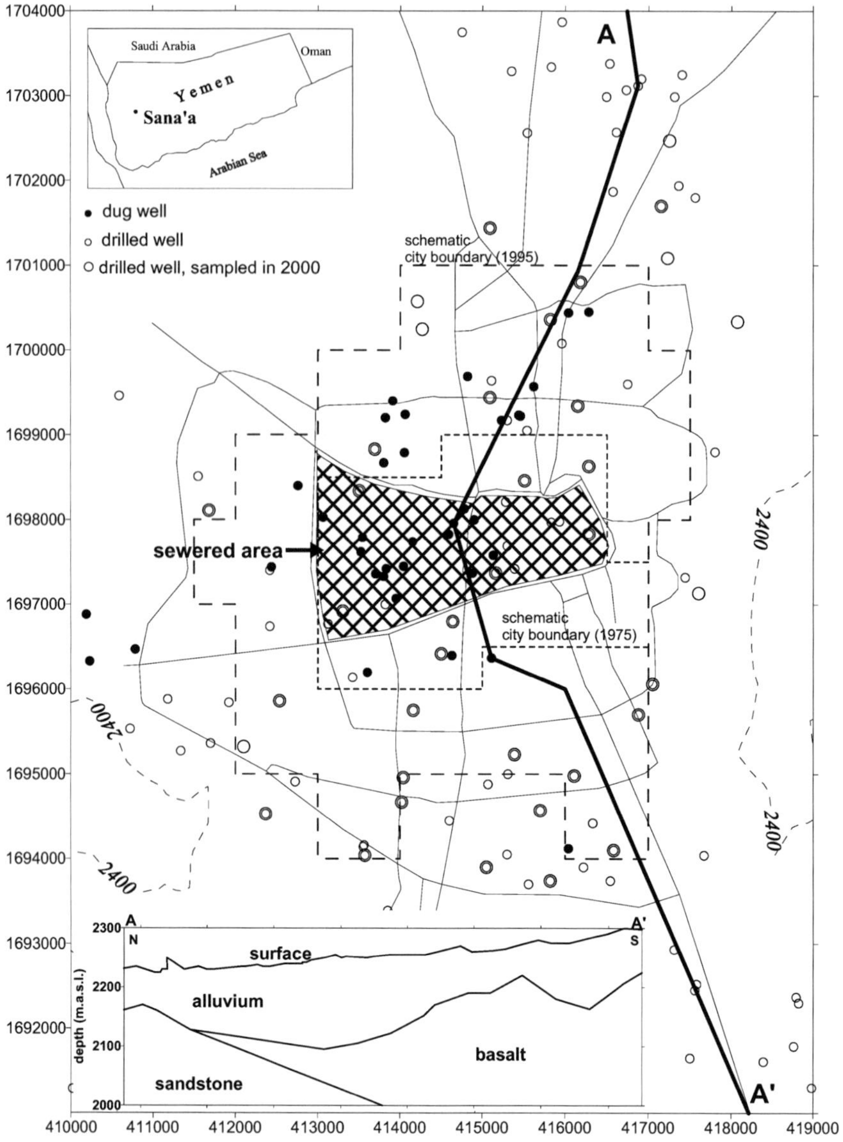

Figure 1: Location map of Sana'a showing the wells inventoried including the wells that were sampled in 2000.

The sandstone, which is the main aquifer throughout the Sana'a Basin, has an average thickness of some 300 m and dips gently southwards. North-east and north-west of Sana'a where it either outcrops or underlies unconsolidated alluvial deposits,

it is heavily exploited. Recent well-inventories in the Sana'a Basin have identified around 4,000 wells, most of which abstract water from the sandstone.
The basalt has a thickness of some 800 m in Sana'a south, but thins out northwards, petering out north of Sana'a. Water is abstracted from the formation but in general it is thought to be a poor aquifer.
The unconsolidated alluvium is the uppermost aquifer and the most important one for this study, since all wastewater infiltrates into this aquifer. This aquifer too has been heavily exploited because of its relatively shallow water table (□ 40 metres below ground level). It has a maximum thickness of 200 m (Shaaban, 1980) and it is poorly permeable. The alluvium mainly consists of basaltic fragments and it is very poorly sorted with extremely varying fractions of clay, silt, sand and pebble sized grains.
Sana'a has had an average annual population growth of around 6.1% during the last four years and around 10% during the last 20 years. The city had 80,000 inhabitants in 1970, but in the most recent census (1994) the total population was estimated to be around 1,000,000. As a consequence, the areal extent of the city has increased tremendously in the last 20 to 30 years. The boundaries of the city in 1975 and 1995 were determined from topographical maps and aerial photographs (Fig. 1).

10.2.2 Water and wastewater

Water supplies have changed rapidly over the past 25 years to cope with this growth. Historically, water was obtained from dug wells and so-called *ghayls* (Jungfer, 1987). Ghayls are underground tunnels tapping a number of rainfed springs, that were present in the mountains south of Sana'a until some 20 years ago. Ghayls can be compared with *qanats* found in Syria, Iraq and Iran. Borehole construction and the introduction of pumps began in the 1960s and increased rapidly and in an uncontrolled fashion from the 1970s onwards. The result has been a drastic decline in aquifer water tables (Charalambous, 1982; Foppen, 1996), causing the springs feeding the *ghayls* and dug wells to dry up. Nowadays, Sana'a relies almost completely on groundwater for both irrigation and urban water supplies. Drinking water is supplied to about 50% of the population by 4 peri-urban wellfields from the National Water and Sanitation Authority (NWSA); private wells supply the other 50% of the population.
With the increasing population and water use the amount of wastewater has increased as well. Night soil used to be collected, dried and used as fertilizer or fuel. Liquid waste was collected in small pit latrines or cesspits with maximum depths of around 3 m (Hamdi, 2000). Nowadays, cesspits with a diameter of 2 to 4 m and depths up to 30 m are dug for disposal of solid and liquid waste. Also, old and abandoned dug wells that have become dry are nowadays used for disposal of solid and liquid waste. This has resulted in areas where drainage of wastewater is seriously hindered: the soil becomes saturated, cesspits overflow and wastewater drains into the streets. Houses on the saturated soil tend to collapse. It was to prevent this that in the 1980s a sewerage system was constructed in the old and densely populated centre of Sana'a (Fig. 1). At present, wastewater from some 15,000 households (about 100,000 people) is collected in sewage ponds a few kilometres north of Sana'a. Coverage of the sewage network is however not complete; pit latrines are still very common, especially in the western part of the sewered area (Fig. 1).

Table 1: *Overview of wastewater quality characteristics at different locations in the world (* : 4 sites in the UK; @: typical domestic sewage characteristics in the USA; #: summary of several studies; ##: calcium and magnesium combined; ###: in brackets: median value; $: raw sewage at Abu Rawash in the Giza area (Egypt); ^: values are averaged*

	Baxter and Clark (1984)*	Canter and Knox (1985)@			Feigin et al (1991)#			Hamad (1993)$	Hamdi (2000) Sana'a raw sewage
		Weak	Medium	Strong	Low	Medium	High		
Biological characteristics									
BOD		100	200	450	100	200	350		400-570
Inorganic contaminants									
PH					7.0	7.2	8.0		7.5-7.9
Cl (mg/l)	50-130				10	150	650		223-558
Ammonia (mg N/l)	13-31	10	25	40	10	25	50		88-203
Nitrate (mg N/l)	1.6-16.5	-	0.5	1.0	0	0.2	1.5		14-38
t-PO4 (mg/l)	5-11	5	15	30	4	10	36		15-26
Sulphate (mg/l)	43								141-180
Alkalinity (mg/l CaCO3)	80-368				50	200	400		460-680
Na (mg/l)	67-74				10	120	460		164-285
K (mg/l)	15-20				5	10	25		38-48
Ca (mg/l)	30-120				25##	80##	150##		72-204
Mg (mg/l)	3-8								29-34
Boron (mg/l)	1.7					<0.123-2.0			1.3-1.7
Trace metals (in mg/l)									
Fe	0.24-1.6								0.63^
Mn	0.012-0.25							0.4-1.0	
Ag	<0.002								
Cd	<0.005					<0.0012-2.1 (0.024)###		0.011-0.06	0.025^
Cr	0.007-0.026					<0.0008-83.3 (0.40)###		1.9-5.9	0.2^
Cu	0.036-0.037					<0.0001-36.5 (0.42)###		0.11-0.61	0.13^
Pb	0.026-0.09					<0.001-11.6 (0.12)###		0.26-2.37	0.25^
Ni	0.03-0.063					<0.002-111.4 (0.23)###		0.17-0.35	
Zn	0.076-0.27					<0.001-28.7 (0.52)###		0.35-2.3	0.19^
As						<0.0003-1.9 (0.085)###			
Hg						<0.0001-3.0 (0.11)###			
Mo						<0.0011-0.9			
Se						<0.002-10.0 (0.041)###			
Co								0.13-0.51	
Trace organic contaminants									
	Over 60 volatile species found								

When compared to wastewater in the United Kingdom (Baxter and Clark, 1984) and the USA (Canter and Knox, 1985), wastewater collected in the sewage ponds in Sana'a is very strong (see Table 1): Biological Oxygen Demand (BOD) is very high and so are all concentrations of major cations and anions. It is not uncommon for a city in a developing country to produce high strength wastewater. This is mainly caused by the limited per capita water use (around 40 to 80 l/cap/day). Trace metal concentrations in Sana'a wastewater are rather low. Hamdi (2000) mentions that this is probably due to the absence of industry in Sana'a, since industrial activity is usually mainly responsible for the presence of trace metals in wastewater. Finally, trace organic contaminants have not (yet) been determined in Sana'a wastewater due to lack of proper laboratory equipment. From Baxter and Clark (1984) it is clear that a large number of trace organic contaminants could be present in wastewater.

10.3 Materials and methods

10.3.1 Well inventory

From April 1995 until January 1996, a well inventory was conducted in Sana'a and its immediate vicinity. A questionnaire was filled out for each well, to obtain information on well details (construction date, total depth, well diameter, type of pump) and water production and consumption (pumping duration, yield, use of the abstracted water). Wells were numbered and located on the map by means of a GPS (Global Positioning System).

10.3.2 Sampling (1995 and 2000)

Together with the well inventory in 1995/1996 a sample of pumped water was taken in a 0.75 l PVC bottle from each well, including some outside the urban area. EC, pH and temperature were measured in the field. The samples were stored cool and transported to the laboratory of the Sana'a branch of NWSA, where they were analysed within 48 hours. Parameters and methods (in brackets) are Na^+ and K^+ (flame photometric); Ca^{2+} and Total Hardness (titration with EDTA indicator); Mg^{2+} (difference between Total Hardness and Ca^{2+}); Cl^- (argentometric); alkalinity (titration with phenolphthalein indicator); SO_4^{2-}, NO_3^- and Fe-Total (colorimetric). The latter three were determined with a HACH DR/2000 spectrophotometer. Also 10 raw sewage samples were taken at the inlet of the sewage ponds.
In 2000 a number of wells was sampled again. Water was collected in a 0.75 l PVC bottle. EC, pH and temperature were measured in the field. The samples were stored cool and transported to the laboratory of the Faculty of Engineering at the Sana'a University, where they were analysed within 48 hours.

10.3.3 Processing the chemical data

Analyses with an ion balance error exceeding ± 5% and analyses exceeding 20% difference between measured and calculated EC were exluded from interpretation.

The Saturation Index for calcite ($SI_{calcite}$) was determined by using PHREEQC (Parkhurst and Appelo, 1999). All results were entered in a spreadsheet for further processing.

10.4 Results and discussion

10.4.1 Well types

In all, 248 wells were inventoried. A total number of 214 wells were used for interpretation after processing the chemical data. The interpretation also drew on chemical data from NWSA production wells. Samples were obtained from both dug and drilled wells. Dug wells are usually older than 50 years, with a diameter of around 3 m and an average depth ranging from 25 to 50 m. Groundwater is present near the bottom of the well (up to 5 m above the well bottom). Because samples from dug wells do not represent groundwater quality in the aquifer owing to degassing of CO_2, precipitation of calcite, oxidation reactions, etc. only Cl^- data were used for further interpretation since this parameter is assumed to behave conservatively.
Drilled wells have a diameter of 0.25 to 0.35 m and are cased where they penetrate the alluvial deposits. In the underlying hard rock the well is usually uncased; depths range mainly from 100 to 400 m and on average from 200 to 300 m. Shallow groundwater from the unconsolidated alluvial deposits was sampled in dug wells and groundwater was sampled from hard rock (sandstone, basalt or both) in drilled wells. The depth of the pump in the well is usually not near the bottom of the well but on average at around half of the total depth.

10.4.2 Natural groundwater composition

The natural groundwater composition was assessed by examining analyses from outside the urban area and can be divided into four types:
The groundwater in the basalt was predominantly of the $NaHCO_3$ type. The EC was usually low (300-600 □S/cm) and the pH high (7.5-9.5). The high pH is probably attributable to silicate weathering, which consumes protons (Appelo and Postma, 1993). However, this assumption cannot be validated, since data on silica were not collected. A good example of the resulting water type is H4, sampled in 1994 and 1999 (Table 2). Some samples were taken in the vicinity of old small volcanoes that are present in the Sana'a Basin. Groundwater quality of these samples seems to differ from the "regular" groundwater quality found in the basaltic aquifer: an example is J/2.13, which contains almost equal amounts of Cl^- , HCO_3^- and SO_4^{2-}.
Groundwater of a $NaHCO_3$ character was also present in the alluvial deposits. The differences between "alluvium" water and "basalt" water are minor, which can be expected since the alluvial deposits mainly consist of basalt fragments. The differences concern the pH of alluvium water (around 7-8) and the less developed dominance of Na^+ and HCO_3^- compared with basaltic groundwater (see W106, sampled in 1995 and 2000 in Table 2).
Groundwater in the sandstone was mainly of the $CaHCO_3$ type (see "Q" sampled in

Table 2: *Results of the analysis of a selection of groundwater samples. Wells E48, E22 and E37 were affected by wastewater but not wells H4, J/2.13, W106, Q and W7*

Well code	date	x-coord UTM	y-coord UTM	EC uS/cm	pH	Temp C	Na^+	K^+	Mg^{2+}	Ca^{2+}	Cl^-	HCO_3^-	SO_4^{2-}	NO_3^-	CO_3^{2-}	iberr %
H4	18-Jun-94	414229	1691788	393	9.30	-	4	0.01	0.08	0.16	0.79	2.48	0.47	0.09	0.23	2.46
H4	13-Mar-99	414229	1691788	392	9.27	-	4.31	0.01	0.04	0.07	0.99	2.82	0.41	0.07	0.57	-1.37
J/2.13	21-Nov-95	419262	1688638	481	9.15	30	4.22	0.01	0.08	0.12	1.3	1.03	0.94	0.04	0.32	2.6
W106	06-Jun-95	413500	1698340	459	8.07	25.1	2.37	0.12	0.11	1.06	1.4	2.4	0.72	0.05	0	-4.74
W106	22-Mar-00	413500	1698340	500	8.01	26.3	3.31	0.06	0.16	0.72	1.81	2.56	0.4	0.27	0	-2.92
Q	27-Sep-94	419956	1703132	721	7.08	-	1.83	0.04	0.86	2.12	1.27	4.16	0.97	0.14	0	2.24
Q	13-Mar-99	419956	1703132	676	6.96	-	1.83	0.05	0.78	2.17	1.3	4.47	0.88	0.09	0	1.06
W7	30-May-95	413730	1702300	556	7.1	24	1.34	0.1	0.83	0.74	1.52	2.56	0.29	0.04	0	-0.94
E48	25-May-95	416880	1695700	731	7.54	25	2.35	0.07	0.25	2.27	3.95	2.07	0.35	0.5	0	1.77
E48	22-Mar-00	416880	1695700	1133	7.56	24	3.91	0.04	0.56	2.64	5.98	3.44	0.49	0.42	0	-2.32
E22	03-Jan-96	415090	1699440	1080	7.65	-	1.91	0.08	1.11	3.72	4.96	4.47	0.73	0.66	0	0.35
E22	22-Mar-00	415090	1699440	1373	7.36	22.4	2.48	0.09	1.31	4.2	5.19	4.88	0.99	0.97	0	2.3
E37	13-Jan-96	416150	1699340	1527	7.7	-	3.65	0.1	1.19	4.79	8.89	2	1.96	0.1	0	2.86
E37	22-Mar-00	416150	1699340	2350	7.22	23	4.35	0.13	1.83	6.88	14.1	2.72	2.6	1.12	0	-5.27

1994 and 1999, Table 2). The EC was usually around 500-750 □S/cm and the pH around 7-7.5. Sometimes, $MgHCO_3$ water types were found, probably associated with the presence of dolomite (W7; Table 2). Also, a few samples seemed to originate from dissolution of gypsum since the dominant ions were Ca^{2+} and SO_4^{2-}. Both dolomite and evaporites like gypsum are known to be (discontinuously) present in the lower part of the sandstone and below it.
Furthermore, the $SI_{calcite}$ in alluvium, basalt and sandstone varied mostly between –0.5 and +0.5. Finally, in all "natural groundwater" samples some NO_3^- was present. This might originate from human activity or from minor NH_3 gas emissions from volcanic activity into the groundwater system where it is oxidised to NO_3^-.

10.4.3 Groundwater from the aquifers below the urban area

In the aquifers below Sana'a, groundwater was characterised by high concentrations of almost all major cations and anions. [NO_3^-] ranged from 1 to 3 mmol/l. The dominant watertype appeared to be $CaCl_2$ (E48, E22 and E37, all sampled in 1995 and 2000, Table 2). [Cl^-] below Sana'a ranged from 3-10 mmol/l (Fig. 2). Fig. 2 shows that [Cl^-] in the centre and northern part of Sana'a is equal to [Cl^-] in wastewater (between 223 and 558 mg/l, on average around 10 mmol/l; see Table 1). This is probably the result of hundreds of years of human settlement and production and infiltration of small amounts of wastewater. This hypothesis is supported by Italconsult (1973) who measured high EC values in this area already in 1972 before the growth of the city. A few samples had [Cl^-] (max. 20 mmol/l) that were higher than the [Cl^-] of wastewater. This is probably attributable to the re-use of abstracted water for irrigation or domestic purposes: in both cases, total ion concentrations will increase either due to evaporation or addition of human waste.
NH_4^+ was absent in all groundwater samples, while [NH_4^+] in Sana'a wastewater is around 10 mmol/l. However, [NO_3^-] in groundwater is around 1 to 3 mmol/l, indicating aerobic conditions in the saturated zone. It is thus very likely that nitrification occurs between the cesspit and the saturated zone. Nitrification is the process whereby NH_4^+ is oxidised first to NO_2^- and then to NO_3^- by *Nitrosomonas* and *Nitrobacter* bacteria (e.g. Sprent, 1987):

$$NH_4^+ + 2O_2 \xrightarrow{Nitrosomonas} NO_2^- \xrightarrow{Nitrobacter} NO_3^- + 2H^+ + H_2O \qquad (1)$$

However, [NO_3^-] is far less than [NH_4^+], indicating that there is a sink for N-species between the cesspit and the saturated zone. Besides the effects of dispersion and diffusion which tends to smear out [NO_3^-], the most likely sink is exchange of NH_4^+. If so, then this would also indicate that infiltration of wastewater occurs to a certain depth under anaerobic conditions and oxidation of NH_4^+ will take place only below this depth. It must be noted that this depth does not necessarily have to coincide with the top of the aerobic saturated zone (= groundwater level). An indication for the presence of an NH_4^+ -sink is that if all NH_4^+ had been oxidised to NO_3^-, this would have lowered the pH, alkalinity and $SI_{calcite}$ tremendously (assuming the absence of buffering minerals like calcite), since oxidising 10 mmol/l of NH_4^+ produces 20 mmol/l of protons, which is in excess of alkalinity (15 mmol/l) present in the wastewater.
Concentrations of major cations were compared with conservative mixing of wastewater and natural groundwater present in the alluvium aquifer. The amount of

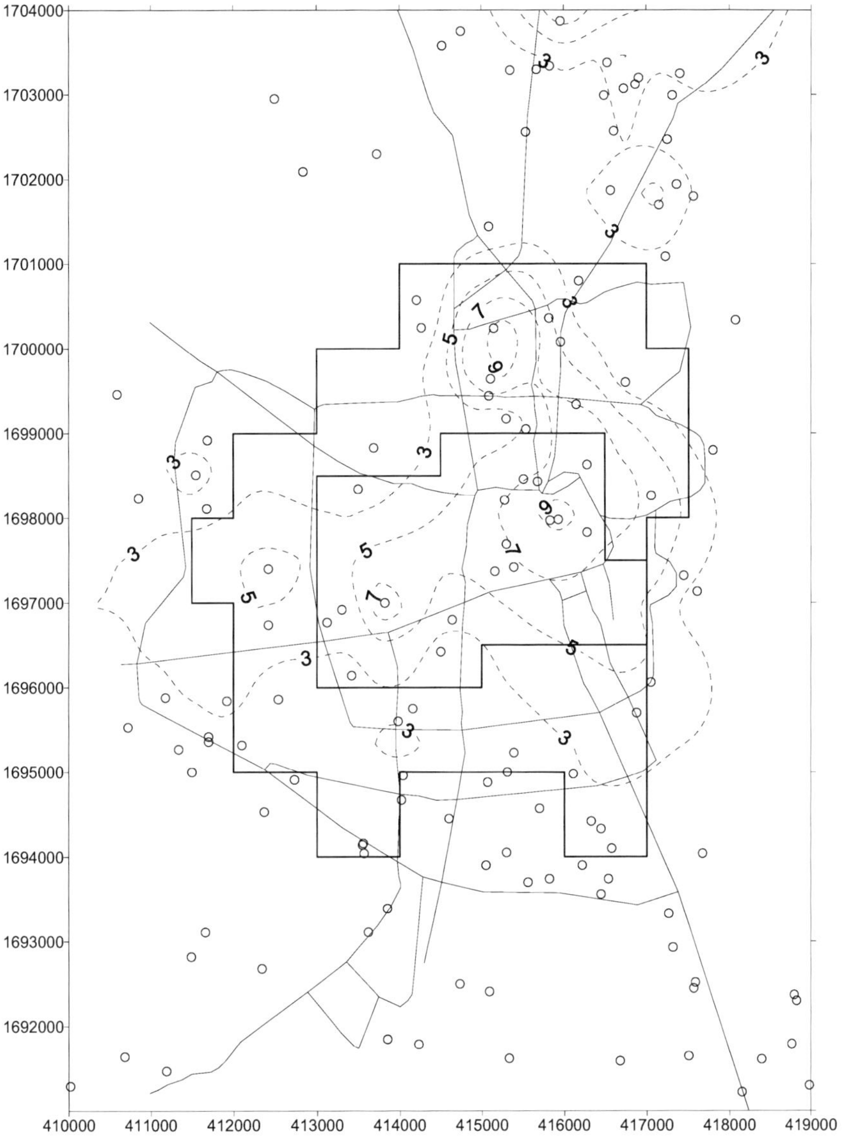

Figure 2: *Contour map of [Cl⁻] (mmol/l) measured in drilled wells. Min. contour = 3 mmol/l; max. contour = 9 mmol/l. Contour line interval = 2 mmol/l*

ion that has undergone reaction(s) other than mixing can be calculated as follows (Appelo and Postma, 1993):

$$m_{i,react} = m_{i,sample} - m_{i,mix} \quad (2)$$

$$m_{i,mix} = f_{wastew} * m_{i,wastew} + (1 - f_{wastew}) * m_{i,aquifer} \quad (3)$$

$$f_{wastew} = \frac{m_{Cl,sample} - m_{Cl,aquifer}}{m_{Cl,wastew} - m_{Cl,aquifer}} \quad (4)$$

where $m_{i,react}$ is the concentration of ion i (mmol/l) as a result of reactions (other than mixing), $m_{i,sample}$ is the concentration of ion i in the sample, $m_{i,mix}$ is the concentration of ion i as a result of mixing, f_{wastew} is the fraction wastewater in the mixed water, $m_{i,wastew}$ is the concentration of ion i in wastewater, $m_{i,aquifer}$ is the concentration of ion i in the aquifer, $m_{Cl,sample}$ is the concentration of chloride in the sample, $m_{Cl,aquifer}$ is the concentration of chloride in the aquifer, and $m_{Cl,wastew}$ is the concentration of chloride in wastewater.

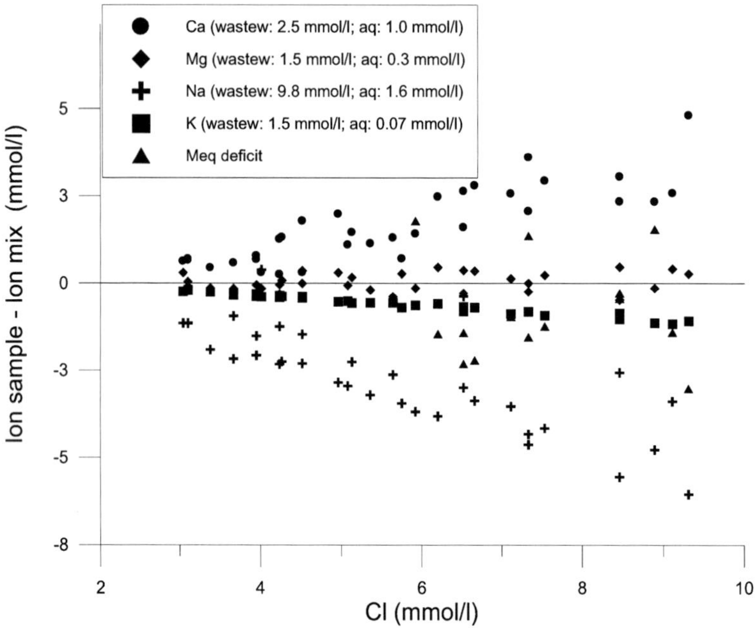

Figure 3: Conservative [Cl⁻] versus concentration of ion i (Ca^{2+}, Mg^{2+}, Na^{+}, K^{+} or NH_4^{+}) as a result of reactions other than conservative mixing. Positive values indicate enrichment, negative values depletion

From Fig. 3 it is clear that compared to conservative mixing of wastewater and natural groundwater in the alluvium aquifer, Na^{+} and K^{+} in groundwater have been depleted and Ca^{2+} has been enriched, while Mg^{2+} shows hardly any signs of exchange. It is also clear that at higher [Cl^{-}] there is an amount of milli-equivalents "missing"; Ca^{2+} in groundwater has been more enriched than Na^{+} and K^{+} have been depleted. It is believed that this is because of exchange of NH_4^{+} that is occurring in the depth interval between the cesspit and the saturated zone. Based on the observed data the overall exchange reaction then becomes:

$$\begin{aligned} &3.9CaX_2 + 0.3MgX_2 + 6.0Na^+ + 1.3K^+ + 1.1NH_4^+ \leftrightarrow \\ &\quad 3.9Ca^{2+} + 0.3Mg^{2+} + 6.0NaX + 1.3KX + 1.1NH_4X \end{aligned} \qquad (5)$$

Fig. 4 shows only the samples collected in a part of the city that has been developed between 1975 and 1995 (see also next section). The boundaries of the city in both 1975 and 1995 are indicated in Fig. 1. From all cation and anion depth profiles it was clear that wastewater has infiltrated until a depth of around 150 m. Below this depth groundwater with a natural composition was found. As was mentioned before, the depth profile of [Cl^-] contains also data points from dug wells, since Cl^- is assumed to behave conservatively. The pH profile shows that samples affected by wastewater (depth < 150 m) have pH values somewhat lower (0.5-1 unit) then pH values of groundwater in the same area that were not affected by wastewater (depth > 150 m) This is in accordance with the nitrification process.

10.4.4 Trends in urban groundwater: comparison between 1995 and 2000 samples

When the results of the samples collected in 2000 are compared with the results of the same wells sampled in 1995, it is clear that concentrations of major cations and anions in general have increased due to continuous wastewater infiltration (Table 2, wells E48, E22 and E37; Fig. 5). As in Fig. 4, Fig. 5 shows only the samples collected in a part of the city that has been developed between 1975 to 1995 and that has been sampled again in 2000. Here, all major cation and anion concentrations have increased while [K^+] and pH values have decreased somewhat. Assuming nitrification is taking place, a lowering of pH would be expected. However, decreased [K^+] cannot be explained.

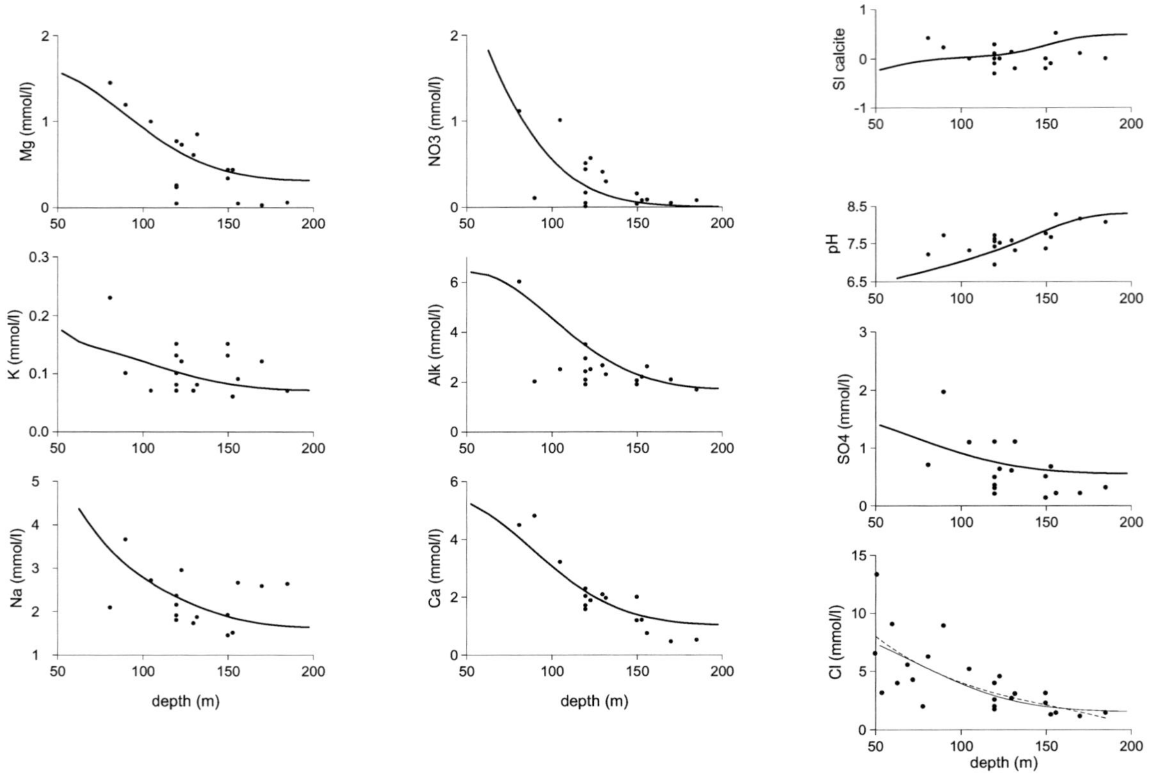

Figure 4:
Depth profiles of major cations and anions together with pH and $SI_{calcite}$ for the data set of 1995. Only the samples collected in a part of the city that has been developed between 1975 and 1995 are shown. Solid lines are model results after 15 years of infiltration of wastewater. Dashed line in the Cl^- depth profile is the depth averaged [Cl^-]

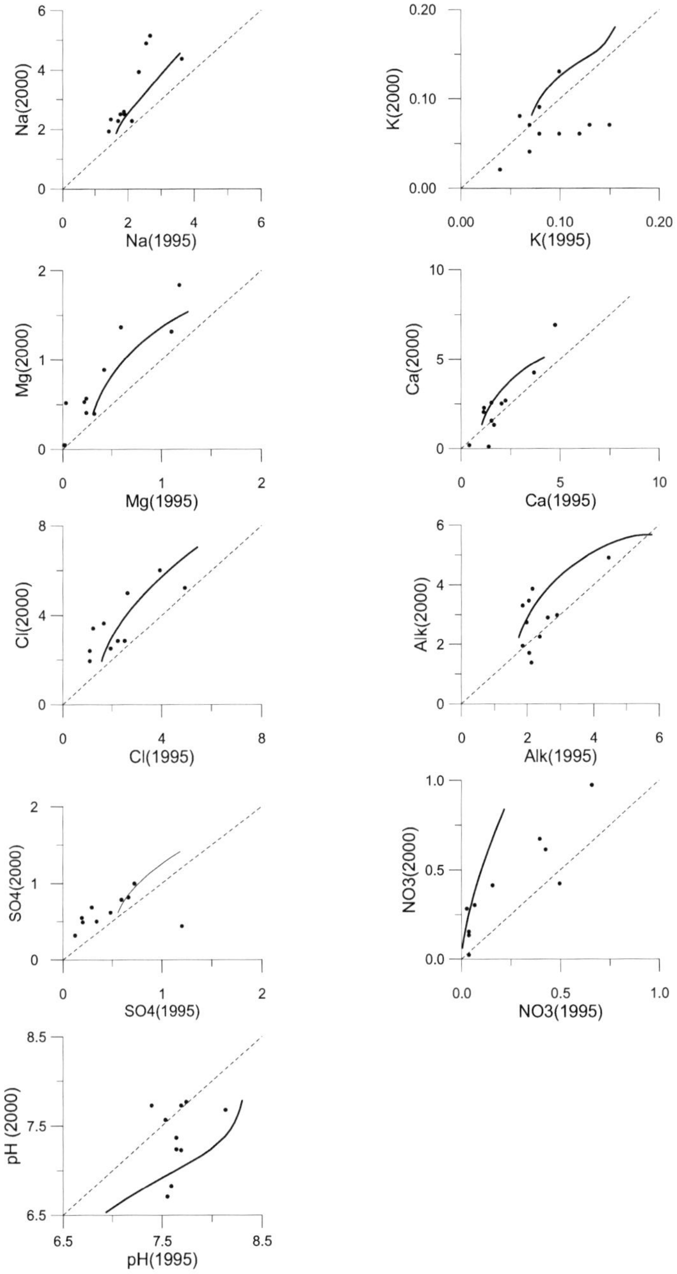

Figure 5: *Concentrations of major cations and anions (mmol/l) and pH measured in 1995 versus concentrations (mmol/l) measured in 2000 in the same wells. Bold solid lines are obtained with the model after 20 years of wastewater infiltration. Dashed: lines of equal values*

10.5 Exploratory modelling of nitrification and cation exchange

To obtain a more quantitative understanding of the relative importance of the processes described above, a one-dimensional transport model consisting of 40 cells of each 5 m was made with PHREEQC version 2 (Parkhurst and Appelo, 1999). The model was used to check whether the processes described above indeed result in the groundwater types found in the aquifers below Sana'a. It is assumed that the cesspit is at the top of the model and the saturated zone is present at a depth of 30 m below the cesspit. Saturated flow of wastewater from the cesspit to the saturated zone is assumed to occur due to the high hydraulic loading of the cesspit (volume of wastewater entering or leaving the cesspit; see below). Based on the well-inventory data and data provided by NWSA, total water production (public and private) is estimated to be $21.1*10^6$ m^3/y. Assuming a loss in the distribution system of 20%, total sewage production is around $16.9*10^6$ m^3/y. Of this, $5.5*10^6$ m^3/y flows to the sewage ponds north of Sana'a, so an estimated $11.4*10^6$ m^3/y infiltrates in to the soil. The area in which this infiltration occurs is around 29 km^2, so the area-averaged hydraulic loading equals a Darcy vertical flow velocity of 0.40 m/year. Assuming an effective porosity of 5-10%, pore water flow velocity was assumed to be 5 m/year. Diffusion at this high pore water flow velocity can be neglected. The dispersivity was assumed to be 10% of the travelled distance. The latter was around 150 m and was estimated from the averaged Cl data (dashed line in Fig. 4, Cl^-- depth profile) so the dispersivity was 15 m.
CEC-values for the alluvium aquifer are around 0.5 meq/100 g soil. These values were derived from a soil study carried out in the Sana'a Basin (Mosgiprovodkhoz, 1986). Ca^{2+} is the dominant base in the exchange complex (around 80%) followed by Mg^{2+} (around 15%), Na^+ and K^+. Mass density of the alluvial deposits is around 2.74 g/cm^3.
Wastewater infiltration was allowed to take place for 15 years and calibrated with the data collected in 1995. The time step used is 1 year. The same model was used to allow for 20 years of wastewater infiltration; results after 20 years of wastewater infiltration were verified with the data collected in 2000. Data for model calibration were obtained from Fig. 4. Since samples were assumed to be taken at a depth equal to the depth of the pump in the well, the model can only be approximate. There are uncertainties in the way the well was constructed and the depth from which a groundwater sample was obtained. The compositions of wastewater and natural groundwater in the alluvium used in the model are given in Table 3. The pe value of raw sewage is assumed to be 2, indicating anaerobic conditions in which NH_4^+ is stable (Sawyer et al., 1994). A pe value of 12 indicates aerobic conditions (Parkhurst and Appelo, 1999) for groundwater in the alluvium. Furthermore it is assumed that at a depth of 30 m and below, P_{O_2} = 0.2 atm since groundwater in the alluvium is aerobic. The P_{CO_2} in the first model cell is fixed at a value of $10^{-2.5}$ atm which is the average of the P_{CO_2} measured in raw sewage (~ $10^{-1.5}$ atm) and atmospheric P_{CO_2} (~ $10^{-3.5}$ atm). Therefore in the first model cell a little CO_2 degassing is allowed to occur from the wastewater used in the model. Only stable nitrogen species with electron valences of –III and +V were allowed in the model. The oxidation reaction from N(-III) to N(+V) does not involve the production or consumption of N_2-gas,

while the intermediate form N(+III) is instable. The exchange coefficients and the database employed in the calculations are the standard values and standard database used in PHREEQC (Parkhurst and Appelo, 1999).

Table 3: Concentrations (mmol/l) of cations and anions for raw sewage and alluvium water used in the PHREEQC calculations

Parameter	Raw sewage	Alluvium water
Temp (°C)	23.0	26.0
pH	7.30	8.35
pe	2	12
Na^+	9.82	1.6
K^+	1.5	0.07
Mg^{2+}	1.5	0.3
Ca^{2+}	2.5	1.0
NH_4^+	10.0	0.0
Cl^-	10.0	1.52
Alkalinity	15.7	1.65
SO_4^{2-}	1.81	0.55

10.5.1 Model results

Modelled concentrations for all major cations and anions after 15 years of infiltration of wastewater are given in Fig. 6 together with modelled pH and $SI_{calcite}$. The modelled [Cl^-] is the result of advection and dispersion. This also applies to modelled [SO_4^{2-}].
At a depth of 2.5 m (the uppermost values in Fig. 6), the original wastewater [NH_4^+] of 10 mmol/l has already decreased to around 6 mmol/l, while [NO_3^-] is still very low. [Na^+] and [K^+] have decreased as well, while [Ca^{2+}] has increased from 2.5 mmol/l in wastewater to more than 4 mmol/l at a depth of 2.5 m.
Fig. 7a shows the composition of the exchange-complex in time at a depth of 22.5 m. Modelling in this figure is extended over a period of 50 years. The major exchange reaction taking place at this depth is NH_4^+ being adsorbed and Ca^{2+} being released into solution. Concentrations of NaX and KX are steadily increasing while there is hardly any change in MgX_2.
It can be calculated that after 15 year of wastewater infiltration at a (Darcy) rate of 0.4 m/year and a [NH_4^+] in wastewater of 10 mmol/l, 60 mol of NH_4^+ has entered the soil. Out of this 60 mol, 8.9 mol is still present as NH_4^+ , 0.3 mol is dissolved ammonia (NH_3), 14.9 mol is oxidised to NO_3^- and 35.8 mol is adsorbed ammonium (NH_4X).
At a depth of 27.5 m, the maximum modelled [NO_3^-] is found. Here acid production is at a maximum and pH, alkalinity and $SI_{calcite}$ are therefore at their minima. Alkalinity is "eaten" away by protons produced from nitrification and $SI_{calcite}$ is low because the activity of carbonate (CO_3^{2-}) is low at low pH.
At a depth of 52.5 m, cation exchange (Fig. 7b) is mainly characterised by the release of Ca^{2+} into solution and an increase of Na^+ on the exchanger (NaX). At a depth of 102.5 m there is hardly any cation-exchange occurring (Fig. 7c) and its significance becomes even less at greater depths.

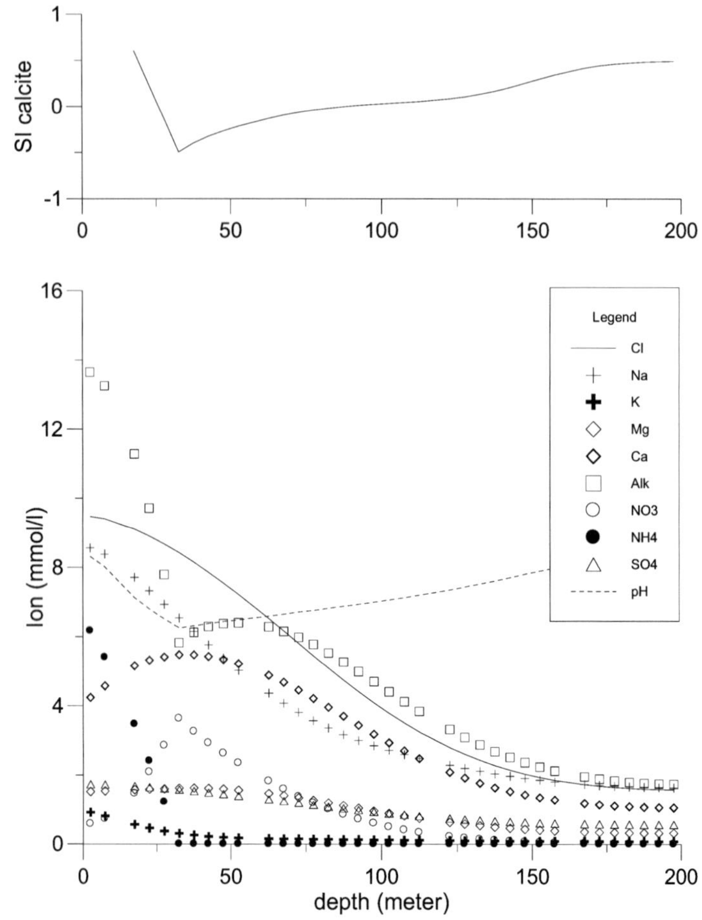

Figure 6: Concentration profiles for all major cations and anions including pH and $SI_{calcite}$ obtained with the PHREEQC model after 15 years of wastewater infiltration

After 50 years of modelling, the modelled pH has decreased dramatically (pH 3) at shallow depth (52.5 m). Alkalinity is the only buffer present in the aquifer and in time NH_4^+ adsorption onto the exchanger becomes less and less (convex curve in Fig. 7a). This means automatically more oxidation of NH_4^+ to NO_3^- and more production of protons. The possibility of exchange of the hydrogen ion exists at low pH. It has not been included in the model due to the selectivity of the hydrogen ion in exchange reactions (Appelo and Postma, 1993).

Correlation between the data collected in 1995 and the model is rather poor (Fig. 4). However, despite the variations in the number of years wastewater has indeed infiltrated into the soil, the population density, the per capita water use, the amount of recharge infiltration, permeability and porosity variations of the subsoil and uncertainties with regard to the depth of sample collection, there still is a clear relationship between the 1995 data and the model.

It should be emphasised that the model is meant to analyse in an exploratory way the process of infiltration and nitrification and cation-exchange in the pit latrine –

groundwater system. Observed data are scattered and can only indicate whether the model produces realistic results.

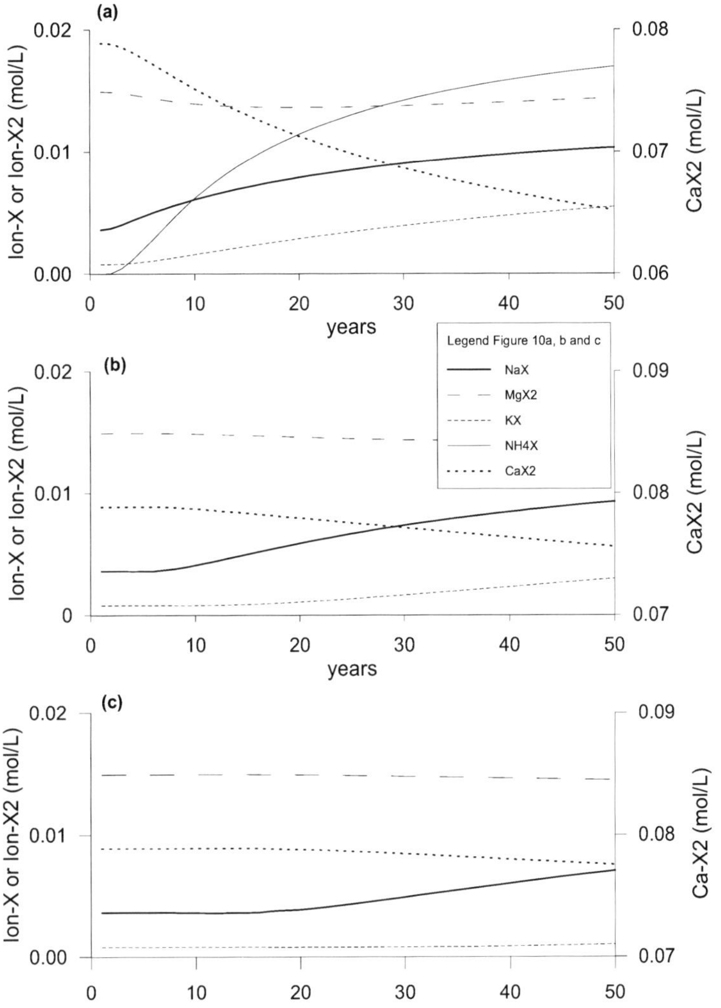

Figure 7: *Distribution of cations (mol/l) on the exchanger in time (a: 22.5 m; b: 52.5 m; c: 102.5 m). Left-hand Y-axis is for ions Na^+, Mg^{2+}, K^+ and NH_4^+; right-hand Y-axis is for Ca^{2+}*

10.5.2 Sensitivity analysis

In the light of the potential uncertainties mentioned above a sensitivity analysis was conducted with the aim of assessing the impact of changes in one or more independent parameters on changes in dependent parameters (major cations and anions, pH and $SI_{calcite}$). Independent parameters chosen to test the model sensitivity are depth to groundwater table, CEC of the soil and losses from total water

production. The choice for these three independent parameters was made since they seem to define the system to a great extent and values used in the model are based upon scarce and/or indirect measurements.
Depth to groundwater table was varied –10 m (indicating a groundwater table rise) and + 10 m (groundwater table decline) by varying the number of constant P_{O_2} cells. CEC of the soil and losses from total water production were both varied –50% and +50%. The latter has an effect on pore water flow velocity of –20% and +20%. Sensitivity coefficients (Table 4) were calculated as

$$s = \left(\frac{P_n - P_0}{P_0}\right) * 100\% \qquad (6)$$

where

s	sensitivity (%)
P_n	new parameter value due to induced change
P_0	original parameter value

Sensitivity coefficients are given only for the most important dependent parameters in Table 4 at three depths: 20 m, 50 m and 100 m (model cell 4-5, 10-11 and 20-21). When the groundwater level is changed –10 m, the parameters NO_3^-, $SI_{calcite}$ and NH_4^+ change 190%, 824% and –100% resp. and are therefore the most sensitive of all dependent parameters to changes in water level. The explanation is that due to a raise in water level, less NH_4^+ will be exchanged and therefore [NO_3^-] will increase almost twofold due to the oxidation of more NH_4^+. [NH_4^+] at a depth of 20 m will be reduced to 0 in the case of a change of –10 m (-100%) due to the forced presence of oxygen. SI values close to 0 explain the high sensitivity of this dependent parameter. The effect on Na^+, K^+ and pH is limited compared to the effect on NO_3^-, NH_4^+ and $SI_{calcite}$. When going deeper (50 m and 100 m), the sensitivity becomes less for all parameters except for $SI_{calcite}$. This is mainly because the original $SI_{calcite}$ values at 50 m and 100 m are closer to 0 than at 20 m thereby causing higher sensitivities.
When the CEC of the soil is decreased by 50%, again NO_3^-, NH_4^+ and $SI_{calcite}$ are the most sensitive dependent parameters. The explanation is that less NH_4^+ is exchanged (increase in [NH_4^+]) and therefore more NH_4^+ will be oxidised (increase in [NO_3^-]). When going deeper (50 m and 100 m), sensitivities become less except for NO_3^- and $SI_{calcite}$. For $SI_{calcite}$ this is already explained above and for NO_3^- the values of [NO_3^-] decrease with depth, but due to increased oxidation, the difference between original and new [NO_3^-] has increased.
When losses from total water production are increased 50%, the resulting pore water flow velocity increases from 5 m/d to 6 m/d. Again NO_3^-, NH_4^+ and $SI_{calcite}$ are the most sensitive dependent model parameters. The explanation is that more NH_4^+ will be absorbed thereby leaving less NH_4^+ for oxidation to NO_3^- (decrease of –42%). Due to the higher pore water flow velocity, the new [NH_4^+] is higher than the original [NH_4^+] (+51%).
Overall, the sensitivity values are rather high, reflecting the small changes in independent parameters within the model required to produce a model that no longer represents the measured data set.

Table 4: *Sensitivities (in italic) of the most important dependent parameters at depths of 20 m, 50 m and 100 m to changes in groundwater level, concentration of exchanger and losses from total water production.*

	Depth: 20 m		Depth: 50 m		Depth: 100 m	
Groundwater level (in m below top of column):	20	40	20	40	20	40
Groundwater level change (in m):	-10	+10	-10	+10	-10	+10
Change in parameter (in %):						
NO_3^-	*190*	*-76*	*56*	*-37*	*82*	*-45*
Na^+	*-2*	*1*	*-2*	*1*	*-2*	*1*
Ca^{2+}	*-11*	*7*	*-9*	*6*	*-5*	*3*
pH	*-18*	*16*	*-6*	*5*	*-5*	*6*
Alk	*-82*	*22*	*-43*	*21*	*-23*	*7*
NH_4^+	*-100*	*25*	-	-	-	-
$SI_{calcite}$	*-824*	*490*	*298*	*-209*	*-2121*	*1942*
	Depth: 20 m		Depth: 50 m		Depth: 100 m	
Concentration of exchanger X (in mol/L):	0.096	0.288	0.096	0.288	0.096	0.288
Change of exchanger X (in %):	-50	+50	-50	+50	-50	+50
Change in parameter (in %):						
NO_3^-	*54*	*-32*	*65*	*-36*	*100*	*-43*
NH_4^+	*33*	*-25*	-	-	-	-
Na^+	*7*	*-5*	*13*	*-8*	*5*	*0*
pH	*-5*	*4*	*-8*	*5*	*-7*	*5*
Ca^{2+}	*-21*	*13*	*-17*	*9*	*-9*	*3*
Alk	*-26*	*8*	*-50*	*20*	*-28*	*6*
$SI_{calcite}$	*-220*	*156*	*397*	*-213*	*-2737*	*1829*
	Depth: 20 m		Depth: 50 m		Depth: 100 m	
Losses from total water production (in 10^6 m^3/y)	2.1	6.3	2.1	6.3	2.1	6.3
Change in losses from total water production (in %):	-50	+50	-50	+50	-50	+50
Resulting pore water flow velocity (in m/day):	4	6	4	6	4	6
Change in parameter (in %):						
NO_3^-	*50*	*-42*	*-12*	*17*	*-47*	*42*
Ca^{2+}	*-4*	*0*	*-11*	*9*	*-30*	*26*
pH	*-6*	*7*	*0*	*-1*	*2*	*-1*
Na^+	*-8*	*8*	*-13*	*16*	*-17*	*16*
Alk	*-33*	*18*	*-16*	*4*	*-30*	*19*
NH_4^+	*-54*	*51*	-	-	-	-
$SI_{calcite}$	*-219*	*226*	*37*	*-8*	*-455*	*405*

10.5.3 Verification of the model with the data from 2000

To verify the model presented above, 20 years (instead of 15 years) of wastewater infiltration was simulated and compared with the (independent) data set collected in 2000. The results are given in Fig. 5 (solid bold lines). Modelled pH in 2000 has decreased due to nitrification; all other modelled cations and modelled anions in 2000 have increased somewhat and are considered to be inside the limits given by the data collected in 2000. Modelled [NO_3^-] and [K^+] in 2000 are considered to be outside these limits. Modelled concentrations of both are too high. For K^+ there is no obvious explanation. For NO_3^- this could be explained as follows (based on the results of the sensitivity analysis):

- The actual distance between the bottom of the cesspit and the aerobic zone is greater than the value used in the model.
- The actual CEC is higher than the value used in the model.
- The actual pore water flow velocity is lower than the value used in the model.

In all, it was concluded that the accuracy and predictive capability of the model are within acceptable limits of error.

10.6 Infiltrated wastewater and use of wells

To ascertain the use of the abstracted contaminated groundwater, wells with a NO_3^- concentration exceeding 25 mg/l were selected from the well inventory. Groundwater is mainly used for irrigation and for ritual ablution in the mosque, but around 20 wells abstract groundwater for sale as drinking water. The groundwater is collected in open or closed reservoirs, which supply tanks of 3-4 m^3 on small trucks that distribute and sell the water. The total measured yield from the wells providing water for sale was around 3700 m^3/day. Assuming an average per capita water consumption of 30-60 l/day for the people dependent on these water vendors, around 60,000-120,000 people are provided with chemically contaminated groundwater for their drinking water supply. This is around 6-12% of the population of Sana'a.

10.7 Concluding remarks

Collection of data concerning nitrification and ammonium exchange taking place between the cesspit and the saturated zone.

The interpretation of the groundwater quality data presented here is based on numerous assumptions with regard to processes taking place between the pit latrine and the saturated zone. The sensitivity analysis has shown that especially [NO_3^-], [NH_4^+] and $SI_{calcite}$ are very sensitive to changes in depth to groundwater table, CEC of the soil and pore water flow velocity. Data on these parameters are needed to make a better verification of the hydrochemical model presented in this paper.

Microbiological contamination.

So far no attention has been paid to microbiological contamination of the alluvium aquifer below Sana'a. Although bacterial and viral pathogen removal percentages are expected to be high (>98%; Baxter and Clark, 1984; Canter and Knox, 1985), enough pathogens will remain present in groundwater at some depths and they pose a potential health hazard. It is evident that bacterial and viral transport mechanisms need to be studied in the aquifers below Sana'a.

Mobility of trace metals when the pH decreases.
Table 1 shows that trace metals are present in Sana'a wastewater. Soil water pH and the redox potential both play a very important role in the retention and mobility of metals. The pH is the controlling factor in many sorption-desorption reactions and many precipitation-solubilisation reactions and anaerobic conditions enhance the mobility of metals. The effect of nitrification is a decrease in pH and an increase in trace metal mobility. Therefore, trace metals will add to the groundwater contamination caused by wastewater infiltration.

Fate of trace organic contaminants.
Although it is very likely that organic contaminants are present in Sana'a wastewater, no data on organic contaminants has been collected and their transport and fate in the alluvium aquifer has not been studied at all. One of the typical problems in this respect is that equipment and technical skills to analyse trace quantities of organic contaminants are not available in Yemen.

The risk of contamination of the peri-urban well fields.
The population is growing at a very high rate, the amount of wastewater that is produced is increasing rapidly and the area of contamination is increasing rapidly as well. It will not take too long before the 4 peri-urban wellfields that are close to the city will become seriously affected. In fact, the wells closest to the city in these well fields already have elevated [NO_3^-] and these values are increasing. It will become increasingly difficult for NWSA in the future to keep on providing drinking water of good quality.

CHAPTER 11:

TRANSPORT OF *ESCHERICHIA COLI* AND SOLUTES DURING WASTE WATER INFILTRATION IN AN URBAN ALLUVIAL AQUIFER

Foppen, J.W.A., M. van Herwerden, M. Kebtie, A. Noman, J.F. Schijven, P.J. Stuyfzand, and S. Uhlenbrook, 2007c. Transport of *Escherichia coli* and solutes during waste water infiltration in an urban alluvial aquifer. J. Contam. Hydrol. (subm.)

Abstract

Recharge of waste water in an unconsolidated poorly sorted alluvial aquifer is a complex process, both physically and hydrochemically. The aim of this paper is to analyse and conceptualise vertical transport mechanisms taking place in an urban area of extensive wastewater infiltration by analysing and combining the water balance, the microbial (*Escherichia coli*) mass balance, and the mass balance for dissolved solutes. For this, data on sediment characteristics (grain size, organic carbon, reactive iron, and calcite), groundwater levels, and concentrations of *E. coli* in groundwater and waste water were collected. In the laboratory, data on *E. coli* decay rate coefficients, and on bacteria retention characteristics of the sediment were collected via column experiments. The results indicated that shallow groundwater, at depths of 50 m below the surface, was contaminated with *E. coli* concentrations as high as 10^6 CFU/100 mL. In general, *E. coli* concentrations decreased only 3 log units from the point of infiltration to shallow groundwater. Concentrations were lower at greater depths in the aquifer. In laboratory columns of disturbed sediments, bacteria removal was 2-5 log units per 0.5 cm column sediment. Because of the relatively high *E. coli* concentrations in the shallow aquifer, transport had likely taken place via a connected network of pores with a diameter large enough to allow bacterial transport instead of via the sediment matrix, which was inaccessible for bacteria, as was clear from the column experiments. The decay rate coefficient was determined from laboratory microcosms to be 0.15 d^{-1}. Assuming that decay in the aquifer was similar to decay in the laboratory, then the pore water flow velocity between the point of infiltration and shallow groundwater, coinciding with a concentration decrease of 3 log units, was 0.38 m/d, and therefore, transport in this connected network of pores was fast. According to the water balance of the alluvial aquifer, determined from transient groundwater modelling, groundwater flow in the aquifer was mainly in vertical downward direction, and therefore, the mass balance for dissolved solutes was simulated using a 1D transport model of a 200 m column of the Quaternary Alluvium aquifer. The model, constructed with PHREEQC, included dual porosity, and was able to adequately simulate removal of *E. coli*, cation-exchange, and nitrification. The added value of the use of *E. coli* in this study was the recognition of relatively fast transport velocities occurring in the aquifer, and the necessity to use the dual porosity concept to investigate vertical transport mechanisms. Therefore, in general and if possible, microbial mass balances should be considered more systematically as an integral part of transport studies.

11.1 Introduction

Many studies exist on the effects of controlled or uncontrolled infiltration of waste water or reclaimed water on groundwater quality. In these studies, focus is either on the hydrochemical processes taking place (e.g. Taylor et al., 2006a; Greskowiak et al., 2005; Quast et al., 2006; Baxter an Clark, 1984; Canter and Knox, 1985; Eiswirth and Hotzl, 1997; Feigin et al., 1991; Gooddy et al., 1997; Hamad, 1993, Klimas, 1997; chapter 10 of this dissertation) or on the determination of the microbiological removal versus the distance of the point of infiltration (Pang et al., 2003; Close et al., 2006; Powell et al., 2003; Sinton et al., 1997; 2000, Hagedorn et al., 1978; Rahe et al., 1978; Reneau, 1978; Reneau and Pettry, 1975). However, case studies combining the hydrochemical mass balance, the microbiological mass balance and the water balance to conceptualise the waste water infiltration process have rarely been documented. One such case is described by Cronin et al (2003). They infer, in a qualitative manner, groundwater flow characteristics in an urban consolidated sandstone aquifer based on depth-specific profiles of faecal organisms and solutes such as nitrate. For the same study area, Taylor et al (2006) concluded that aquifer penetration rates of faecally derived microorganisms were orders of magnitude more rapid than aquifer penetration rates suggested by anthropogenic solutes.

Sana'a, the capital of Yemen (Fig. 1), is in the center of the Sana'a Basin, an intermontane plain of some 3,200 km^2 in the central Yemen Highlands. During the last 20 years Sana'a has grown at an annual rate of around 10%: there were 80,000 inhabitants in 1970, 1,000,000 in 1994 and the current estimate is around 2,000,000 inhabitants. As a consequence, the area of the city has grown tremendously in the last 20 to 30 years, and so has the total water consumption and wastewater production. While total water consumption in 1970 was estimated to be 0.3 million m^3/y, present total water consumption is estimated to be 50-75 million m^3/y (Chapter 9). Until some 5 years ago, waste water was mainly disposed of on-site via pit latrines or cesspits, and infiltration of waste water took place below the city in the unconsolidated aquifer, consisting of layers of poorly sorted alluvial gravels, sands and clays (Quaternary Alluvium; Chapter 10). In the last 5 years, in response to unacceptable health risks, odor nuisance and road damage associated with sewage continuously flooding the streets in some areas of Sana'a (Chapter 9), entire city neighbourhoods have been connected to the sewerage and a major sewage treatment plant has been built. In the recent past, hydrochemical processes occurring in the aquifer upon infiltration of wastewater were studied and modelled using a one-dimensional hydrochemical transport model (Chapter 10). In addition, a calibrated transient groundwater model of the entire Sana'a Basin covering the period from 1970-2000 was constructed (Chapter 9). However, combining the results of these studies and the microbiological mass balance to better understand the waste water infiltration process has not been carried out. Therefore, the aim of this paper is to conceptualise the vertical transport mechanisms taking place in the urban area of Sana'a, in which extensive wastewater infiltration has been taking place, by determining and combining the transient water balance, the transient *E. coli* mass balance, and the transient dissolved solutes mass balance. In addition, distinctly different transport mechanisms (e.g. single porosity, dual porosity) will be evaluated in terms of their occurrence in the aquifer.

11.2 Water balance and hydrochemistry

The water balance of the urban area (Fig. 1) for the year 1995, prior to the sewerage construction, as determined from a calibrated transient groundwater model (Chapter 9), indicates that the main supplier of water of the Quaternary Alluvium is infiltration from waste water (6352 m^2/day) via cesspits. The main component flowing out of the Quaternary Alluvium is discharge into the underlying aquifers. Therefore, groundwater flow in the Alluvium is mainly in a vertically downward direction, while the recharge from precipitation, and lateral in- and outflow fluxes are relatively unimportant. Waste water is infiltrating at high rates, causing the groundwater table to rise (814 m^3/day into storage) in some parts of the city, while in other parts waste water is flooding the streets (500 m^3/day waste water flooding).

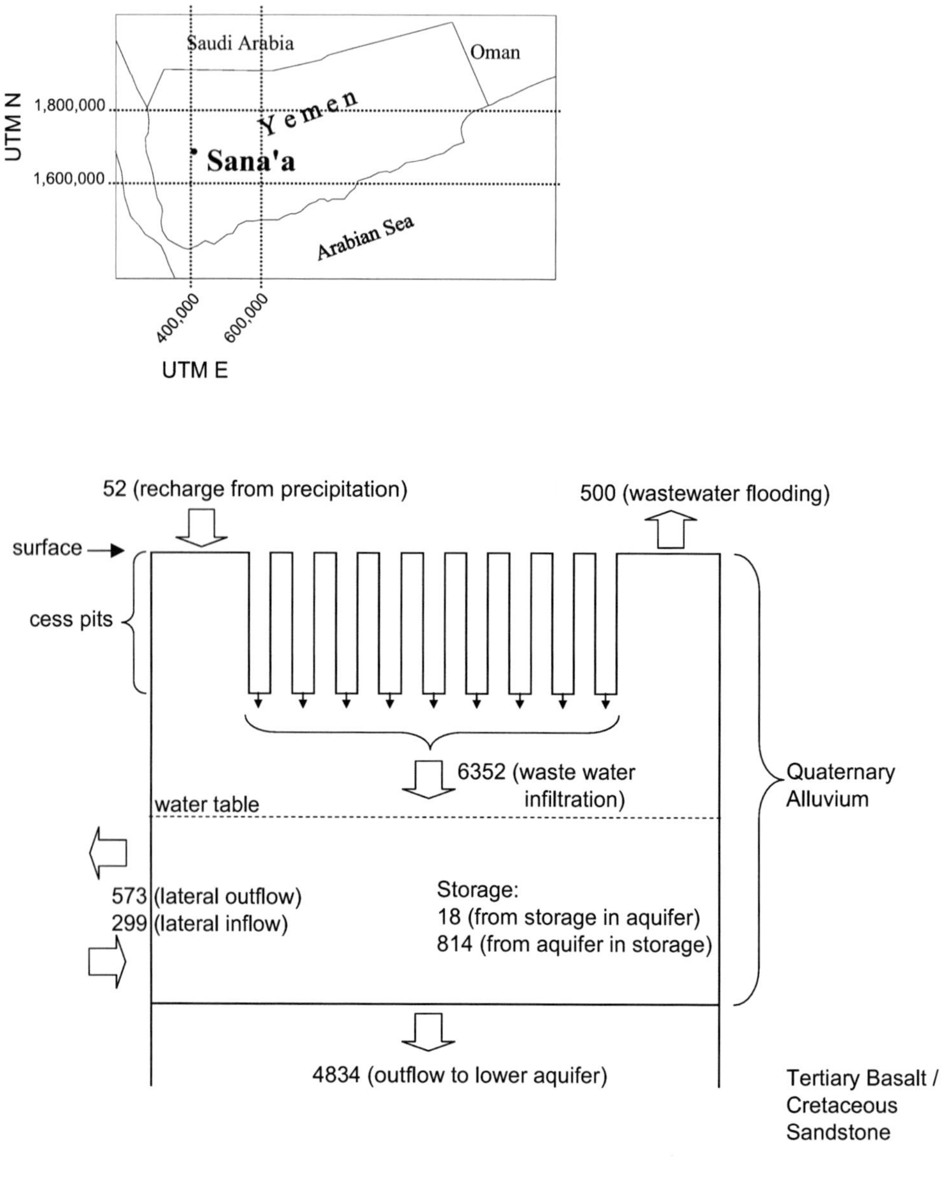

Figure 1: Water balance of the Quaternary Alluvium below a part of the city centre in 1995 (2.3 by 2.3 km; values are in m^3/day), determined with a calibrated transient groundwater model of the area (Chapter 9; drawing is not to scale). Upper left: location of Sana'a

As a result of the infiltration of wastewater, groundwater quality in the Quaternary Alluvium is characterised by high concentrations of almost all major cations and anions (Chapter 10). In Table 1, a number of typical analyses of groundwater affected by wastewater is given. Usually, NH_4^+ is absent in groundwater samples, while [NH_4^+] in Sana'a wastewater is around 10 mmol/l. However, [NO_3^-] in groundwater is around 1 to 3 mmol/l, indicating aerobic conditions in the saturated zone. It is thus very likely that nitrification occurs between the cesspit and the saturated zone. Nitrification is the process whereby NH_4^+ is oxidised to NO_3^-:

$$NH_4^+ + 2O_2 \quad \rightarrow \quad NO_3^- + 2H^+ + H_2O \qquad (1)$$

However, [NO_3^-] is far less than [NH_4^+], indicating that there is a sink for N-species between the cesspit and the saturated zone. Besides the effects of dispersion and diffusion which tends to reduce [NO_3^-], the most likely sink is cation-exchange of NH_4^+. When comparing Na^+, K^+ , Ca^{2+} and Mg^{2+} in the Quaternary Alluvium aquifer to conservative mixing of wastewater and natural groundwater, it can be seen that Na^+ and K^+ in groundwater are depleted, Ca^{2+} is enriched, while Mg^{2+} shows hardly any signs of exchange, according to (Chapter 10):

$$\begin{aligned} &3.9CaX_2 + 0.3MgX_2 + 6.0Na^+ + 1.3K^+ + 1.1NH_4^+ \leftrightarrow \\ &\qquad 3.9Ca^{2+} + 0.3Mg^{2+} + 6.0NaX + 1.3KX + 1.1NH_4X \end{aligned} \qquad (2)$$

where X indicates the soil exchanger. This process of cation-exchange is typical for a salinising environment, which in this case is caused by the infiltration of high strength sewage. The net result is a strong increase in total hardness of the groundwater (Chapter 10).

Table 1: Results of the analysis of a selection of groundwater samples affected by infiltrated wastewater (ion concentrations are in mmol/L; Chapter 10). Well locations are given in Figure 2. Ww: average waste water composition, based on 10 samples taken at the inlet of the sewage ponds; All: average groundwater composition, based on 12 samples taken from the Quaternary Alluvium outside Sana'a, and not affected by wastewater

Well code	Date	EC µS/cm	pH	Temp °C	Na^+	K^+	Mg^{2+}	Ca^{2+}	Cl^-	HCO_3^-	SO_4^{2-}	NO_3^-	NH_4^+
E90	14-Nov-95	1433	7.2	24.0	6.1	0.05	0.7	3.6	6.5	2.0	2.3	1.8	0.0
E97	10-Jan-96	1720	7.3	21.1	2.9	0.08	1.6	6.4	7.5	4.4	2.2	2.1	0.0
E22	03-Jan-96	1080	7.7	-	1.9	0.08	1.1	3.7	4.9	4.4	0.7	0.7	0.0
E22	22-Mar-00	1373	7.4	22.4	2.5	0.09	1.3	4.2	5.2	4.9	1.0	1.0	0.0
E37	13-Jan-96	1527	7.7	-	3.7	0.10	1.2	4.8	8.9	2.0	2.0	0.1	0.0
E37	22-Mar-00	2350	7.2	23.0	4.4	0.13	1.8	6.9	14.1	2.7	2.6	1.1	0.0
Ww			7.3	23.0	9.8	1.50	1.5	2.5	10.0	15.7	1.8	0.0	10.0
All			8.4	26.0	1.6	0.07	0.3	1.0	1.5	1.7	0.6	0.0	0.0

11.3 Materials and Methods

11.3.1 Sediment characteristics

In two large diameter wells (SW and SW2) of approximately 2 m diameter, located in the centre of the town (see Fig. 2), a person was lowered into the well to collect soil samples of about 0.5 kg from freshly carved surfaces of the wall of the well at various depth intervals, to a total depth of around 50 m below the surface. Since the sample locations were not representative for undisturbed aquifer conditions, the samples could only give an indication of sediment characteristics within the aquifer. Samples were stored cool and airtight in a refrigerator. In each soil sample, grain size distribution, hydroxide bound iron, organic carbon, and calcite were determined, as described below.

11.3.2 Grain size distribution

Oven-dried samples (100-300 g) were passed through a series of sieves (45-560 μm) via shaking, and the relative weight contribution of each fraction was determined.

11.3.3 Reactive iron

Oven-dried samples (1 g) were digested in steps via evaporation on a hot plate with concentrated HNO_3, $HClO_4$ and 50% HF (APHA, 2005). Total concentrations of Al and Fe (in % dry weight, or % d.w.) were determined (ICP; Perkin Elmer Plasma 1000), and the content of reactive iron (Fe_{react}) was calculated as

$$Fe_{react} = Fe_{total} - Fe_{clay} \tag{3}$$

whereby Fe_{clay} was taken as (Huisman and Kiden, 1997)

$$Fe_{clay} = \frac{Al_{total}}{4} \tag{4}$$

11.3.4 Organic carbon (C_{org} ; in % d.w.)

Soil Organic Material (SOM) was determined by the samples' loss on ignition from 105 to 550 °C (SOM_{550} , in % d.w.; Rabenhorst, 1988). Then, a correction for the loss of structurally bound water was made (Hieltjes and Breeuwsma, 1983)

$$SOM_{corr} = SOM_{550} - 0.067L - 0.12Fe_{react} \tag{5}$$

where L is the fraction (% d.w.) particles < 2 μm. C_{org} was derived from SOM_{corr} , according to (Hieltjes and Breeuwsma, 1983)

$$C_{org} = \frac{SOM_{corr}}{2} \qquad (6)$$

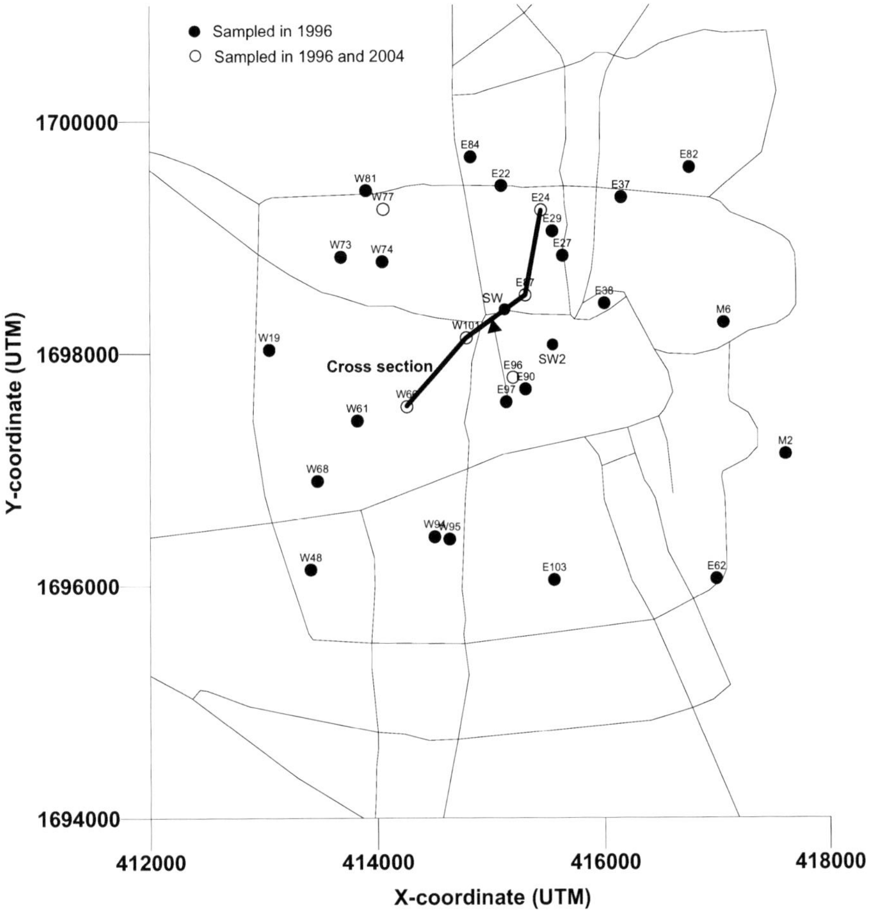

Fig. 2: Location map of sampled wells and geological cross-section

11.3.5 Calcite (in % d.w.)

Calcite was determined by back titrating an excess of 1 mol/L HCl added to 5 g of sample with 0.5 mol/L NaOH (Allison and Moodie, 1965).

11.3.6 Hydrogeological schematization

In 1996, total depths of some 30 cesspits in the centre of the city were determined with a measuring tape. In 2004, in six large diameter wells, a commercially available digital video camera (Sony DCR-TRV22E) with a 3W video light (Sony HVL-S3D), attached to the tape of a water level sounding device, was lowered into the well to obtain an impression of sediments exposed to the wall of the well, zones of exfiltration of groundwater, and to observe the static water table depth.

11.3.7 Microbiological sampling

In 1996, during the period of massive and uncontrolled wastewater infiltration, 25 duplicate samples were collected from wells in the city (Fig. 2) and from a small sewage treatment plant north of the city in sterile flasks and transported in a cooling box (4 °C) to the Microbiological Laboratory of the Ministry of Health in Sana'a. Assays were begun on the day the samples were taken. Total coliform (TC) and thermotolerant coliform concentrations were determined according to the multiple-tube fermentation technique (APHA, 1992). This used MacConkey broth (48 h at 37 °C for TC and 24 h at 44 °C for thermotolerant coliform), with positive tubes confirmed for TC in brilliant green lactose bile broth 2% (48 h at 37 °C) and for thermotolerant coliform (24 h at 44 °C). In the thermotolerant coliform tubes, the presence of *Escherichia coli* was confirmed with an indole test. In all tests, blanks were included (demineralised water). All growth media were prepared by the Laboratory of the Ministry of Health.

In 2004, almost after completion of the sewerage construction, TC and *E. coli* concentrations for 5 wells with high *E. coli* concentrations in 1996 (Fig. 2) were determined via membrane filtration (24 h at 37 °C) using Chromocult agar. On this agar, coliform colonies appear pink, while *E. coli* colonies appear deep purple. In addition, *Shigella* species appear as dark blue colonies. In the 2004 tests, blanks were included (demineralised water), and in most cases results were replicated by determining counts of various sample dilutions. The reason for sampling in 2004 those parts of the shallow Quaternary Alluvium aquifer that were most polluted in 1996, was to establish whether *E. coli* concentrations had indeed reduced due to the sewerage construction.

11.3.8 Representative nature of microbiological data: chlorination of wells

To obtain an idea of whether the microbiological parameters of the samples collected from the wells were representative for the aquifer, and were not caused by contamination from the surface due to poor wellhead protection, four wells, that were in use every day for a few hours for local water supply, were chlorinated to remove all bacteria. If, after replacing the water volume in the well a number of times by pumping, bacteria concentrations would reach pre-chlorination values, we interpreted the time series of the bacteria concentrations before and after chlorination as being representative for aquifer conditions. Chlorination of the wells was carried out by dissolving solid calcium hypochlorite ($Ca(OCl)_2$ 70%) to a free chlorine concentration of 50 mg/l well water (Driscoll, 1986). Small containers filled with solid $Ca(OCl)_2$ were lowered into the wells and pulled up and down in the water column to allow for dissolving and homogeneous mixing. After complete dissolution of the added $Ca(OCl)_2$, a sample was taken to determine microbiological parameters and free chlorine concentration with a portable chlorine set (model CN-66 from HACH). If necessary, additional $Ca(OCl)_2$ was added to reach the required free chlorine concentration in the well water. Every 24 h after completion of the disinfection, and after abstracting several well volumes from the well, a sample was taken and analysed for TC and *E. coli.*

11.3.9 Decay of E. coli

One litre polyethylene bottles (or microcosms; in duplicate) were cleaned with a 0.01 M HCl solution, and then rinsed with demineralised water until the electrical conductivity of the rinse water was less than 1 μS/cm. Then, the bottles were autoclaved, dried, wrapped in black plastic, and kept in a temperature controlled room at 20 °C, which is close to aquifer conditions (Table 1). The bottles were filled with water from the Quaternary Alluvium, obtained from mixing several fresh groundwater samples from the centre of the city, and spiked with a wild *E. coli* strain (10^9 cells/mL), which was harvested and isolated from the Quaternary Alluvium (well # E24). The strain was serotyped as O2:K1 (Diagnostic Laboratory for Infectious Diseases and Perinatal Screening of the National Institute of Public Health and the Environment, The Netherlands). For comparison, various microcosms were prepared consisting of well water (as mentioned above), autoclaved or sterile well water, tap water, and autoclaved tap water. The *E. coli* O2:K1 concentrations were determined immediately after the preparation of each microcosm on plates with Chromocult agar, and then on a weekly basis for 7 weeks (in duplicate). The decay rate coefficient obtained from the concentration versus time plots gave an indication of the *E. coli* decay occurring in the Quaternary Alluvium aquifer.

11.3.10 Column experiments

In order to assess the bacterial removal during transport of *E. coli* O2:K1, column experiments were conducted in borosilicate glass columns with an inner diameter of 2.5 cm (Omnifit, Cambridge, U.K.) with polyethylene frits (25 μm pore diameter) and one adjustable end-piece. The columns were packed wet with sediment samples from the Quaternary Alluvium, obtained from the two large diameter wells, SW and SW2 (Fig. 2). Prior to packing, the column sediment was sieved (sieve size: 250 μm) to remove coarse material. Column sediment length varied between 0.5 and 1.0 cm, and the pore water flow velocity varied between 0.36 m/d ± 0.16. Columns were allowed to equilibrate with artificial wastewater (AW) overnight, and then spiked with AW suspensions of *E. coli* O2:K1 (10^5 cells/mL). The AW was prepared by dissolving 443.9 mg/L $CaCl_2$, 87.7 mg/L NaCl, 7.5 mg/L KCl, 167.8 mg/L $NaHCO_3$, 30 mg/L Na-humic acid, and 85 mg/L $NaNO_3$, and was assumed to represent the average hydrochemical wastewater conditions in Sana'a (Table 1). *E. coli* O2:K1 was obtained by inoculation in 50 mL of nutrient broth (Oxoid CM001) for 24 h at 37 °C, yielding *E. coli* concentrations of 10^9 cells/mL. Bacteria were washed and centrifuged (3000 rpm) three times in AW and then diluted with AW to 10^5 cells/mL. Effluent *E. coli* concentrations were determined by plating various effluent dilutions with Chromocult. Duration of *E. coli* spiking was 6-10 pore volumes. Then, the column was flushed for some 5 pore volumes with *E. coli* free AW in order to observe *E. coli* detachment characteristics. The removal of *E. coli* in the columns due to a combination of attachment and straining (Chapter 2) gave an indication of the bacteria removal processes taking place in the Quaternary Alluvium aquifer.

11.3.11 E. coli mass balance formulation

The mass balance of *E. coli* suspended in groundwater due to infiltration of waste water via cesspits was considered to be a steady state one-dimensional vertical flow process with first order decay, and neglecting detachment, according to

$$\frac{d^2C}{dz^2} - v\frac{dC}{dz} - \mu C = 0 \qquad (7)$$

where

C *E. coli* concentration [number of colonies/100 ml]

z depth [m]

v pore water flow velocity [m/d]

μ first order decay rate constant [1/d]

The first order decay rate constant, μ, was considered to be the combined rate coefficient of decay and removal due to attachment and straining. With boundary conditions

$$C(0) = C_0 \text{ and } \frac{dC}{dz}(\infty) = 0, \qquad (8)$$

Equation (7) can be solved analytically according to (Van Genuchten, 1981)

$$C = C_0 \exp\left\{ \frac{(v - v(1 + \frac{4\mu D}{v^2})^{\frac{1}{2}})z}{2D} \right\} \qquad (9)$$

where D is the longitudinal dispersion [m^2/d]. Equation (9) was used in combination with the chemical mass balance (see below) for determining vertical transport velocities in the Quaternary Alluvium.

11.3.12 Modelling the hydrochemical mass balance

The dissolved solutes mass balance was assessed via modelling a 1D vertical column with PHREEQC (Parkhurst and Appelo, 1999) consisting of 40 cells from the bottom of the average cesspit depth until the lower boundary of the Quaternary Alluvium, covering an aquifer thickness of 200 m, similar to Foppen (2002; Chapter 10). In this model, waste water, with a composition as given in Table 1, was flushed through the cells, whereby cation-exchange could take place throughout the column. Saturated flow of wastewater from the cesspit to the saturated zone is assumed to occur due to the high hydraulic loading of the cesspit. Initially, prior to wastewater infiltration, groundwater composition in the cells was similar to the alluvium type water given in Table 1. The pe-value of the waste water was assumed to be 2, indicating anaerobic conditions in which NH_4^+ is stable (Chapter 10). These anaerobic conditions were assumed to prevail in the upper 6 cells of the model. In reality, the thickness of this anaerobic zone was unknown, because of a lack of depth-specific measurements. Then, from cell 7 to 40, the pe-value of the alluvium water was assumed to be 12, indicating aerobic conditions, which was in accordance with the high nitrate concentrations found in the aquifer. In this part of the model, oxidation (nitrification) could take place when solutes entered the aerobic part of the groundwater flow system, which was schematised in the model with a fixed P_{O_2} = 0.2 atm. In the model, only stable nitrogen species with electron valences of –III and +V were allowed, preventing the production of N_2-gas and instable N(+III) species upon oxidation of NH_4^+ to NO_3^-.

11.4 Results

11.4.1 Sediment characteristics and hydrogeological schematization

Samples were very poorly sorted (Fig. 3), with grain size fractions ranging from < 45 µm to > 560 µm. For the samples in the upper 35 m of SW and SW2, more than 50% d.w. of each sample was in the grain size range > 74 µm, while grains were smaller than 4.75 mm. In addition, these samples contained more than 12% fines, defined as grains < 74 µm, according to Unified Soil Classification System (USCS; ASTM, 1985), and was in our case defined as the averages of the 45 µm and 106 µm sieve fractions. These samples classified as poorly sorted silty-clayey sands, according to the USCS. For the samples below 35 m of SW and SW2, around 90% d.w. was larger than 560 µm. Since various rounded pebbles of several centimeters were frequently present, these samples were classified as poorly sorted coarse sands to sandy gravels. Hydroxide bound iron (Fe_{react}) was around 50 mg/g for SW and 30 mg/g for SW2, while C_{org} was 4-5% for SW and 1-4% for SW2. Calcite was 5-20% in the upper parts of SW and SW2, and reduced to 4-6% d.w. in the lower parts of the sampled wells. We concluded that the sediments were relatively rich in iron-oxi-hydroxides, organic material, and in calcite.

Two types of sediment (silty-clayey sands and coarse sand to sandy gravels) also appeared from visual inspection of the wall of the well (video logs ; Fig. 4). Since the gravels in the lowest parts of the logged well depths were all very well rounded, layered, and sometimes with pebble imbrications, this pointed towards a fluviatile origin of the sediments. The static water table was located in the gravels at a depth of 40-50 m below the surface, and thus the large diameter wells were tapping from this aquifer. In some wells, exfiltration of groundwater was evident from the presence of travertine, covering the surface of the wall of the well in those sections where gravel was exposed. The average cesspit depth was ± 15-20 m below the surface, so the average vertical transport distance of wastewater to groundwater in the gravels was on average 30 m.

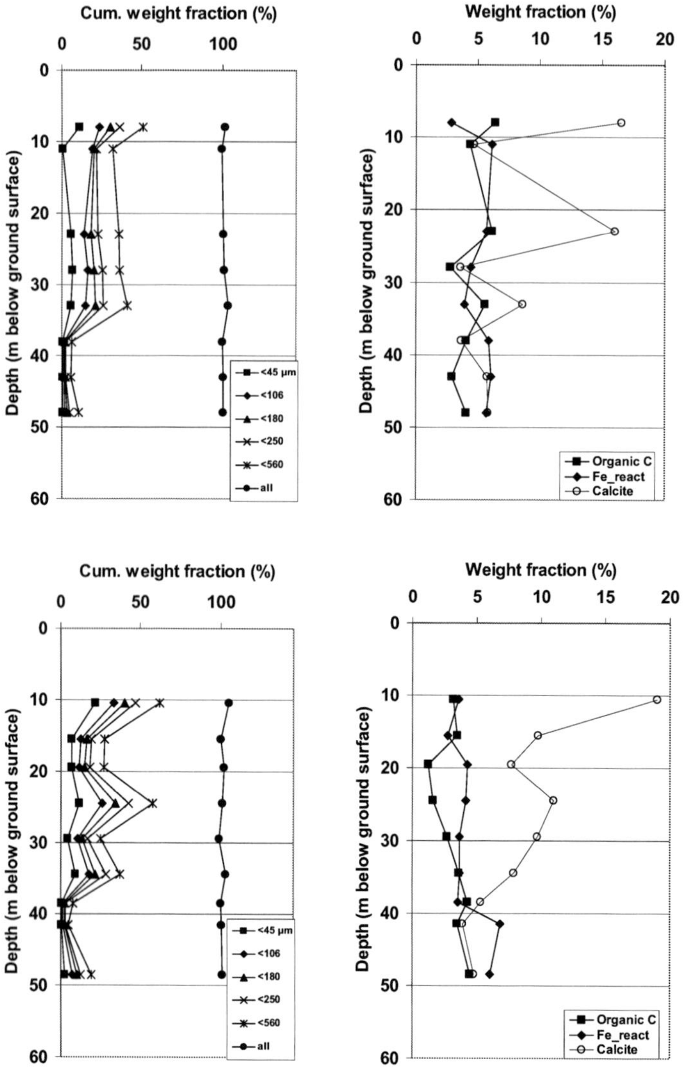

Figure 3: Grain size distribution, reactive iron concentration, organic carbon, and calcite content of wells SW (above) and SW2 (below)

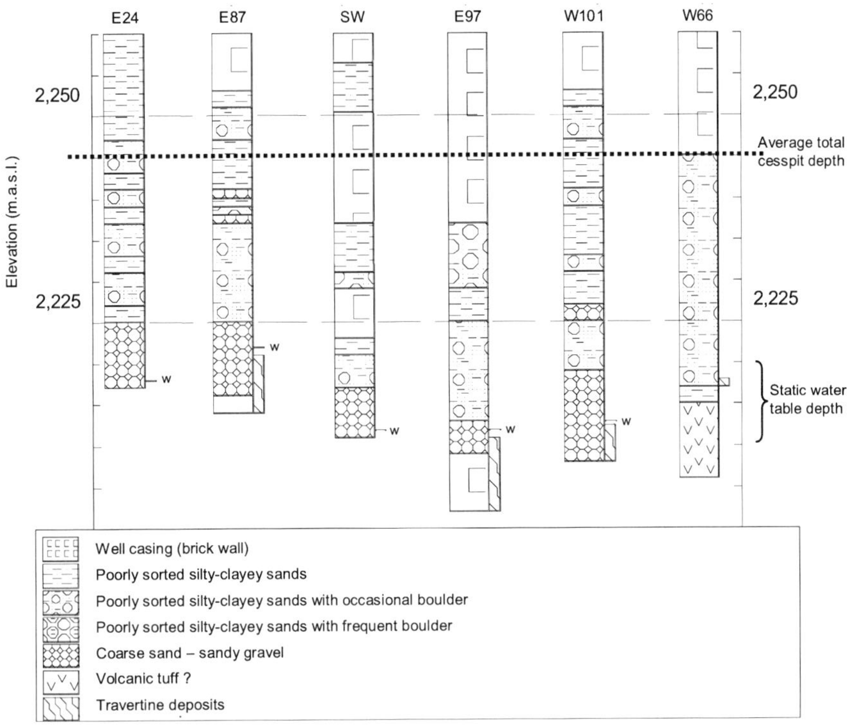

Fig. 4: Lithological description of 6 large diameter wells obtained with a video camera. Elevation of all wells was estimated to be around 2260 m.a.s.l. Video-taped well depths varied from 42 m to 57 m. In some wells, travertine deposits covered the wall of the well, indicating zones of exfiltration of groundwater and degassing of CO_2. The static water table (May 2004) was at a depth of 40-50 m (indicated with a 'w'), while the average total cess pit depth was at ± 15-20 m.

11.4.2 Microbiological sampling

In 1996, thermotolerant coliform concentrations in the centre of town were as high as 920,000 CFU/100 mL, and concentrations rapidly decreased from the centre of the town to the periphery (Fig. 5). Since *E. coli* was confirmed, thermotolerant coliform concentrations were interpreted as *E. coli* concentrations. Most of the wells sampled were shallow large diameter wells (with an average depth of 40-60 m), tapping the water table in the sandy gravels, while the 0 CFU/100 mL values from the eastern and southern samples were from deep tube wells (200-400 m depth). This window of faecally polluted shallow groundwater was probably a result of sampling bias, caused by the presence of shallow large diameter wells in the centre of the city and their absence in the periphery. Faecal contamination of deep well samples was usually absent. *E. coli* concentrations of raw sewage were 2-3*10^8 CFU/100 mL and were high, reflecting high strength waste water conditions (Chapter 10). Since *E. coli* concentrations in the shallow groundwater were between 10^2 and 10^6 CFU/100 mL, reduction of the *E. coli* concentration was between 2-6 log units over an estimated average transport distance of 30 m (distance between cesspit bottom and static groundwater table; see previous section). Deeper in the aquifer, *E. coli* in the tube well samples was usually absent.

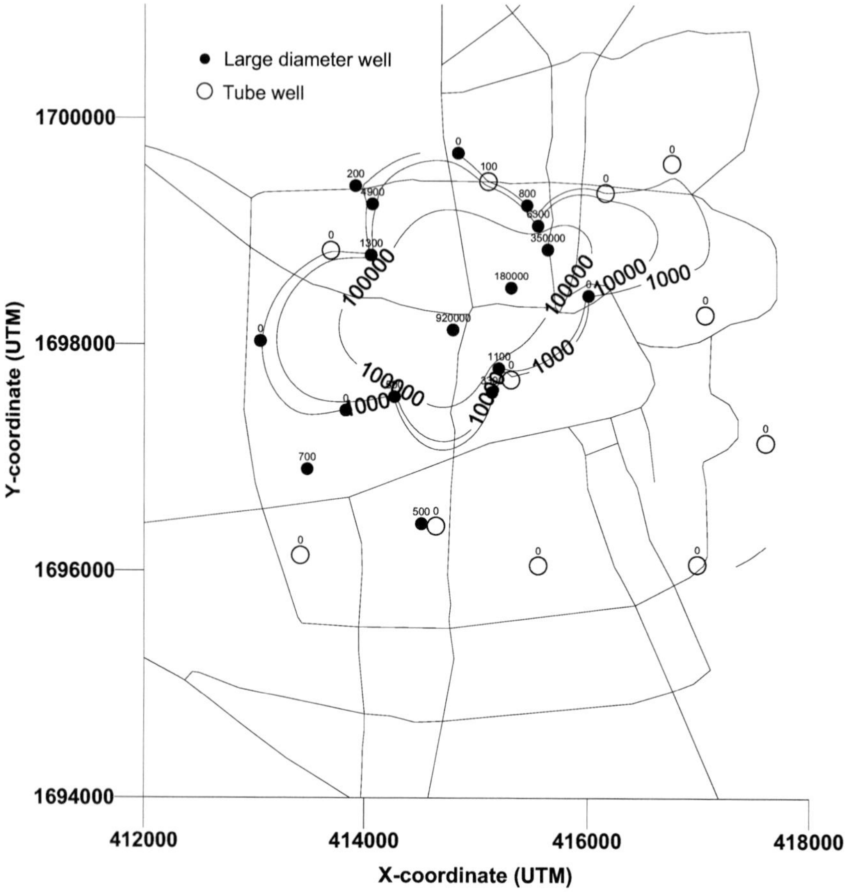

Figure 5: E coli concentrations in 1996 (in CFU/100 mL). Average E. coli concentration in raw sewage was 2-3x10^8 CFU/100 mL

In 2004, after construction of the sewerage, *E. coli* usually was absent in those wells that were most polluted in 1996 (Fig. 6). However, occasionally, concentrations were as high as 1-10 CFU/100 mL. Compared to 1996, concentrations had reduced 3 to 6 log units. In 4 out of 5 wells, *Shigella ssp.* were frequently found and sometimes at higher concentrations than *E. coli* (well W66). TC (Total Coliform) concentrations in 2004 ranged from 10^3 to 10^6 CFU/100 mL, and, compared to 1996, TC were not reduced significantly. Concentrations of bacteria before and after chlorination were in the same log range in 3 out of 4 wells (except W66), and from this we concluded that bacteria concentrations in the wells were representative of concentrations in shallow groundwater in the Quaternary Alluvium aquifer.

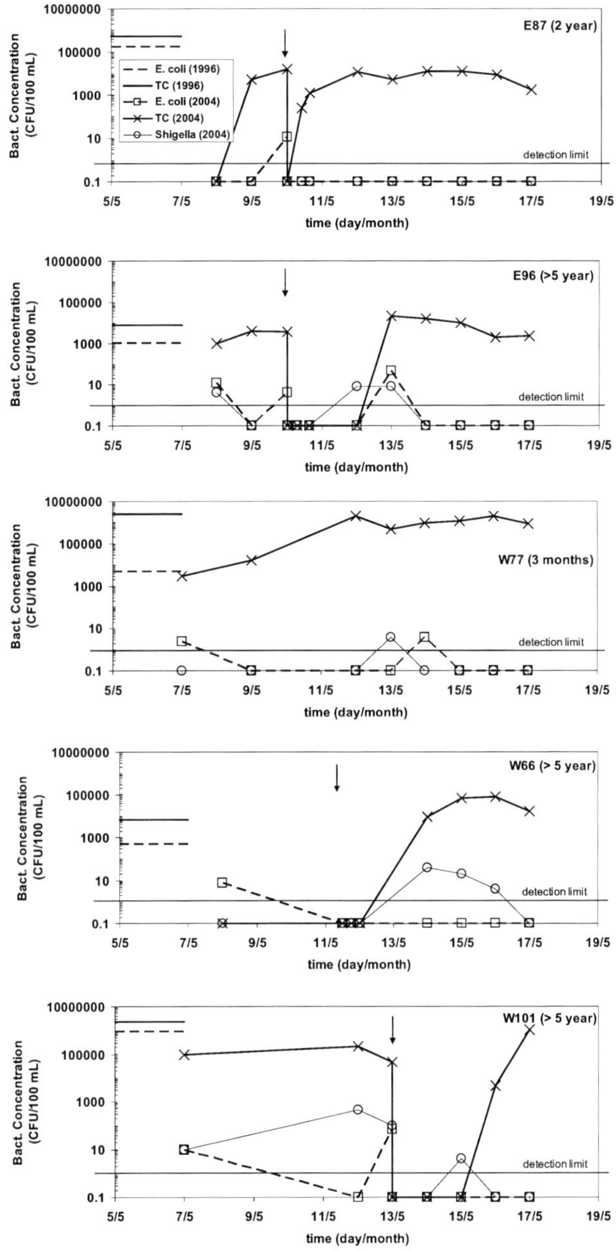

Figure 6: Time series of microbiological parameters collected in 2004. Duration (in brackets behind the well code) indicates the time elapsed since the sewerage was constructed in the vicinity of the well. Arrows in the time series indicate start of disinfection.

11.4.3 Decay of E. coli O2:K1

The time dependent decrease of *E. coli* O2:K1 concentrations in autoclaved, sterile, well water and tap water was around 1 log unit in 49 days, while in non-sterile waters the decrease was around 3 log units (Fig. 7). The decay rate coefficient for the sterile cases was 0.03 and for the non-sterile cases 0.15. The difference between the two decay rates was probably caused by the presence of other organisms in the non-sterile microcosms. After week 5, the growth of Enterobacteriaceae other than *E. coli* (white colonies on Chromocult) became visible. For the tap water, these appeared to be *Enterobacter aerogenes* (determined with an API test). For the well water, *Citrobacter ssp.* were identified. In addition, other flagellate-type organisms were observed microscopically in a counting chamber. They ranged in size from 6-19 μm, while estimated concentrations were in the order of 10^5 to 10^6 cells/mL. These flagellates might have been feeding on *E. coli.*

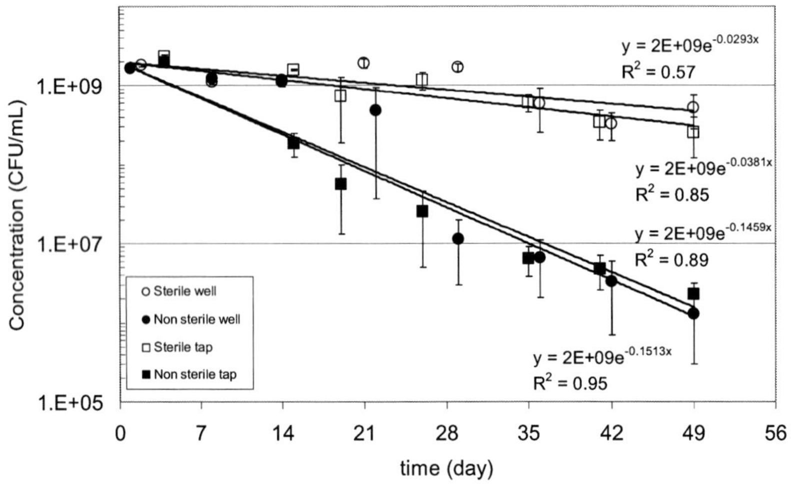

Figure 7: Die-off of E. coli O2:K1 in sterile and non-sterile tap and well water

11.4.4 Column experiments

For a number of cases, removal was 1-3 log units for 0.4 cm column heights (Fig. 8). In those cases, breakthrough rapidly increased to a constant C/C_0-value after 1-2 pore volumes, and remained constant for the entire duration of the *E. coli* O2:K1 spiking period. Then, during deloading (flushing of the columns with bacteria free artificial wastewater), C/C_0-values decreased around 1 log unit within 2-4 pore volumes after deloading had started, indicating that some detachment of bacteria from the columns was occurring. The log removal of *E. coli*, obtained from the plateau phases of the breakthrough curve was determined for each soil sample of SW and SW2. The results, given in Figure 9, indicate that *E. coli* O2:K1 concentrations in all sediment columns decreased 2 to more than 5 log units over column heights ranging between 0.4 and 0.9 cm. So, removal of *E. coli* by the sediment was very high. From the shape of the breakthrough curves, it was impossible to distinguish attachment or straining as being more important for removal.

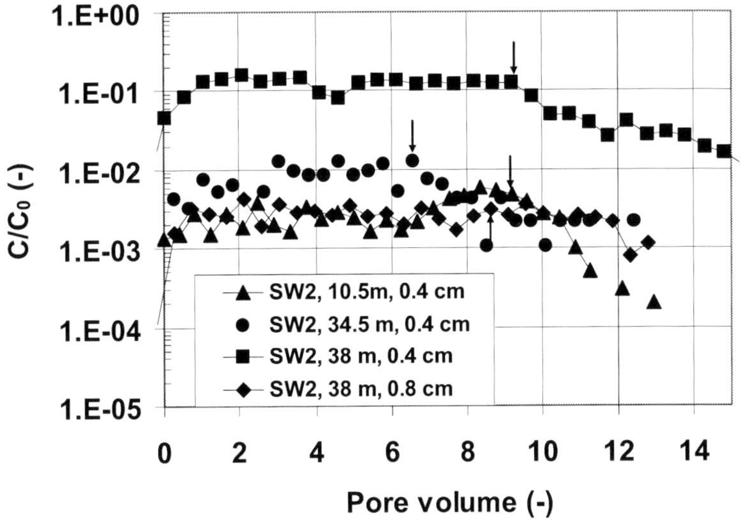

Figure 8: Examples of breakthrough curves of E. coli O2:K1 isolated from Sana'a groundwater in columns of 0.4-0.8 cm unconsolidated sediment from the Quaternary Alluvium aquifer (arrows indicate start of flushing with bacteria-free AW)

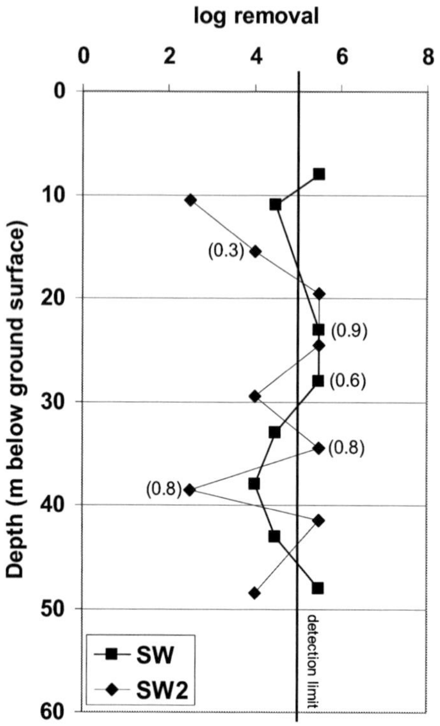

Figure 9: Removal of E. coli O2:K1 (expressed in log-units) in sediment columns of 0.4 cm column height. Numbers in brackets: column heights (in cm) other than 0.4 cm

11.5 Modeling vertical transport of mass

11.5.1 Transport model

Removal of *E. coli* in the column experiments was high (2 to >5 log units per 0.4-0.9 cm of sediment), while in 1996 the decrease of *E. coli* concentration in the aquifer was low (2-6 log units over an estimated average transport distance of 30 m; see section above). In addition, because decay of *E. coli* O2:K1 in the microcosms was around 3 log units per 49 days, and assuming that the laboratory inactivation rate applied to field conditions, transport of (at least) this wild *E. coli* strain to shallow groundwater must have been relatively fast. We conclude that, due to bacterial retention processes (mainly attachment and straining), parts of the Quaternary Alluvium aquifer were inaccessible for *E. coli*. Transport had likely taken place via a connected network of pores with a diameter large enough to allow bacterial transport within the Quaternary Alluvium. Transport in such network of pores was fast, when taking the decay of *E. coli* in this apparently hostile environment into consideration. These two findings (large pore network, fast transport) were used to quantitatively explore whether vertical flow in the aquifer could be described by a dual pore water flow velocity system. Such system was considered realistic if the three relevant mass balances, i.e. the water balance of the area, the *E. coli* mass balance formulation (eq. 9), and the mass balance of dissolved solutes, could be simulated adequately. Although various approaches for modeling preferential flow in aquifers exist (Simunek et al., 2003), we used the relatively simple dual-porosity approach, offered in PHREEQC (Parkhurst and Appelo, 1999). In this approach, the waste water infiltration process was schematised into a connected network of pores in which transport was relatively fast, and a sediment matrix with stagnant water in which diffusion of solutes took place (Fig. 10, left side). The dissolved solutes mass balance was modelled with a 1D vertical column in PHREEQC, consisting of 40 cells for the entire Quaternary Alluvium aquifer thickness of 200 m from the average cesspit depth to the lower boundary of the aquifer, in which water was allowed to flow (mobile water). Each mobile column cell was attached to a stagnant cell in which only diffusion could take place. Exchange between mobile and stagnant cells was possible via an exchange factor. The exchange factor is an empirical rate coefficient which determines the diffusional exchange between the mobile and stagnant regions (Van Genuchten, 1985). Infiltration of waste water was allowed to take place for 15 years, similar to Foppen (2002; Chapter 10). Cation exchange took place throughout the column, and the CEC (in meq/kg dry weight) was calculated according to Appelo and Postma (2005), with the pH-dependence of the CEC of organic matter according to Scheffer and Schachtschabel (2002):

$$CEC = 7L + 5.1(pH - 1.16)C_{org} \qquad (10)$$

where pH is the pH of the groundwater. Assuming constant values for L, C_{org}, and pH of 5 %, 4.5 %, and 7.5 resp., the CEC was 180.5 meq/kg (Table 2). In addition, nitrification occurred when ammonium entered the aerobic part of the groundwater system, which was schematised in the model from cell 7-40 with a fixed $P_{O_2} = 0.2$ atm. From cell 1-6, because of the high pH-value, and due to cation exchange, the calculated saturation index of calcite was above 0, indicating supersaturation. In these model cells, we did not adjust for the presence of calcite via the saturation index. From cell 7-40, calcite was simulated assuming

$$SI_{calcite} = 0 \qquad (11)$$

where $SI_{calcite}$ is the calcite saturation index.

Removal of *E. coli* was assessed with eq. (9), where we assumed that *E. coli* was present in the mobile part of the aquifer only. Removal was 3 log units due to decay (decay rate coefficient = 0.15 d^{-1}) over a vertical distance of 30 m (Fig. 10, far right), which was the distance between the average cesspit bottom and the shallow groundwater depth (Fig. 4). In the mobile region with large pores, attachment and straining were neglected. The dispersivity was assumed to be 10% of the travelled distance (= 3 m; Appelo and Postma, 2005; Gelhar et al., 1985). With these parameter values, the pore water flow velocity was 0.38 m/d. From the water balance of Fig. 1, the rate of waste water infiltration was determined (~ 0.42 m/y) and together with the pore water flow velocity, the mobile porosity was calculated to be 0.3 %. The parameter values used are summarised in Table 2. Although the single porosity, low velocity 1D transport model was not realistic in terms of process

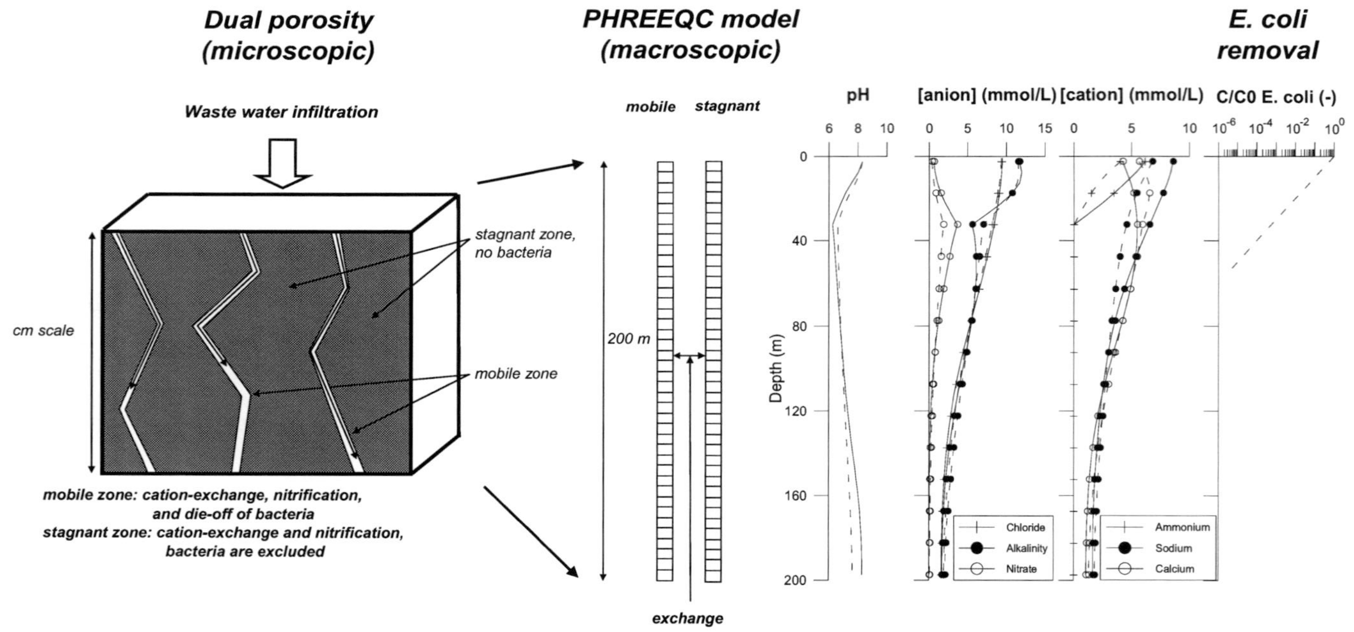

Figure 10: Left: conceptual representation of microscopic groundwater flow velocities (adapted from Taylor et al., 2004). Right: Results of the 1D transport model for selected parameters, determined with PHREEQC (solid line: single porosity; dashed line: dual porosity). Far right: Concentration of E. coli (C) relative to waste water concentration (C_0) as a function of depth, determined with the E. coli mass balance (eq. (9))

Table 2: Vertical transport mechanisms: values used for the single porosity case (Foppen, 2002, see Chapter 10) and the dual porosity case

Transport mechanism	Single porosity, low velocity (Foppen, 2002)	Dual porosity mobile column	stagnant column
Dissolved solutes mass balance			
No. of cells	40	40	40
Cation exchange	cell 1-40	cell 1-40	cell 1-40
CEC (meq/kg)	5	180.5	180.5
Nitrification	cell 7-40	cell 7-40	cell 7-40
$SI_{calcite}$	-	cell 7-40	cell 7-40
Exchange factor	-	2E-9	
E. coli mass balance			
Decay rate (1/d)	-	0.15	-
Retention rate (1/d)	-	0	-
E. coli removal (log-unit/10m)	(>>1)	1	-
Water balance			
porosity (%)	8	0.3	8
Infiltration (m/y)	0.42	0.42	-
v (m/d)	0.014	0.384	-
Sample mix	-	0.04	0.96

understanding, dual porosity was considered realistic, if calculated concentrations of solutes would be close to values modelled by Foppen (2002; Chapter 10) for the single porosity, low velocity situation, since latter model values were the best fit to measured hydrochemical data in the area. Best results were obtained with an exchange factor between mobile and stagnant phase of $2x10^{-9}$ (1/s). The exchange factor can be related to specific geometries of the stagnant zone, but we did not attempt to do so, because of a lack of information. As stated before, our aim was just to explore fast component vertical transport mechanisms and to demonstrate their likelihood in the Quaternary Alluvium, when taking all mass balances into consideration. The modelled [Cl^-] in Fig. 10 was the result of advection and dispersion only. At a depth of 2.5 m (the uppermost values in Fig. 10), the original wastewater [NH_4^+] of 10 mmol/l had already decreased to around ~ 5 mmol/l due to cation exchange, while [NO_3^-] was still low. Also due to cation exchange, [Na^+] decreased as well, while [Ca^{2+}] increased from 2.5 mmol/l in wastewater to ~ 5 mmol/l. At a depth of 30 m, the maximum modelled [NO_3^-] was found. Here acid production due to nitrification is at a maximum and pH, and alkalinity were therefore at their minima. Alkalinity was "eaten" away by protons produced from nitrification. At a depth of 50 m, cation exchange was mainly characterised by the release of Ca^{2+} into solution and an increase of Na^+ on the exchanger (NaX). At larger depths, there was hardly any cation-exchange occurring, and its significance became even less at greater depths. We concluded that, in general, the distribution of dissolved solutes along the column length (Fig. 10) was similar to concentration patterns obtained with the single porosity, low velocity model from Chapter 10, which indicated that, within the water balance boundary conditions of Fig. 1, both the *E. coli* mass balance and the dissolved solutes mass balance in the Quaternary Alluvium could be adequately simulated using the dual porosity approach. However, it should be emphasized that the dual flow model we employed had at least two serious limitations: (1) depth-specific data were missing, and therefore the depth-dependent conceptualisation of hydrochemical processes might have been doubtful, (2) the lack of retention due to attachment and straining in large pore structure was unrealistic. Therefore, the estimated pore water flow velocity in the mobile pore matrix might have been too low.

11.5.2 Size exclusion or dual porosity?

In the literature, there are other field studies of sewage contamination that provide evidence of groundwater flow paths that permit the transport of microorganisms by rapid velocities (Taylor et al., 2004). In media where microorganisms are excluded from parts of the sediment, the disparity between the average linear velocity of groundwater flow ($\bar{v}$) and flow velocities transporting microorganisms ($v_{colloid}$) might be intensified. Measured ratios of $\frac{v_{colloid}}{\bar{v}}$ in such media may range from 1-2 in alluvial sands and gravels to more than 300 in fractured sandstone aquifers (Taylor et al., 2006). This phenomenon is generally also known as size exclusion or differential advection (Taylor et al., 2004). In the Sana'a case, based on the pore water flow velocities calculated for the dual porosity case and the single porosity case (Table 2), this ratio was $\frac{0.38}{0.014} \approx 27$, which was relatively high compared to measured ratios in other alluvial sediments. However, in Sana'a, not only were micro-organisms excluded from parts of the aquifer. Solute transport could also have been excluded from parts of the aquifer, because their transport was well described using the dual porosity concept. In such cases, where also solutes are transported in a dual flow type of transport mechanism, the average linear velocity of groundwater flow, calculated from observed penetration depths of conservative components like chloride, while neglecting dual advection, is not correct, and might severely underestimate the real groundwater flow velocity. In fact, for dual porosity cases with mobile and stagnant zones, the ratio $\frac{v_{colloid}}{\bar{v}}$ can also be written as:

$$\frac{v_{colloid}}{\bar{v}} = \frac{\dfrac{q}{\theta_{mobile}}}{\dfrac{q}{\theta_{total}}} = \frac{\theta_{total}}{\theta_{mobile}} \qquad (12)$$

where q is the average groundwater discharge (m/d), θ_{mobile} is the porosity of the mobile zone of the aquifer, and θ_{total} is the total porosity of the aquifer. So, if the porosity of the mobile parts of the aquifer is low, then the ratio $\frac{\theta_{total}}{\theta_{mobile}}$ is high, and the $\frac{v_{colloid}}{\bar{v}}$ ratio is then indicating porosity variations in the aquifer instead of indicating transport velocity variations between colloids and solutes. Therefore, more in general, we concluded that it is more accurate to use high $\frac{v_{colloid}}{\bar{v}}$ ratios as an indicator of the occurrence of dual flow type of transport mechanisms in aquifers, which may affect both dissolved solutes and micro-organisms, than to use the ratio as an indicator of size exclusion or real pore water flow velocity variations.

11.6 Conclusions

Knowledge about the components determining the microbiological mass balance in aquifers provided valuable information on the occurrence and magnitude of fast transport mechanisms. Combining the water balance, the dissolved solutes mass balance, and the microbiological mass balance yielded a realistic description of the vertical transport mechanism taking place in this unconsolidated alluvial aquifer during infiltration of waste water. In the Sana'a case, because of the high removal in the column experiments, ranging from 2 to more than 5 log units per 0.4-0.9 cm of sediment, we concluded that parts of the Quaternary Alluvium aquifer were inaccessible for *E. coli*. Because of the relatively high *E. coli* concentrations in the shallow aquifer, up to 10^6 CFU/100 mL, transport in the aquifer from cesspit to groundwater had likely taken place via a connected network of pores with a diameter large enough to allow bacterial transport. In addition, transport in such a network of pores was fast, when taking the decay rate coefficient of *E. coli* of 0.15 per day, measured in laboratory microcosms, into consideration. A 1D transport model including dual porosity constructed with PHREEQC was able to adequately simulate microbiological removal processes and hydrochemical processes in the Quaternary Alluvium. The pore water flow velocity in this model was 0.38 m/d and was estimated from the removal of *E. coli* in the aquifer, inferred from *E. coli* concentrations in the shallow aquifer and the decay rate coefficient determined from the laboratory microcosms. Based on the water balance of the city area, determined from a calibrated groundwater model of the entire Sana'a Basin, the porosity available for flow of water was only 0.3% of the total sediment volume. By fitting the calculated depth dependent concentrations of solutes with best fits with measured data, obtained from a single porosity transport model, the porosity of the stagnant zone was determined to be 8% of the total sediment volume, while the exchange factor between mobile and stagnant zones was 2×10^{-9} per second.
In combination with laboratory experiments, knowledge on the transport of microorganisms can add valuable information to mechanisms of transport in aquifers. More generally, we believe that there is added value in using microbial tracers for the conceptualization of such transport mechanisms. Therefore, in general and if possible, microbial mass balances should be considered more systematically as an integral part of transport studies.

CHAPTER 12:

SUMMARY, CONCLUSIONS AND RECOMMENDATIONS

12.1 Summary

12.1.1 Interpreting retention of *E. coli* with the classical CFT

Under fully controlled laboratory conditions, with constant chemical groundwater quality, one type of collector (quartz grains), a range of Darcy flow velocities, a range of grain sizes, and in columns of limited height, the *E. coli* concentration decreases logarithmically with the transport distance. Also, there is reasonable agreement between the measured filter coefficients from breakthrough curves and the filter coefficients calculated using the single collector contact efficiency, and assuming a constant sticking efficiency for the *E. coli* – quartz system. Because the *E. coli* – quartz grain system behaves according to the classical CFT, as defined in chapter 1 of this thesis, the theory provides a valuable framework for predicting the transport of *E. coli* in porous media under saturated conditions. This is one of the main outcomes of the work carried out by Matthess et al. (1991a, 1991b) on an extensive set of column experiments. However, based on the research presented in this thesis, this outcome needs to be modified for the following reasons:

- The conclusions of Matthess et al. (1991a, 1991b), and many other workers, are based on effluent concentrations from column experiments only. As was concluded by Bradford et al (2002, 2003), Tufenkji and Elimelech (2004a, 2005), Li et al. (2004), Tong and Johnson (2007), and by our work (Chapter 8), such type of experiment does not give information on the transport distance dependent retention processes taking place in the column. Therefore, more types of experiments are required. In literature, column extrusion experiments have been mentioned a number of times, and we added a new type of experiment, the so called flow reversal experiment, in which the occurrence of non-adhesive retention in column experiments is directly assessed;
- We have demonstrated in columns of spherical glass beads with various diameters, that, when the beads were large, or column heights were too limited, dual mode deposition, as a result of attachment heterogeneity among members of the *E. coli* population, was completely masked, because in such cases just one type of (fast) deposition dominated the retained profiles, resulting in log-linear retained bacteria masses. Therefore, the extrapolation of experiments, where masking of dual mode deposition has taken place, to predict removal of *E. coli* at a larger scale will almost certainly lead to overestimation of *E. coli* removal;
- The CFT does not include straining. In columns of ultra-pure angular quartz sand, from eluted concentrations, column extrusion experiments, flow reversal experiments, and model fitting of eluted concentrations, we found that straining could be as much as 25-30% of the total bacteria mass applied to the column, while the maximum attached fraction was 30-35%. Since straining occurred in all grain size fractions we used, we are convinced that a good deal of literature on colloid retention processes, using coulmns of angular collectors, and neglecting straining or non-adhesive retention, have come to erroneous conclusions with regard to the nature of the processes causing retention in their experiments. In these cases, as was mentioned above, the extrapolation of such results to predict

removal of *E. coli* at a larger scale will almost certainly lead to overestimation of *E. coli* removal;
- The CFT does not include geochemical heterogeneity as a result of the presence of various minerals in the sediment. Since these minerals can have different surface charges as a result of pore water chemistry and/or lattice imperfections, and since *E. coli* is also (negatively) charged, mineralogical variations in the sediment can have an influence on attachment. We found that in columns of limited height (2 cm), due to the presence of positively charged goethite or calcite, the sticking efficiency, as determined from eluted *E. coli* concentration curves increased from 0.2 to 0.9. Also, the sticking efficiency of PRD1, the model virus that we used, calculated in the same way as described above, increased 100%, from 0.05 to 0.1, when goethite coated quartz grains were added to the sediment. Characterization of the surface of the collector is therefore of utmost importance, especially if various mineral assemblages are considered or for comparison and interpretation of sticking efficiency values in literature;
- In cases of dual porosity aquifers, like the shallow unconsolidated alluvial aquifer below Sana'a, where we proved that certain parts of the aquifer were inaccessible for microorganisms, the importance of CFT was limited. Here, decay rate factor, pore water flow velocity, and pore volume available for fast flow in the aquifer were the most important factors determining the transport of *E. coli*. Relatively small volumes of fast flowing polluted water in aquifers can give rise to considerable microbiological contamination, while hydrochemical patterns may be hardly affected. More in general, we have demonstrated the added value of using microbial tracers for the conceptualization of contaminant transport processes in aquifers.

12.1.2 Characteristics of straining of *E. coli*

We found that the mass balance for strained bacteria (S) in angular ultra-pure quartz sand could be described by

$$\frac{\partial S}{\partial t} = \frac{\theta}{\rho_{bulk}}\frac{a}{x}(C - C^{1-b}x^{b}a_c^{bd}S) \qquad (1)$$

whereby in our case $a = 0.05$ $b = 0.717$ and $d = 1.20$. This newly derived mass balance for strained bacteria, which is describing straining as a Langmuirian type of site saturation, was dependent on the concentration in the fluid, C, transport distance, x, grain diameter, a_c, and 3 constants. The values of these constants were found to be applicable for a wide range of porous medium sizes, transport distances, and input concentrations, and were obtained from a combination of column experiments, column extrusion experiments, flow reversal experiments, and model fitting.

Straining was also studied in columns consisting of mixtures of various quartz grain sizes. In both studies there was a maximum pore space available for straining, which was rapidly filled at very high bacteria concentrations. The maximum pore space in the study of the ultra-pure angluar quartz sand was 0.01% of the total column volume, while this was 0.21-0.35% in the study with the quartz sand with mixtures of various grain sizes. The latter values corresponded well with values calculated with a formula based on purely geometrical considerations and also with values calculated with a pore size density function. Because of the relation between pore size density and the occurrence of straining, and because pore size density functions can be determined with the commonly used Van Genuchten soil parameters, the potential for straining at field scale can be tentatively indicated.

12.1.2 The relative importance of mineral assemblages in aquifers

As was already described in section 12.1.1, we found that the presence of various minerals in the sediment affected the retention of *E. coli* and PRD1. However, often (bio)colloids travel through soils and aquifers in a plume of wastewater, in which dissolved organic material (DOM) is present, like humic materials, polysaccharides, polyphenols, proteins, lipids, and heterogeneous organic molecules (Fujita *et al.*, 1996; Ma *et al.*, 2001; Imai *et al.*, 2002; Ilani *et al.*, 2005) with concentrations ranging from 1-100 mg/L, dependent on types of waste water treatment. We found that due to the presence of DOM, attachment of *E. coli* was 2 to 80 times less than in cases without DOM, dependent on the chemical composition of the *E. coli* suspension. The cause for the removal reduction due to sorption of DOM was a combination of:

- Alteration of the positive surface charge at the mineral surface due to negatively charged DOM;
- Site competition between DOM and divalent ions that were required for cation bridging between *E. coli* and the mineral surface, and
- Steric hindrance whereby DOM sorption reduced the short range attractive van der Waals force between collector surface and *E. coli.*

In our experiments there was proof of the occurrence of all three mechanisms. Therefore, our results indicated that in the presence of DOM, the concept of geochemical heterogeneity in explaining attachment of biocolloids has limited relevance.

A similar conclusion was drawn from analyzing column experiments of sediment mixtures of quartz sand with goethite-coated sand and PRD1, which is often used as a model virus. We used PRD1 to assess the attachment characteristics of a model virus, and to evaluate the justification of making conservative assumptions about the attachment of viruses when calculating protection zones fro groundwater wells (Schijven et al., 2006). Due to the presence of DOM, effluent concentrations of PRD1 increased by as much as 5 log units, thereby almost completely eliminating attachment to the quartz grains and to the goethite coated grain surface. Our results implied that in the presence of dissolved organic matter (DOM), viruses can be transported for long distances due to poor attachment, because DOM has occupied favorable sites for attachment. Therefore, also in the case of virus transport, the concept of geochemical heterogeneity in explaining attachment for biocolloids had limited relevance.

12.1.3 Population heterogeneity

We found that the sticking efficiency within a population of *E. coli* bacteria may vary at least 100-fold, resulting in different deposition patterns. In a series of column experiments with glass beads of various sizes, attachment of *E. coli* decreased hyper-exponentially, or, on logarithmic scale in a bimodal way, as a function of the transported distance from the column inlet. From modeling, we concluded that 60% of the bacteria population attached fast, while 40% attached 100 fold slower. To confirm this modeling result, obtained from fitting retained profiles, column experiments with *E. coli* subpopulations consisting of slow attachers with low sticking efficiencies were carried out, and eluted concentrations were significantly higher than for *E. coli* total populations. So, heterogeneity among members of the bacteria population indeed caused deposition variations within the column. If there is steady state and if the influence of hydrodynamic dispersion is negligible, for a continuous particle injection at concentration C_0 (at $x = 0$) and time period t_0, the mass balances for bacteria with two distinct sticking efficiencies in the fluid and attached to the solid phase may be obtained from the general mass balance equation for transport of mass in porous media (De Marsily, 1986; Johnson and Li, 2005) according to:

$$C(x) = C_0\left(f\,exp\left[-\frac{k_{fast}}{v}x \right] + (1-f)\,exp\left[-\frac{k_{slow}}{v}x \right] \right) \tag{2}$$

and

$$S(x) = \frac{t\theta C_0}{\rho_{bulk}}\left(k_{fast} f\,exp\left[-\frac{k_{fast}}{v}x \right] + (1-f)k_{slow}\,exp\left[-\frac{k_{slow}}{v}x \right] \right) \tag{3}$$

whereby f is the population fraction responsible for fast removal, $(1-f)$ is the fraction for slow removal, and each fraction has a different attachment rate coefficient, k_{fast} and k_{slow}.

12.1.4 Contaminant transport in the aquifers below Sana'a

In Sana'a, the capital of the Yemen Arab Republic, due to rapid and uncontrolled waste water infiltration into the aquifers via cesspits, groundwater below the urban area was characterised by high concentrations of almost all major cations and anions. The most important cation and anion of the watertype that was predominantly found in Sana'a were Ca and Cl. The Cl^--concentration ranged from 3-10 mmol/l and NO_3^--concentration ranged from 1-3 mmol/l while NH_4^+ was absent in all samples. Furthermore, from comparing conservative mixing of wastewater and uncontaminated groundwater in the Quaternary Alluvium, Ca^{2+} in groundwater was enriched, while Na^+, K^+ and NH_4^+ were depleted. We concluded that nitrification of ammonium, present in wastewater, and cation-exchange had taken place. Over the period between two surveys, one in 1995 and one in 2000, concentrations of almost all major anions and cations increased, while pH decreased, both owing to the continuous infiltration of wastewater. An exploratory one-dimensional transport model of a 200 m column of the aquifer underlying Sana'a showed that around 30% of the NH_4^+ present in raw sewage would oxidise to NO_3^- thereby producing acidity and some 70 % would be adsorbed. Recharge of waste water in the poorly sorted alluvial aquifer below Sana'a is a complex process, both physically and hydrochemically. By analysing and determining components of the microbial (*Escherichia coli*) mass balance, vertical transport mechanisms were analysed in a conceptual way. The outcomes were used to explore transport mechanisms in a one-dimensional vertical hydrochemical model. The results indicated that shallow groundwater (0-100 me depth) was fecally contaminated with *E. coli* concentrations as high as 10^5 CFU/100 mL. Concentrations reduced at greater depths (100-300 m) in the aquifer. However, in laboratory columns, bacteria removal was 2-5 log units per 0.5 cm column sediment. We concluded that because of the relatively high *E. coli* concentrations in the shallow aquifer, transport in the aquifer had likely taken place via a connected network of pores with a diameter large enough to allow bacterial transport instead of via the sediment matrix. Transport in such network of pores must have been fast, because of the considerable decay rate coefficient of *E. coli*, which was determined from laboratory experiments to be 0.15 d^{-1}. Combined fast microbial transport and infiltration of waste water, including cation exchange and nitrification, was accurately simulated with a one-dimensional transport model of a 200 m column of the aquifer with dual porosity.

In the period from 2001-2005, management of waste water has changed dramatically and entire city neighborhoods have been connected to a conventional sewer system. The effects of this measure on long-term groundwater quality development of the aquifers underlying Sana'a and, more specifically, on water quality of the public supply peri-urban wellfields were analyzed. The results, obtained with a transient groundwater model, indicated that by 2020 the construction of the sewerage will have considerably reduced the area that is hydrochemically polluted by groundwater, but the process is slow. Furthermore, although construction of the sewerage had already reduced *E. coli* concentrations to very low values of less than 10 CFU/mL in 2004, the effect on chemical groundwater quality near the public supply well-fields is expected to be limited, since flow is not directed towards most of the production wells. An adverse side effect of the sewerage construction is the lowering of the groundwater table due to the decrease of waste water recharge, and the drying out of wells. One of the major water user groups in the city is agriculture, and many people, especially the elder ones depend on this agriculture for their food supply. There is a good chance that a part of the agricultural areas inside the city will disappear due to the sewerage construction.

12.1.5 Improved transport model

Based on the results of the laboratory column studies, and the Sana'a case, a new model approach for the transport of *E. coli* in saturated porous media was developed that includes the new description for straining, and the descriptions for population heterogeneity (Johnson and Li, 2005), and geochemical heterogeneity (Johnson et al., 1996; Elimelech et al., 2000), that were validated for *E. coli* in this research. The one-dimensional transport may be generally described by the advection-dispersion-sorption (ADS) equation (Pang et al., 2003; Powelson and Mills, 2001; Matthess and Pekdeger, 1981 and 1985; Matthess et al., 1985, 1988; Chapters 2 and 5 of this study) along with terms for attachment, detachment, straining, and inactivation or die-off:

$$\frac{\partial C}{\partial t} = D\frac{\partial^2 C}{\partial x^2} - v\frac{\partial C}{\partial x} - \frac{\rho_{bulk}}{\theta}\frac{\partial S}{\partial t} - k_i C \quad (4)$$

where C is the concentration of suspended bacteria in the aqueous phase (cells/cm^3), D is the hydrodynamic dispersion coefficient and includes the effects of random motility and chemotaxis (cm^{2}s^{-1}), v is the pore water flow velocity (cm/s), ρ_{bulk} is the bulk density of the porous medium (g/cm^3), θ is the effective porosity, S is the dimensionless retained bacteria concentration (g/g), k_i is the inactivation or die-off rate coefficient of bacteria in the fluid (s^{-1}), t is time (s), and x is the distance traveled (cm). The mass balance equation for retained bacteria can be expressed as

$$\frac{\partial S}{\partial t} = \frac{\partial S_{att}}{\partial t} + \frac{\partial S_{str}}{\partial t} - k_{is} S \quad (5)$$

$$\frac{\partial S_{att}}{\partial t} = \frac{\partial S_{att,fast}}{\partial t} + \frac{\partial S_{att,slow}}{\partial t} \quad (6)$$

$$\frac{\partial S_{str}}{\partial t} = \frac{\theta}{\rho_{bulk}}\frac{a}{x}(C - C^{1-b} x^b a_c^{bd} S_{str}) \quad (7)$$

where S_{att} is the bacteria mass retained due to attachment, $S_{att,fast}$ is the attached bacteria mass due to the presence of fast attaching bacteria in the total bacteria population, $S_{att,slow}$ is the attached bacteria mass due to the presence of slow attaching bacteria in the total bacteria population, S_{str} is the bacteria mass retained due to straining, k_{is} is the inactivation or decay rate coefficient of bacteria on the solid matrix (s^{-1}), k_{str} is the straining rate coefficient (cm^{-1}), C_0 is the influent concentration (cells/cm^3), a_c is the grain diameter (cm), and values for parameters a, b, and c were taken from the ultrapure angular quartz sand case: $a = 0.05$ $b = 0.717$ and $c = 1.20$. Kinetic fast and slow attachment can be expressed as

$$\frac{\partial S_{att,fast}}{\partial t} = k_{fast} C_{fast} \quad (8)$$

$$\frac{\partial S_{att,slow}}{\partial t} = k_{slow} C_{slow} \quad (9)$$

where k_{fast} and k_{low} are the attachment rate coefficients (s^{-1}) for fast and slow attachment, respectively. At any point in the fluid $C = C_{fast} + C_{slow}$, and because attachment differs for the two population fractions, the ratio $\frac{C_{fast}}{C_{slow}}$ is not constant, and reduces with increasing transport distance. When geochemical heterogeneity is included, the attachment rate coefficient can be defined as:

$$k_a = \frac{3(1-\theta)}{2a_c} v\eta_0 [\lambda\alpha_f + (1-\lambda)\alpha_u] \qquad (10)$$

where η_0 is the single collector contact efficiency (SCCE), determined from physical considerations (e.g. Tufenkji and Elimelech, 2004), α_f and α_u are the sticking efficiencies to favorable and unfavorable attachment sites respectively, and λ is a dimensionless heterogeneity parameter describing the fraction of aquifer grains composed of minerals favorable for attachment or grains coated with favorable attachment patches (Chapter 5). The single collector contact efficiency, η_0, is defined as (Yao et al., 1971; Tufenkji and Elimelech, 2004b)

$$\eta_0 = \eta_D + \eta_I + \eta_G \qquad (11)$$

where η_D, η_I and η_G, represent theoretical values for the single-collector contact efficiency when the sole transport mechanisms are diffusion, interception, and sedimentation, respectively. In case of fast attachment ($k_{a,fast}$): $\alpha_f = \alpha_u = 1$, and eq. (10) reduces to:

$$k_{fast} = \frac{3(1-\theta)}{2a_c} v\eta_0 \qquad (12)$$

In case of slow attachment $\alpha_f = 1$ and $\alpha_u << 1$, and eq. (10) is written as:

$$k_{slow} = \frac{3(1-\theta)}{2a_c} v\eta_0 [\lambda\alpha_f + (1-\lambda)\alpha_u] \qquad (13)$$

12.1.6 Discussion of an example case

To demonstrate the importance of processes, the combined mass balances of bacteria in the fluid (eq. 4) and retained by the sediment (eq. 5) for one type of *E. coli* (C_{fast} or C_{slow}) were solved numerically with an explicit finite difference scheme that is forward in time, central in space for dispersion and upwind for advective transport. With each time step, first advective transport is calculated, then rate reactions and finally dispersive transport. Rates must be integrated over a time interval, which involves calculating the changes in bacteria concentrations both in the fluid and on the solid matrix. The scheme was prepared in a spreadsheet and is the same as that used in PHREEQC version 2 (Parkhurst and Appelo, 1999), and was used in determining the importance of straining in the study on straining in quartz sand, consisting of mixtures of various grain sizes (Chapter 5). The spreadsheet is flexible, and therefore the newly formulated mass balance for strained bacteria, which is not available in other model codes, could be easily implemented. The example cases are two columns, one with a height of 50 cm and one with a height of 50 m. Both were spiked with a 10^3 cells/mL *E. coli* suspension for 5 pore volumes. Pore water flow velocity is constant at 0.3 m/d. The columns consist of quartz grains with a diameter of 250 μm, while the porosity is 0.3. The decay rate coefficient was taken as 0.1 d^{-1}, which was relatively low, when compared with values from literature (Chapter 2). The straining rate coefficient was taken as 0.05 cm^{-1} , determined from the straining experiments in the columns of pure quartz grains, and reduced with increasing transport distance in the aquifer. Based on the dual mode deposition experiments, the attachment rate coefficient was calculated assuming a sticking efficiency of 0.01 for the fraction slowly attaching bacteria. The fraction fast attaching bacteria (assumed to be 0.4) was not included in the calculations, assuming that transport of bacteria in the aquifer and associated retention processes were mainly determined by the fraction slow attaching bacteria. Other parameters are mentioned in Table 1.

Table 1: Overview of parameters and parameter values used for the example case (Fig. 1)

GENERAL FLUID CONDITIONS		**STRAINING PARAMETERS**	
Temperature	293 K	a	0.717 -
Density of water	1 g/cm^3	b	1.2 -
TRANSPORT PARAMETERS		**SINGLE COLLECTOR CONTACT EFFICIENCY**	
Total distance	50 cm	Hamaker constant	6.50E-21 J
Total time	144000 s		
Velocity (pore water)	3.47E-03 cm/s	**STICKING EFFICIENCY**	
Dispersivity	5 cm	Geochem. heterog. param.	0 -
Diffusion coeff.	1.00E-05 cm^2/s	Sticking eff. (unfav.)	1.00E-02 -
		Sticking eff. (fav.)	1 -
SEDIMENT PARAMETERS			
Porosity	0.3 -	**RATE COEFFICIENTS**	
Bulk density	1.86 g/cm^3	Decay rate coeff.	1.16E-06 1/s
Grain diameter	250 micron	Attachment rate coeff.	2.54E-06 1/s
		Straining rate coeff.	5.00E-02 1/cm
BACTERIA PARAMETERS			
Influent C	1.00E+03 $cells/cm^3$		
Fast fraction	0.4		
Density of bact.	1.05 g/cm^3		
Diameter	1 micron		

Breakthrough of non-decaying, non-attaching and non-straining *E. coli* ("slow *E. coli*"; Fig. 1a) in the 50 cm column was 0.6 after flushing for 2 pore volumes. Breakthrough did not increase to 1, because the fast attaching bacteria had already attached near the column inlet. When transport of *E. coli* through the column was subject to decay only ("decay" in Fig. 1a), then normalized concentrations were a fraction lower than for the "slow *E. coli*" case, indicating that the effect of decay was limited. Similarly, the effect of attachment due to a sticking efficiency of 0.01 was limited. Normalized breakthrough concentrations were close to 0.6 after two pore volumes. When straining was included, then the shape of the breakthrough curves changed significantly. Initially, breakthrough was almost zero, and normalized concentrations started to rise only after two pore volumes, due to the filling of pore space with strained bacteria cells. The effect of 5% surface area of positively charged mineral surface, giving rise to favorable attachment of the negatively charged *E. coli*, further reduced the breakthrough of *E. coli* significantly.

Breakthrough of non-decaying, non-attaching and non-straining *E. coli* ("slow *E. coli*"; Fig. 1a) in the 50 m column was, like the 50 cm column, 0.6 after flushing for

2 pore volumes. Again, breakthrough did not increase to 1, because the fast attaching bacteria had already attached near the column inlet. When transport of *E. coli* through the column was subject to decay only ("decay" in Fig. 1b), then normalized concentrations reduced almost 1 log-unit. Due to slow attachment, normalized concentrations reduced 6 log-units. When straining was included, the breakthrough curve hardly changed, because of the inverse relationship between straining rate parameter and transport distance: the effect of straining for such large distances was negligible. The 5% surface area consisting of positively charged mineral surface, and giving rise to favorable attachment of the negatively charged *E. coli*, reduced the normalized concentration to 10^{-24} (values are not given in Fig. 1b).

From these two example cases, two main conclusions can be drawn:

- In columns of limited heights, the effect of decay of bacteria on their breakthrough was limited, provided that the decay rate coefficient was not too large. In addition, the effect of slow attachment was also negligible. Straining and favorable attachment were important processes in columns of limited height. Straining can be easily overlooked in column experiments when the grain size or pore size is large, and if so, then the interpretation of breakthrough curves might lead to wrong conclusions with regard to the type of retention processes occurring in the column. For instance, the breakthrough curves due to straining in Fig. 1a could be easily interpreted as being the result of a combination of equilibrium sorption, resulting in retardation of 2 pore volumes before the breakthrough curve started to rise, plus attachment with reduction of the attachment rate coefficient due to site saturation.
- In the large 50 m column, the dominant removal processes were unfavorable and favorable attachment. Decay and straining were relatively unimportant. However, because patches of ironoxides, assumed to be present in the sediment, and responsible for favorable attachment, are usually likely to have been in contact with organic molecules from DOM, in aquifers the role of favorable attachment probably is limited. The presence of a preferential flow or dual porosity system in the 50 m column would have increased normalized concentrations. For instance, if we would have assumed that pore water of Fig. 1b for the case of decay + unfavorable attachment + straining was affected by 0.1% pore water from preferential flow, in which no decay and no attachment had taken place, then normalized concentrations would have become

$$0.001x10^3 + \frac{0.999x10^3}{10^7\ (= removal)} = 1\ cell / cm^3$$

So, as a result of preferential flow, the *E. coli* concentration increased 7 log-units! This example clearly demonstrates the relative importance of preferential flow.

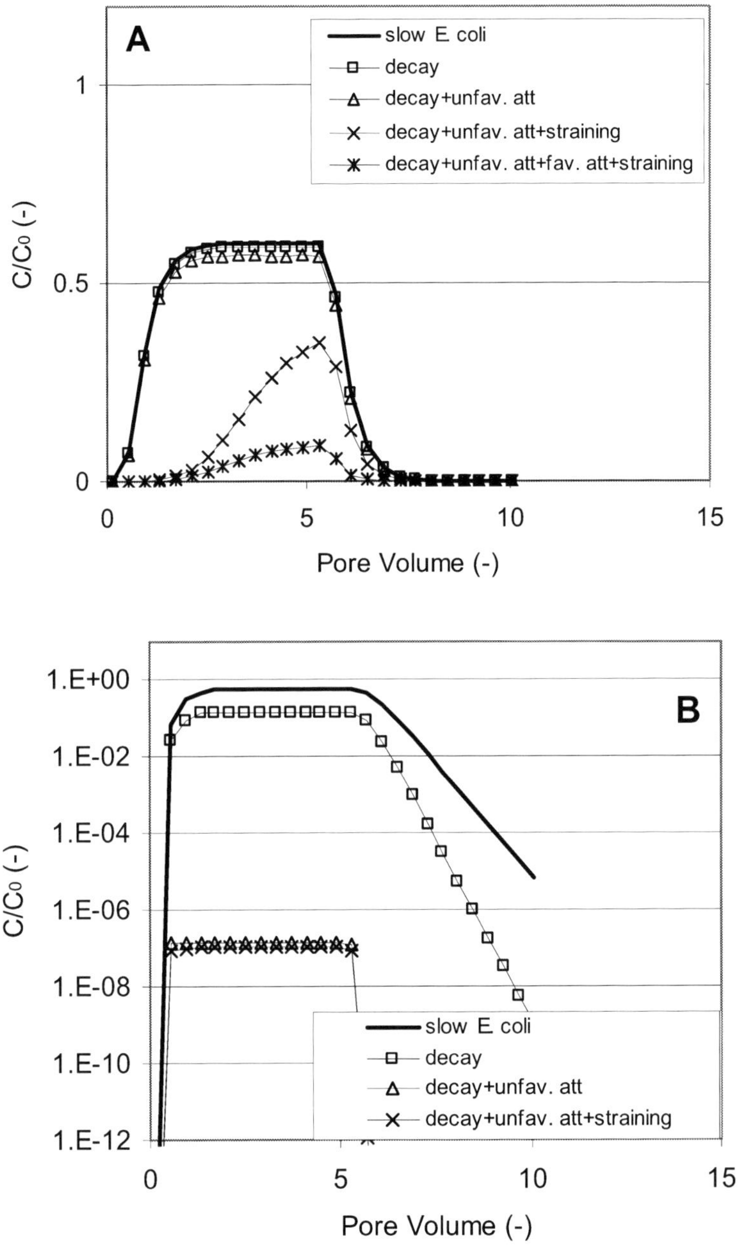

Fig. 1: Removal of E. coli after application of a pulse of 5 pore volumes of a suspension containing 10^3 cells/mL E. coli, as a function of pore volume for a 50 cm column (a) and a 50 m column (b)

12.2 Future research directions

12.2.1 Straining and other types of non-adhesive retention

Although the shape of the pore structure has been recognized as an important parameter in determining the retention of microorganisms, the topic has not been subject of systematic research. Up to date, straining, and, more in general, non-adhesive retention of microorganisms is not included in the colloid filtration theory, and therefore the interpretation of laboratory breakthrough curves, which has provided the experimental basis for the theory, might have resulted and will continue to result in inaccurate interpretations of processes occurring in columns.
In recent years, micromodels have been increasingly employed to study the fate and transport of colloids and specifically biocolloids at the pore scale. Micromodels are transparent physical models of porous media, with a pore size in the range of 10-100 μm, etched in glass, silicate wafers, or polymer substrates. The main purpose of microscale experiments has been to visualize biocolloid transport processes at the dimensions of a pore or collection of pores, validating or negating hypotheses that have been put forward with regard to processes that had not been adequately observed. A second objective has been to quantify the importance of these processes. Significant advances have been made in the understanding of biocolloid transport with pore scale visualizations (Keller and Auset, 2006). The process of size exclusion and early arrival of biocolloids in relation to their size is now better understood, there is evidence that interception of biocolloids by the collector surface depends on fluid velocity, and furthermore, there are now clear indications that within the pore structure there are regions of stagnant pore water that can lead to storage of biocolloids.
To improve the accuracy of our interpretations of colloid transport phenomena, the use of micromodels of complex pore structures, and the developmeny of physically based simulation models, with the aim to quantify retention patterns of biocolloids, might lead to new insights in the transport of colloids in porous media.

12.2.2 Population heterogeneity and sticking efficiency

In column experiments with glass beads of various sizes, we found that attachment of *E. coli* decreased hyper-exponentially, or, on logarithmic scale in a bimodal way, as a function of the transported distance from the column inlet. From data fitting of the retained bacteria concentration profiles, the sticking efficiency of 40% of the *E. coli* population was high ($\alpha = 1$), while the sticking efficiency of 60% was low ($\alpha = 0.01$). In an homogeneous isotropic aquifer consisting of grains with a diameter of 250 μm with steady state groundwater flow conditions, resulting in a pore water flow velocity field of 1.0 m/d, a sticking efficiency of 10^{-2} leads to a removal of 5 log-units within 400 m. However, if a small fraction of a population of bacteria consists of cells with a much lower sticking efficiency of 10^{-4}, then removal of that fraction is less than 1 log unit in 1000 m (Fig. 2), and this low removal may significantly influence the total bacteria concentration.

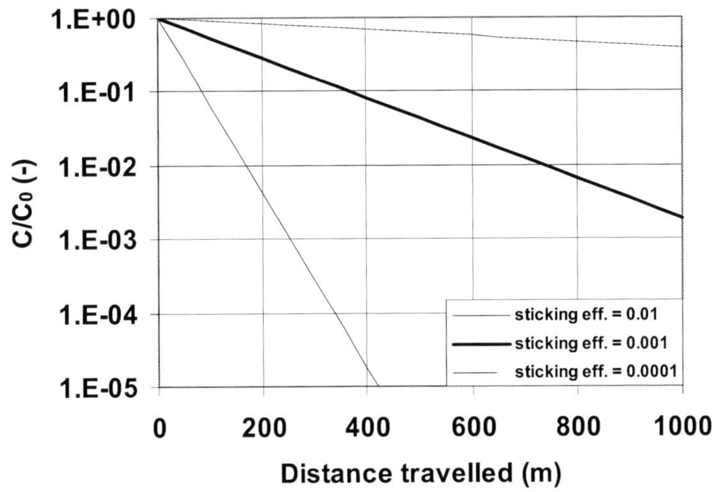

Fig. 2: Normalized concentration as a function of travel distance for various sticking efficiency values

Therefore, immediate, and very relevant questions for better assessment of environmental health risks associated with the transport of biocolloids, are:

- What is the environmentally relevant minimum sticking efficiency a biocolloid population can have?
- What is the fraction of the total population belonging to this minimum sticking efficiency?
- Is survival of such fraction of the population different from survival of the total population?

12.2.3 Large organic molecules (DOM)

E. coli or thermotolerant coliforms usually travel in a plume of wastewater, consisting of many organic and inorganic compounds, all with highly variable concentrations. These compounds may block attachment sites, thereby leading to enhanced transport of *E. coli*. Laboratory experiments are usually carried out with solutions that do not reflect wastewater compositions, which might be the cause for the observed discrepancy between laboratory and field sticking efficiencies.
The magnitude of blocking was investigated for the system *E. coli* – goethite coated sand – humic acid (HA), as a surrogate for organic compounds present in wastewater (Chapter 6). One of the main conclusions of this work was that removal in 0.25 cm columns consisting of goethite coated due to the presence of humic acid attached to the goethite patches, decreased 2-100 times. In similar work with the system PRD1 - goethite coated sand - dissolved organic carbon (Chapter 7), removal in a 7 cm column consisting of goethite coated sand was 5 log units. In the presence of HA, C/C_0-values of PRD1 increased by as much as 5 log units, thereby almost completely eliminating retention in the column. A second result was that the inactivation rate coefficient of PRD1 was much lower in the presence of dissolved organic carbon.

DOM in groundwater is a complex mixture of acidic, neutral, and basic organic poly-molecules (Thurman, 1985). Concentrations vary widely and may be as high as 100-200 mg/L near the water table to less than 0.5 mg/L in deeper groundwater (Thurman, 1985). Often pathogens travel through soils and aquifers in a plume of wastewater. DOM in wastewater is composed of humic materials, polysaccharides, polyphenols, proteins, lipids, and heterogeneous organic molecules (Fujita *et al.* , 1996; Ma *et al.* , 2001; Imai *et al.* , 2002; Ilani *et al.* , 2005) with concentrations ranging from 1-100 mg/L, dependent on types of treatment. The presence of (large) organic molecules may significantly reduce the attachment of biocolloids while traveling through an aquifer and, in addition, these organic molucules may reduce the inactivation of biocolloids. Central theme for future research would therefore be to assess the influence of the presence of organic compounds on the retention of microorganisms.

12.2.4 Aquifer geometry and upscaling

The upscaling of data obtained from controlled experiments in the laboratory to (controlled) field conditions needs due attention. Concepts, that have now been quantified in laboratory set-ups, like straining, dual mode deposition and minimum sticking efficiency, and the effect of organic molecules on the retention of microorganisms, need to be investigated in controlled field set-ups. In addition, if preferential flow occurs, how significant is preferential flow, which factors are most important in determining *E. coli* transport behavior, how to quantify preferential flow at different scales, and how should preferential flow characteristics be incorporated into the conceptualization of an *E. coli* transport model. Then, at a larger scale, one of the most relevant questions is how can the most important biocolloid transport parameters be conceptualized or described with the commonly present data sets on geology, climate, soils, topography, landuse, etc.?

LIST OF SYMBOLS

A_S	Happel's cell model constant	
Al_{total}	total content of aluminum	(% d.w.)
B	fitting parameter	(M/cell)
C	number or mass concentration of suspended bacteria in the aqueous phase	(L^{-3}), (M/L^3), ($cells/L^3$), (CFU/100 ml)
C_0	influent concentration	
C/C_0	column outlet normalised bacteria concentration	
C_{org}	organic carbon content	(% d.w.)
$B(S_{att})$	dynamic blocking function	
D	hydrodynamic dispersion coefficient	(L^2/T)
D'	coefficient of mechanical dispersion	(L^2/T)
D_B	coefficient of diffusion due to Brownian motion.	(L^2/T)
EM	electrophoretic mobility	((μm/s)/(V/cm))
Fe_{react}	content of reactive iron	(% d.w.)
Fe_{total}	total content of reactive iron	(% d.w.)
Fe_{clay}	content of reactive iron in clay	(% d.w.)
H	Hamaker constant between the collector and colloid	(J)
K_D	empirical distribution coefficient	(L)
K	empirical constant	
L	filter medium packed length	(L)
$L_{<2}$	fraction particles < 2 μm	(% d.w.)
N	coordination number or the number of contact points between grains	
N_A	attraction number representing the combined influence of the van der Waals attraction forces and fluid velocity on particle deposition rate due to interception	
N_G	gravity number for sedimentation; ratio of Stokes particle settling velocity to approach	
N_{Pe}	Peclet number characterizing the ratio of convective transport to diffusive transport	
N_R	aspect ratio	
N_{vdW}	van der Waals number characterizing the ratio of van der Waals interaction energy to the velocity of the fluid	
P_n	new parameter value due to induced change	
P_0	original parameter value	
R_F	retardation factor due to equilibrium sorption	
S	fractional surface coverage with bacteria, total retained bacteria concentration	(-), (# of cells/M)

S_{att}	fraction available for kinetic attachment, bacteria mass retained due to kinetic attachment	(-), (# of cells/M)
$S_{att,fast}$	attached bacteria mass due to the presence of fast attaching bacteria	
$S_{att,slow}$	attached bacteria mass due to the presence of slow attaching bacteria	
S_{str}	fraction available for straining, retained bacteria concentration due to straining	(-) , (# of cells/M)
S_{eq}	fraction available for equilibrium sorption	
S_{DEP}	bacteria concentration due to straining in dead end pores	
S_{TEMP}	bacteria concentration due to temporary retention	
$S_{MAX,DEP}$	maximum bacteria concentration that can be retained due to DEP straining	
SOM_{corr}	soil organic matter corrected for lutum and reactive iron concentration	(% d.w.)
SOM_{550}	soil organic matter minus loss due to ignition at 550 °C	(% d.w.)
T	temperature	(K)
$T-value$	test statistic	
U	uniformity of the sediment	
W	weight	(M)
Z	charge number or valence of the ion	
a , b , c	fitting parameters	
a_c	collector diameter , median of the grain size number distribution	(L)
$a_{c,60}$, $a_{c,10}$	60th and 10th percentile values of the log-normal grain size number distribution curve	(L)
a_i	coefficient	
a_p	diameter of the colloid	(L)
e	charge of the electron	(C)
f	specific surface area of the porous medium	(L^{-1})
f	fraction fast attaching bacteria cells	
$f(r)$	pore size density function	
f_{wastew}	fraction wastewater in the mixed water	
g	gravitational accelaration constant	(L/T^2)
h	empirical constant	
k	Boltzmann constant	(J/K)
k_a	attachment rate coefficient	(L/T), (T^{-1})
k_{fast}	attachment rate coefficients for fast attachment	(T^{-1})
k_{slow}	attachment rate coefficients for slow attachment	(T^{-1})
k_r	detachment or release rate coefficient	(T^{-1})
k_{str}	straining rate coefficient	(T^{-1}), (L^{-1})
k_i	inactivation or die-off rate coefficient of bacteria in the fluid	(T^{-1})
k_{is}	inactivation or die-off rate coefficient of bacteria on the solid matrix	(T^{-1})
k_{total}	total retention rate coefficient	(T^{-1})
k_{DEP}	rate coefficient for DEP straining	(T^{-1})

k_{ret}	rate coefficient for temporary retention	(T^{-1})
$m_{i,react}$	concentration of ion i	(mmol/l)
$m_{i,sample}$	concentration of ion i in the sample	(mmol/l)
$m_{i,mix}$	concentration of ion i as a result of mixing	(mmol/l)
$m_{i,wastew}$	concentration of ion i in wastewater	(mmol/l)
$m_{i,aquifer}$	concentration of ion i in the aquifer	(mmol/l)
$m_{Cl,sample}$	concentration of chloride in the sample	(mmol/l)
$m_{Cl,aquifer}$	concentration of chloride in the aquifer	(mmol/l)
$m_{Cl,wastew}$	concentration of chloride in wastewater	(mmol/l)
m_v, n_v, α_v	Van Genuchten soil parameters	
pH_s	pH at the bacterial surface	
pH_b	pH of the bulk solution	
r	pore-radius	(L)
s	sensitivity (%)	
$s_{Y'}$	standard error of estimate	
t	time	(T)
t_0	time needed to flush one pore volume from a unit volume	(T)
v	pore water flow velocity	(L/T)
v_w	mean groundwater flow velocity (at pore scale)	(L/T)
v_m	mean transport velocity of microorganisms (at pore scale)	(L/T)
$v_{colloid}$	velocity of the colloid	(L/T)
x	distance traveled	(L)
z_{pg}	separation distance between colloidal particle and collector grain	(L)
z	depth	(L)
α	sticking efficiency of the colloid	
α_f	sticking efficiency to favorable attachment sites	
α_{str}	straining correction factor to account for fluid velocity variations at pore level, grain geometry variations, and filter bed porosity	
α_{total}	overall sticking efficiency	
α_u	sticking efficiency to unfavorable attachment sites	
β	filter coefficient; filter effect with respect to a certain flow length	(L^{-1})
β_{aws}	contact angle formed in the air–water–solid system	
δ_{aw}	interfacial tension between air and water phases	(Nm^{-1})
δ	geometrical suffusion security	
ε	permittivity of the medium	(C/(VL)
η	single collector removal efficiency	
η_0	single collector contact efficiency	
η_{total}	overall single collector removal efficiency	
η_D	theoretical value for the single-collector contact efficiency when the sole transport mechanism is diffusion	
η_I	theoretical value for the single-collector contact efficiency when the sole transport mechanism is interception	

η_G	theoretical value for the single-collector contact efficiency when the sole transport mechanism is gravitational sedimentation particle's thermal energy	
η_S	straining contact efficiency	
κ	Debye-Huckel length	(L^{-1})
λ	heterogeneity parameter	
μ	dynamic viscosity	(M/(LT))
μ_1	first order decay rate constant	(T^{-1})
v	kinematic viscosity	(L^2/T)
ρ	specific weight	(M/L^3)
ρ_p	particle or colloid density	(M/L^3)
ρ_{fl}	fluid density	(M/L^3)
ρ_{bulk}	bulk density of the porous medium	(M/L^3)
σ	volume of spherical colloids that could be retained as part of the total sediment volume	
σ_c	collision diameter	(L)
θ	effective porosity	
θ_b	bed porosity	
θ_p	internal or particle porosity	
ψ	mean	
ψ_1, ψ_2	electrical surface potentials	(V)

Abbreviations

CFT	Colloid Filtration Theory
ADS	Advection-Dispersion-Sorption
DOC	Dissolved Organic Carbon
VIP	Ventilated Improved Pit latrine.
RT-PCR	Reverse Transcription - Polymerase Chain Reaction. A laboratory technique for amplifying a defined piece of a ribonucleic acid (RNA) molecule.
LPS	Lipopolysaccharide
OS	Oligosaccharide
KDO	2-keto-3-deoxyoctulasonic acid
VBNC	Viable but non culturable
NCTC	National collection of type cultures
ATCC	American type culture collection
SCRE	single collector removal efficiency
SCCE	single collector contact efficiency
DLVO	theory named after Darjaguin, Landau, Verwey and Overbeek. Describes the force between charged surfaces interacting through a liquid medium. It combines the effects of the van der Waals attraction and the electrostatic repulsion/attraction due to the so called double-layer of counterions.
TC	Total Coliform
FC	Faecal Coliform
DEP	dead end pore; a pore in which the exit pore throat diameter equals zero

REFERENCES

Al-Degs, Y., M.A.M. Khraisheh, S.J. Allen, and M.N. Ahmad, 2000. Effect of carbon surface chemistry on the removal of reactive dyes from textile effluent. Wat. Res. Vol. 34, No. 3, p. 927-935.

Alderwish, A.M., and J. Dottridge, 1998. Recharge components in a semi-arid area: the Sana'a Basin, Yemen. In: Groundwater pollution, aquifer recharge and vulnerability. British Geological Society Special Publications No. 130, edited by N.S. Robins, p. 169-177.

Alexander, I., and K.P. Seiler, 1983. Lebensdauer und Transport von Bakterien in typischen Grundwasserleitern. Münchener Schotterebene. DVGW-Schr., 35, p. 113-125 (ZfGW, Franfurt/Main).

Allison, L.E. and C.D. Moodie, 1965. Carbonate. In: C.A. Black et al. (ed.) Methods of Soil Analysis. Part 2. Chemical and microbiological properties, ASA, Medison, WI, USA.

Althaus, H., K.D. Jung, G. Matthess and A. Pekdeger, 1982. Lebensdauer von Bacterien und Viren in Grundwasserleitern. Umweltbundesamt, Materialien 1/82. Schmidt-Verlag, Berlin.

American Public Health Association (APHA), 1992. Standard methods for the examination of water and wastewater, 18th edition. American Public Health Association, Washington D.C.

American Public Health Association (APHA), 2005. Standard methods for the examination of water and wastewater, 21st edition. American Public Health Association, Washington D.C.

American Society for Testing and Materials (ASTM), 1985. Classification of Soils for Engineering Purposes: Annual Book of ASTM Standards. Vol. 04.08.

Amirbahman, A. and T.M. Olson, 1993. Transport of humic matter-coated hematite in packed beds. Environ. Sci. Technol., 1993, Vol. 27, p. 2807-2813.

Amor, K., D.E. Heinrichs, E. Frirdich, K. Ziebell, R.P. Johnson and C. Whitfield, 2000. Distribution of core oligosaccharide types in lipopolysaccharides from *Escherichia coli*. Infection and Immunity, Vol. 68, No.3, p. 1116-1124.

Appelo, C.A.J. and Postma, D., 2005. Geochemistry, groundwater and pollution, second edition. A.A.Balkema, Rotterdam.

Arana, I., C. Seco, K. Epelde, A. Muela, A. Fernandez-Astorga and I. Barcina, 2004. Relationships between *Escherichia coli* cells and the surrounding medium during survival processes. Antonie van Leeuwenhoek, Vol. 86, p. 189-199.

Artz, R.R.E. and K. Killham, 2002. Survival of *Escherichia coli* O157:H7 in private drinking water wells: influences of protozoan grazing and elevated copper concentrations. FEMS Microbiol. Lett. (216), p. 117-122.

Aulenbach, D.B. and T.J. Tofflemire, 1975. Thirty-five years of continuous discharge of secundary treated effluent onto sand beds. Ground Water, 1975, 13(2), 161-166.

AWWA, 1999. Waterborne Pathogens. American Water Works Association Manual of Water Supply Practices AWWA M48. ISBN 1-58321-022-9.

Barcina, I., P. Lebaron and J. Vives-Rego, 1997. Survival of allochthonous bacteria in aquatic systems: a biological approach. FEMS Microbiol. Ecol. (23), p. 1-9.

Baxter, K.M. and L. Clark, 1984. Effluent recharge. The effects of effluent recharge on grondwater quality. Technical Report 199. Water Research Centre. United Kingdom.

Baygents, J.C., J.R. Glynn Jr., O. Albinger, B.K. Biesemeyer, K.L. Ogden, and R.G. Arnold, 1998. Variation of surface charge density in monoclonal bacterial populations: Implications for transport through porous media. Environ. Sci. Technol. Vol. 32, p. 1596-1603.

Bedbur, E., 1989. Laboruntersuchungen zum Einfluss sedimentologischer und hydraulischer Parameter auf die Filterwirkung gleichförmiger Sande. Verichte-Reports, Geol.-Paläont. Inst. Univ. Kiel 31 (1989), p1-77.

Bell, F.G., 1987. Ground engineer's reference book. Butterworth and Co Ltd. ISBN 0-408-01173-4.

Bengtsson, G. and L. Ekere, 2001. Predicting sorption of groundwater bacteria from size distribution, surface area, and magnetic susceptibility of soil particles. Water Resour. Res. 37, 1795-1812.

Bengtsson, G. and R. Lindqvist, 1995. Transport of soil bacteria controlled by density-dependent sorption kinetics. Water Resour. Res. 31, 1247-1256.

Bergey, D.H., J.G. Holt, and N.R. Krieg, 1984. Bergey's manual of systematic bacteriology. Williams & Wilkins, Baltimore, USA. ISBN 0-683-04108-8

Bhattacharjee, S., J.N. Ryan and M. Elimelech, 2002. Virus transport in physically and geochemically heterogeneously subsurface porous media. J. Cont. Hydrol. 57 (2002), p. 161-187.
Bitton, G. and R.W. Harvey, 1992. Transport of pathogens through soils and aquifers. In: Environmental Microbiology, edited by R. Mitchell, ISBN 0-471-50647-8, p. 103-124.
Bitton, G., S.R. Farrah, R.H. Ruskin, J. Butner and Y.J. Chou, 1983. Survival of pathogenic and indicator organisms in ground water. Ground Water (21), p. 405-410.
Blanford, W.J., M.L. Brusseau, T.C. Jim Yeh, C.P. Gerba, and R. Harvey, 2005. Influence of water chemistry and travel distance on bacteriophage PRD-1 transport in a sandy aquifer. Wat. Res., Vol. 39, p. 2345-2357.
Bogosian, G., L.E. Sammons, P.J.L. Morris, J.P. O'Neill, M.A. Heitkamp and D.B. Webber, 1996. Death of *Escherichia coli* K-12 strain W3110 in soil and water. Appl. Environ. Microbiol. (62), p. 4114-4120.
Bouwer, H., J.C. Lange and M.S. Riggs, 1974a. High-rate land treatment I: Infiltration and hydraulic aspects of the Flushing Meadows project. J. Wat. Poll. Cont. Fed. (46), p. 834-843.
Bouwer, H., J.C. Lange and M.S. Riggs, 1974b. High-rate land treatment II: Water quality and economic aspects of the Flushing Meadows project. J. Wat. Poll. Cont. Fed. (46), p. 844-859.
Bradford, S.A. and M.Bettahar, 2005. Straining, attachment, and detachment of Cryptosporidium Oocysts in saturated porous media. J. Environ. Qual. (34), p. 469-478.
Bradford, S.A., and M. Bettahar, 2006. Concentration dependent transport of colloids in saturated porous media. J. Contam. Hydrol. 82, p. 99-117.
Bradford, S.A., S.R. Yates, M.Bettahar, and J. Simunek, 2002. Physical factors affecting the transport and fate of colloids in saturated porous media. Water Resour. Res., 38(12), 1327, doi:10.1029/2002WR001340.
Bradford, S.A., J.Simunek, M. Bettahar, M.Th. van Genuchten, and S.R. Yates, 2003. Modeling colloid attachment, straining, and exclusion in saturated porous media. Environ. Sci. Technol. 37, 2242-2250.
Bradford, S.A., J. Simunek, M. Bettahar, Y.F. Tadassa, M.T. van Genuchten, and S.R. Yates, 2005. Straining of colloids at textural surfaces. Water Resour. Res., 41, W10404, doi: 10.1029/2004WR003675.
Brown, D.G., J.R. Stencel and P.R. Jaffe, 2002. Effects of porous media preparation on bacteria transport through laboratory columns. Water Res. 36 (2002), 105-114.
Buol, S.W., F.D. Hole and R.J. McCracken, 1989. Soil genesis and classification. Iowa State University Press. ISBN 0-8138-1462-6.
Butler, R.G., G.T. Orlob, and P.H. McGauhey, 1954. Underground movement of bacterial and chemical pollutants. J. Am. Wat. Works Assoc. (46), p. 97-111.
Caldwell, E.L., 1937. Pollution flow from pit latrines when an impervious stratum closely underlies the flow. J. Infect. Dis. (61), p. 270-288.
Caldwell, E.L. and L.W. Parr, 1937. Groundwater pollution and the bored hole latrine. Am. J. Pub. Health 23:467, p. 149-183.
Caldwell, B.A., C. Ye, R.P. Griffiths, C.L. Moyer and R.Y. Morita, 1989. Plasmid expression and maintenance during long-term starvation-survival of bacteria in well water. Appl. Environ. Microbiol. (55), p. 1860-1864.
Cameron, D.R. and A. Klute, 1977. Convective-dispersive solute transport with a combined equilibrium and kinetic adsorption model. Water Resour. Res. 13, 183-188.
Canter, L.W. and R.C. Knox, 1985. Septic tank system effects on groundwater quality. Lewis Publishers, Inc., Chelsea, Michigan USA. ISBN 0-87371-012-6.
Champ, D.R. and J. Schroeter, 1988. Bacterial transport in fractured rock – A field-scale tracer test at the Chalk River Nuclear Laboratories. Water Sci. Technol. 20, p. 81-87.
Chao, W.L. and R.L. Feng, 1990. Survival of genetically engineered *Escherichia coli* in natural soil and river water. J. Appl. Bacteriol. (68), p. 319-325.
Charalambous, A.N., 1982. Problems of groundwater development in the Sana'a Basin, Yemen Arab Republic. Proceedings of the *Symposium "Improvements of Methods of Long Term Prediction of Variations in Groundwater Resources and Regimes Due to Human Activity"*. IAHS Publication no. 136.
Chen, Y. ,C.W. Young, T.K. Jan and N. Rotaghi, 1974. Trace metals in wastewater effluent. Journal of the Water Pollution Control Federation, 1974, 46(12), 2663-2675.
Chi, F-H and G.L. Amy, 2004. Kinetic study on the sorption of dissolved natural organic matter onto different aquifer materials: the effect of hydrophobiciy and functional groups. J. Colloid Interface Sci., Vol. 274, p. 380-391.
Chu, Y., Y. Jin, M. Flury and M. V. Yates, 2001. Mechanisms of virus removal during transport in unsaturated porous media. Wat. Resour. Res. Vol. 37, No. 2, p. 253-263.
Close, M.E., L. Pang, M.J. Flintoft, and W. Sinton, 2006. Distance and flow effects on microsphere transport in a large gravel column. J. Environ. Qual., Vol 35, p. 1204 – 1212.

Corapcioglu, M. Y. and A. Haridas, 1984. Transport and fate of microorganisms in porous media: a theoretical investigation. J. Hydrol. 72, 149-169.
Corapcioglu, M. Y. and A. Haridas, 1985. Microbial transport in soils and groundwater: a numerical model. Adv. Water Resour. 8, 188-200.
Cornell, R.M. and U. Schwertmann, 1992. The iron oxides. Structure, properties, reactions, occurrence and uses. ISBN 3-527-28576-8. Weinheim: VCH.
Crane, S.R. and J. Moore, 1984. Bacterial pollution of groundwater: a review. Water, Air, Soil Pollution (22), p. 67-83.
Daniel C. and F.S. Wood, 1971. Fitting equations to data. Wiley-Interscience, New York, 1971.
Daubner, I., 1975. Changes in the Properties of *E. coli* under the influence of water environments. Internationale Vereinigung fur Limnologie Verhandlungen, Vol. 19, p. 2650-2657.
Davis, J.A., 1982. Adsorption of natural dissolved organic matter at the oxide/water interface. Geochim. Cosmochim. Acta, 46, p. 2381-2393.
De Kerchove, A.J. and M. Elimelech, 2005. Relevance of electrokinetic theory for 'soft' particles to bacterial cells: implications for bacterial adhesion. Langmuir, Vol. 21, p. 6462-6472
De Marsily, G., 1986. Quantitative Hydrology, Academic Press, San Diego. ISBN 0-12-208916-2.
Dizer, H., A. Nasser, and J.M. Lopez, 1984. Penetration of different human pathogenic viruses into sand columns percolated with distilled water, groundwater or wastewater. Appl. Environ. Microbiol.. Vol. 47, p. 409-415.
Dong, H., 2002. Significance of electrophoretic mobility distribution to bacterial transport in granular porous media. J. Microbiol. Methods 51, p. 83-93.
Dong, H., T.C. Onstott, M. F. Deflaun, M.E. Fuller, T.D. Scheibe, S.H. Streger, R.K. Rothmel, and B.J. Mailloux, 2002. Relative dominance of physical versus chemical effects on the transport of adhesion-deficient bacteria in intact cores from South Oyster, Virginia. Environ. Sci. Technol., 2002, 36, 891-900.
Driscoll, F.G., 1986. Groundwater and wells, 2nd edition. Published by Johnson Division, St. Paul, Minnesota 55112. ISBN 0-9616456-0-1.
Dullien, F.A.L., 1991. Porous media. Fluid transport and pore structure. Second Edition. ISBN 0-12-223651-3. Academic Press, Inc. San Diego, California 92101.
Dyer, B.R. and T.R. Bhaskaran, 1943. Investigation of groundwater pollution, Part I: Determination of the direction and the velocity of flow of groundwater. Indian Journal of Medical Research (31), p. 231-243.
Dyer, B.R. and T.R. Bhaskaran, 1945. Investigation of groundwater pollution, Part II: Soil characteristics in West-Bengal, India, at the site of groundwater pollution investigations. Indian Journal of Medical Research (33), p. 17-22.
Dyer, B.R. and T.R. Bhaskaran, 1945. Investigation of groundwater pollution, Part III: Groundwater pollution in West Bengal, India. Indian Journal of Medical Research (33), p. 23-62.
Eiswirth, M. and H. Hötzl. The impact of leaking sewers on urban groundwater. In: Chilton, J. (Ed.), Groundwater in the Urban Environment. Proc. of the XXVII IAH Congress on Groundwater in the Urban Environment, Nottingham, UK, 21-27 Sept., Vol. 1, pp. 399-404. . Balkema, Rotterdam, The Netherlands, ISBN 9054108371.
Elimelech, M and C.R. O'Melia, 1990. Effect of particle size and collision efficiency in the deposition of Brownian particles with electrostatic energy barriers. Langmuir 1990, 6, p. 1153-1163.
Elimelech, M., M. Nagai, C-H. Ko, and J. Ryan, 2000. Relative insignificance of mineral zeta potential to colloid transport in geochemically heterogeneous porous media. Environ. Sci. Technol. 2000, Vol. 34, p. 2143-2148.
Feigin A., I. Ravina and J. Shalhevet, 1991. Irrigation with treated sewage effluent: management for environmental protection. ISBN 3-540-50804-X. Springer Verlag Berlin Heidelberg New York.
Feigin A., I. Ravina and J. Shalhevet, 1991. Irrigation with treated sewage effluent: management for environmental protection. ISBN 3-540-50804-X. Springer Verlag Berlin Heidelberg New York.
Filip, Z., H. Dizer, D. Kaddu-Mulinda, M. Kiper, J.M. Lopez-Pila, G. Milde, A. Nasser, K. Seidel, 1986. Untersuchungen über das Verhalten pathogener und anderer Mikroorganismen und Viren im Grundwasser im Hinblick auf die Bemessung von Wasserschutzzonen. Institut für Wasser-, Boden- und Lufthygiene des Bundesgesundheitsamtes. Berlin, 1986 (WaBoLu-Hefte 3/1986).
Foppen, J.W.A., 1996. Availability of data and conceptualisation of the groundwater flow system. SAWAS Technical Report No. 5, Volume II. NWSA, Sana'a, Yemen and NITG-TNO. Delft, The Netherlands
Foppen, J.W.A., 1996. Evaluation of the effects of groundwater use on groundwater availability in the Sana'a Basin. SAWAS Technical Report No. 5, Vol. I-V. NWSA, Sana'a, Yemen and NITG TNO, Delft, The Netherlands.

Foppen, J.W.A., 2002. Impact of high-strength wastewater infiltration on groundwater quality and drinking water supply: the case of Sana'a. Yemen. J. Hydrol. 263 (2002), 198-216.

Foppen, J.W.A. and J. Griffioen, 1995. Contribution of groundwater outflow to the phosphate balance of ditches in a Dutch polder. Man's influence on fresh water ecosystems and water use. IAHS Publ. No. 230, 177-184.

Foppen, J.W.A. and J.F. Schijven, 2005. Transport of *Escherichia coli* in columns of geochemically heterogeneous sediment. Wat. Res. 39, p. 3082-3088.

Foppen, J.W.A and J.F. Schijven, 2006. Transport and survival of *Escherichia coli* and thermotolerant coliforms in aquifers under saturated conditions – a review. Wat. Res. Vol. 40, p. 401-426.

Foppen, J.W.A., A. Mporokoso and J.F. Schijven, 2005. Determining straining of *Escherichia coli* from breakthrough curves. J. Contam. Hydrol., Vol. 76, p. 191-210.

Foppen, J.W.A., M. Naaman, and J.F. Schijven, 2005a. Managing urban water under stress : the case of Sana'a, Yemen. WIT Transactions Ecol. Environ. (2005), p.101-110;

Foppen, J.W.A., M. Naaman, and J.F. Schijven, 2005b. Managing urban water under stress in Sana'a, Yemen. Arabian Journal for Science and Engineering, Vol 30, No. 2C, p. 69-83, 2005.

Foppen, J.W.A., S. Okletey, and J.F. Schijven, 2006. Effect of goethite and humic acid on the transport of bacteriophage PRD1 in columns of saturated sand. J. Contam. Hydrol. Vol. 76, p. 287-301.

Foppen, J.W.A., M. Van Herwerden, and J.F. Schijven, 2006a. Characteristics of straining of *Escherichia coli* in saturated porous media. In: Gimbel, R., N.J.D. Graham, and M. Robin Collins (Ed.). Recent progress in slow sand and alternative biofiltration processes, IWA publishing, p. 113-122, ISBN 9781843391203.

Foppen, J.W.A., M. van Herwerden, and J.F. Schijven, 2007. Transport of *Escherichia coli* in saturated porous media: dual mode deposition and intra-population heterogeneity. Wat. Res. DOI: 10.1016/j.watres.2006.12.041.

Foppen, J.W.A., M. van Herwerden, and J.F. Schijven, 2007a. Measuring and modeling straining of *Escherichia coli* in saturated porous media. J. Contam. Hydrol. DOI: 10.1016/j.jconhyd.2007.03.001.

Foppen, J.W.A., Y. Liem, and J.F. Schijven, 2007b. The effect of humic acid on the attachment of *Escherichia coli* in columns of goethite coated sand. Wat. Res. (subm.).

Foppen, J.W.A., M. van Herwerden, M. Kebtie, A. Noman, J.F. Schijven, P.J. Stuyfzand, and S. Uhlenbrook, 2007c. Transport of *Escherichia coli* and solutes during waste water infiltration in an urban alluvial aquifer. J. Contam. Hydrol. (in press).

Foster, S.S.D., 2000. Groundwater resources at the turn of the millennium: taking stock and looking forward. In: *Groundwater: past achievements and future challenges, Sillo et al. (eds.),* Balkema, Rotterdam, The Netherlands, ISBN 90 5809 159 7.

Franchi, A., and C. R. O'Melia, 2003. Effects of natual organic matter and solution chemistry on the deposition and reentrainment of colloids in porous media. Environ. Sci. Technol., Vol. 37, p. 1122-1129.

Fujita, Y., W-H Ding and M. Reinhard, 1996. Identification of wastewater dissolved organic carbon characteristics in reclaimed wastewater and recharged groundwater. Water Environ. Res., Vol. 68, p. 867-876.

Fujita, Y., W-H Ding and M. Reinhard, 1996. Identification of wastewater dissolved organic carbon characteristics in reclaimed wastewater and recharged groundwater. Water Environ. Res., Vol. 68, p. 867-876.

Gaboriaud, F. and J-J. Ehrhardt, 2003. Effects of different crystal faces on the surface charge of colloidal goethite (a-FeOOH) particles: an experimental and modeling study. Geochim. et Cosmochim. Acta, Vol. 67, No. 5, p. 967-983

Gelhar, L.W., A. Mantoglou, C. Welty, and K.R. Rehfeldt, 1985. A review of field-scale physical solute transport processes in saturated and unsaturated porous media. EPRI, Palo Alto, CA 94303.

Gerba, C.P., 1984. Applied and theoretical aspects of virus adsorption to surfaces. Adv. Appl. Microbiol., Vol. 30, p. 133-168.

Gilbert, P., D.J. Evans, E. Evans, I.G. Duguid and M.R.W. Brown, 1991. Surface characteristics and adhesion of *Escherichia coli* and Staphylococcus epidermis. J. Appl. Bacteriol. (71), p. 72-77.

Gimbel, R.D. and H. Sontheimer, 1980. Einfluss der Oberflächenstruktur von Filtermaterialien auf die Partikelabscheidung in Tieffiltern. Vom Wasser 55 (1980), p. 131-147.

Goldshmid, J., D. Zohar, Y. Argamon and Y. Kott, 1973. Effects of dissolved salts on the filtration of coliform bacteria in sand dunes. In: Advances in water pollution research, edited by S.H. Jenkins. New York: Pergamon Press, p. 147-155.

Gonzalez, J.M., J. Iribirri, L. Egea and I. Barcina, 1992. Characterization of culturability, protistan grazing, and death of enteric bacteria in aquatic ecosystems. Appl. Environ. Microbiol. (58), p. 998-1004.

Gooddy, D.C., A.R. Lawrence, B.L. Morris and P.J. Chilton, 1997. Chemical transformation of groundwater beneath unsewered cities. In: Chilton, J. (Ed.), Groundwater in the Urban Environment. Proc. of the XXVII IAH Congress on Groundwater in the Urban Environment, Nottingham, UK, 21-27 Sept., Vol. 1, pp. 405-410. Balkema, Rotterdam, The Netherlands, ISBN 9054108371.

Graton, L.C. and H.J. Fraser, 1935. Systematic packing of spheres with particular relation to porosity and permeability. J. Geol. 8(43), pp. 785-909.

Greskowiak, J., H. Prommer, G. Massmann, C.D. Johnston, C.D., G. Nutzmann, and A. Pekdeger, 2005. The impact of variably saturated conditions on hydrogeochemical changes during artificial recharge of groundwater. Applied Geochemistry (20), Issue 7, Jul 2005, p. 1409-1426.

Griffioen, J., 1994. Uptake of phosphate by iron hydroxide during seepage in relation to development of groundwater composition in coastal areas. Environ. Sci. Technol. 28, 675-681.

Gu, B., J. Schmitt, Z. Chen, L. Liang, and J.F. McCarthy, 1994. Adsorption and desorption of natural organic matter in iron oxide: mechanisms and models. Environ. Sci. Technol., Vol. 28, p. 38-46.

Gu, B., T.L. Mehlhorn, L. Liang, and J.F. McCarthy, 1996. Competitive adsorption, displacement, and transport of organic matter on iron oxide: II Displacement and transport. Geochim. et Cosmochim. Acta., Vol. 60, No. 16, p. 2977-2992.

Hagedorn, C., E.L. McCoy and T.M. Rahe, 1981. The potential for ground water contamination from septic effluents. J. Environ. Qual. (10), p. 1-8.

Hagedorn. C., D.T. Hansen and G.H. Simonson, 1978. Survival and movement of fecal indicator bacteria in soil under conditions of saturated flow, J. Environ. Qual. (7), p. 55-59.

Hahn, M.W., and C.R. O'Melia, 2004. Deposition and reentrainment of Brownian particles in porous media under unfavorable chemical conditions: some concepts and applications. Environ. Sci. Technol., Vol. 38, p. 210-220

Hall, W.A., 1957. An analysis of sand filtration. J. Sanitary Engineering Division: Proceedings of the Am. Soc. Civ. Eng. (83) SA 3, p. 1276/1-1276/9.

Hamad, T.M.H., 1993. Sewage effluent used for irrigation and its impact on soil environment in some developing African countries. Ph.D. thesis. University of Cairo, Institute of African Research and Studies, Department of Natural Resources.

Hamdi, M., 2000. Competition for scarce groundwater in the Sana'a Plain, Yemen. A study on the incentive systems for urban and agricultural water use. IHE PhD thesis. Balkema Publishers. ISBN 90 5410 426 0.

Harvey, R.W. and J.N. Ryan, 2004. MiniReview: Use of PRD1 bacteriophage in groundwater viral transport, inactivation, and attachment studies. FEMS Microbiol. Ecol., Vol. 49, p. 3-16.

Harvey, R.W. and S.P. Garabedian, 1991. Use of colloid filtration theory in modeling movement of bacteria through a contaminated sandy aquifer. Environ. Sci. Technol. Vol. 25, p. 178-185.

Harvey, R.W., 1997. Microorganisms as tracers in groundwater injection and recovery experiments: a review. FEMS Microbiol Rev. (20), p. 461-472.

Havemeister, G., and R. Riemer, 1985. Laborversuche zum Transport-verhalten von Bakterien. Umweltbundesambt Materialien, 2/85, p. 27-32.

Hendry, M.J., J.R. Lawrence and P. Maloszewski, 1997. The role of sorption in the transport of Klebsiella oxytoca through saturated silica sand. Ground Water Vol. 35, No. 4, p. 574-584.

Hendry, M.J., J.R. Lawrence and P. Maloszewski, 1999. Effects of velocity on the transport of two bacteria through saturated sand. Ground Water Vol. 37, No. 1, p. 103-112.

Herzig, J.P., D.M. Leclerc, and P. Le Golf, 1970. Flow of suspensions through porous media – Application to deep infiltration. In: Flow Through Porous Media, Am. Chem. Soc., Washington, D.C., pp. 129-157.

Hieltjes, A.H.M. and A. Breeuwsma, 1983. Chemische bodemonderzoekmethoden voor bodemkenmerken en anorganische stoffen. Bodembescherming 21, Staatsuitgeverij, 384p.

Hijnen, W.A.M., J.F. Schijven, P. Bonne, A. Visser, and G.J. Medema, 2004. Elimination of viruses, bacteria and protozoan oocysts by slow sand filtration. Wat. Sci. Technol., Vol. 50, No. 1, p. 147-154.

Hirtzel, C.S. and R. Rajagopalan, 1985. Colloidal phenomena – advanced topics. Noyes Publications, Park Ridge, N.J., USA.

Huisman, D.J., and P. Kiden, 1997. A geochemical record of Late Cenozoic sedimentation history in the Southern Netherlands. Geologie en Mijnbouw, Vol. 76, No. 4, p. 277-291.

Hull, S., 1997. *Escherichia coli* lipopolysaccharide in pathogenesis and virulence, p. 145-167. In: M. Sussman (ed.), *Escherichia coli*: mechanisms of virulence. Cambridge University Press, Cambridge, United Kingdom.

Ilani, T., E. Schulz, and B. Chefetz, 2005. Interactions of organic compounds with wastewater dissolved organic matter: role of hydrophobic fractions. J. Environ, Qual., Vol. 34, p. 552-562.

Imai, A., T. Fukushima, K. Matsuhige, Y-H. Kim, and K. Choi, 2002. Characterization of dissolved organic matter in effluents from wastewater treatment plants. Wat. Res., Vol. 36, p. 859-870.

Iriberri, J., I. Azua, A. Labirua-Iturburu, I. Artolozaga and I. Barcina, 1994. Differential elimination of enteric bacteria by protists in a freshwater system. J. Appl. Microbiol. (77), p. 476-483.

ISO (international Organization for Standrdization), 1995. Water Quality – Detection and enumeration of F-specific RNA bacteriophages – Part 1: Method by incubation with a host strain., ISO 10705-1, Geneva.

Italconsult, 1973. Watersupply for Sana'a and Hodeidah. Sana'a Basin Groundwater Studies, Volume 1, 2 and 3. UNDP Project YEM 507 and WHO Project Yemen 3202, Rome.

Iwasaki, T., 1937. Some notes on sand filtration. J. Am. Wat. Works Ass., Vol. 29, No. 10, p. 1591-1602

Jacks. G., F. Sefe, M. Carling, M. Hammar and P. Letsamao, 1999. Tentative nitrogen budget for pit latrines – eastern Botswana. Environmental Geology 38 (3), p. 199-203.

James, A.M., 1957. The electrochemistry of the bacterial surface. Progress in biophysics and biophysical chemistry, Vol. 8, p. 95-142.

JinY, Y. Chu and Y. Li, 2000. Virus removal and transport in saturated and unsaturated sand columns. J. Contam. Hydrol., 43, p. 111-128

Johnson, P.R. and M. Elimelech, 1995. Dynamics of colloid deposition in porous media: blocking based on random sequential adsorption. Langmuir 1995, 11, 802-812.

Johnson, P.R., N. Sun and M. Elimelech, 1996. Colloid transport in geochemically heterogeneous porous media: modelling and measurements. Environ. Sci. Technol. 1996, 30, 3284-3293.

Johnson, W.P., and X. Li, 2005. Comment on breakdown of colloid filtration theory: role of secondary energy minimum and surface charge heterogeneities. Langmuir 2005, Vol. 21, p. 10895-10895.

Johnson, W.P., K.A. Blue, B.E. Logan, and R.G. Arnold, 1995. Modeling bacterial detachment during transport through porous media as a residence-time-dependent process. Wat. Resour. Res., Vol. 31, No. 11, p. 2649-2658.

Jungfer, E.V., 1987. Zur Frage der Grundwasserneubilding in Trockengebieten. Fallstudien aus der Arabischen Republik Jemen und dem Königreich Marokko. Erlanger Geographische Arbeiten. Sonderband 18. Selbstverlag der Fränkischen Geografischen Gesellschaft.

Kaiser, K., 2003. Sorption of natural organic matter fractions to goethite (α-FeOOH): the effect of chemical composition as revealed by liquid state 13C NMR and wet-chemical analysis. Organic Geochemistry 34, p. 1569-1579.

Kaper, J.B., J.P. Nataro and H.L.T. Mobley, 2004. Pathogenic *Escherichia coli*. Nature Reviews Microbiology, Vol. 2, p. 123-140.

Karim, M.R., F.D. Manshadi, M.M. Karpiscak and C.P. Gerba, 2004b. The persistence and removal of enteric pathogens in constructed wetlands. Wat. Res. (38), p. 1831-1837.

Karim, M.R., M. W. LeChevallier, M. Abbaszadegan, A. Alum, J. Sobrinho, and J. Rosen, 2004a. Field testing of USEPA Methods 1601 and 1602 for coliphage in groundwater. AWWA Research Foundation. ISBN 1-58321-348-1.

Kastowksy, M., A. Sabisch, T. Gutberlet and H. Bradaczek, 1991. Molecular modelling of bacterial deep rough mutant lipopolysaccharide of *Escherichia coli*. Eur. J. Biochem. 197, p. 707-716.

Kastowksy, M., T. Gutberlet, H. Bradaczek, 1993. Comparison of X-ray powder diffraction data of various bacterial lipopolysaccharide structures with theoretical model conformations. Eur. J. Biochem. 217, p. 771-779.

Keller, A.A., and M. Auset, 2006. A review of visualization techniques of biocolloid transport processesat the pore scale under saturated and unsaturated conditions. Adv. Water Resour., doi 10.1016/j.advwatres.2006.05.013.

Kerr, M., M. Fitzgerald, J.J. Sheridan, D.A. McDowell and I.S. Blair, 1999. Survival of *Escherichia coli* O157:H7 in bottled natural mineral water. J. Appl. Microbiol., Vol. 87, p. 833-841.

Keswick, B.H,. C.P. Gerba, S.L. Secor and I. Cech, 1982. Survival of enteric viruses and indicator bacteria in groundwater. J. Environ. Sci. Health, A17(6), p. 903-912.

Kim, S. and M. Y. Corapcioglu, 1996. A kinetic approach to modeling mobile bacteria-facilitated groundwater contaminant transport. Wat. Resour. Res., Vol. 32, No. 2, p. 321-331.

Kjelleberg, S. and M. Hermansson, 1984. Starvation-induced effects on bacterial surface characteristics. Appl. Environ. Microbiol. (48), p. 497-503.

Klimas, A., 1997. Impact of urbanization on shallow groundwater in Lithuania. In: Chilton, J. (Ed.), Groundwater in the Urban Environment. Proc. of the XXVII IAH Congress on Groundwater in the Urban Environment, Nottingham, UK, 21-27 Sept., Vol. 1, pp. 463-468. Balkema, Rotterdam, The Netherlands, ISBN 9054108371.

Ko, C.H., S. Bhattacharjee, and M. Elimelech, 2000. Coupled influence of colloidal and hydrodynamic interactions on the RSA blocking function for particle deposition onto packed spherical collectors. J. Colloid INterface Sci. 229, p. 554-567.

Korhonen, L. K., and P.J. Martikainen, 1991. Survival of *Escherichia coli* and Campylobacter jejuni in untreated and filtered lake water. J. Appl. Bacteriol. (71), p. 379-382.

Kostka, J.L. and G.W. Luther III, 1994. Partitioning and speciation of solid phase iron in saltmarsh sediments. Geochim. Cosmochim. Acta 58, 1701-1710.

Kudryavtseva, B.M., 1972. An experimental approach to the establishment of zones of hygienic protection of underground water sources on the basis of sanitary-bacteriological indices. J. Hyg. Epidemiol. Microbiol. Immun. (16), p. 503-511.

Lee, J.W., W.G. Shim, J.Y. Ko, and H. Moon, 2004. Adsorption equilibria, kinetics, and column dynamics of chlorophenols on a non-ionic polymeric sorbent, XAD-1600. Separation Sci. Technol., Vol. 39, No. 9, p. 2041-2065.

Lewis, W.J., Foster, S.S.D. and Drasar, B.S., 1980. The risk of groundwater pollution by on-site sanitation in developing countries. IRCWD Report No. 01/82. International Reference Centre fr Waste Disposal (IRCWD), Duebendorf, Switzerland.

Li, X., T.D. Scheibe, and W.P. Johnson, 2004. Apparent decreases in colloid deposition rate coefficients with distance of transport under unfavorable deposition conditions: a general phenomenon. Environ. Sci. Technol., Vol. 38, p. 5616-5625.

Lindqvist, R. and G. Bengtsson, 1991. Dispersal dynamics of groundwater bacteria. Microb. Ecol. Vol. 21, p. 49-72.

Lindqvist, R., J.S. Cho and C.G. Enfield, 1994. A kinetic model for cell density dependent bacterial transport in porous media. Wat. Res. Research., Vol. 30, No. 12, p. 3291-3299.

Liu, D., P.R. Johnson and M. Elimelech, 1995. Colloid deposition dynamics in flow through porous media: role of electrolyte concentration. Environ. Sci. Technol. 1995, 29, 2963-2973.

Logan, B.E., D.G. Jewett, R.G. Arnold, E.J. Bouwer, and C.R. O'Melia, 1995. Clarification of clean-bed filtration models. J. Environ. Eng. Dec. 1995, p. 869-873.

Loveland, J.P., J.N. Ryan, G.L. Amy, and R.W. Harvey, 1996. The reversibility of virus attachment to mineral surfaces. Colloids and Surfaces A: Physicochemical and Engineering Aspects, Vol. 107, p. 205-221.

Ma, H., H.E. Allen, and Y. Yin, 2001. Characterization of isolated fractions of dissolved organic matter from natural waters and a wastewater effluent. Wat. Res., Vol. 35, p. 985-996.

Macler, B.A. and J.C. Merkle, 2000. Current knowledge on groundwater microbial pathogens and their control. Hydrogeol. J. (2000), vol. 8, no. 1, pp. 29-40.

Mann, J.F., 1976. Wastewater in the vadose zone of arid regions: Hydrologic interactions. Ground Water, 1976, 14(6), 367-373.

Martin, R.E., E.J. Bouwer, and L.M. Hanna, 1992. Application of clean-bed filtration theory to bacteria deposition in porous media. Environ. Sci. Technol. 26(5), p. 1053-1058.

Matthess, G. and A. Pekdeger, 1981. Concepts of a survival and transport model of pathogenic bacteria and viruses in groundwater. In: Quality of Groundwater, Proceedings of an international symposium, edited by Van Duijvenbooden, W., P. Glasbergen and H. van Lelyveld, p. 427-437.

Matthess, G. and A.Pekdeger, 1988. Survival and transport of pathogenic bacteria and viruses in groundwater. In Groundwater Quality, edited by C.H. Ward, W. Giger, and P.L. McCarty, pp. 472-482, John Wiley, New York, 1985.

Matthess, G., 1982. The properties of groundwater. ISBN 0-471-08513-8, John Wiley & Sons, Inc., New York.

Matthess, G., A. Pekdeger and J. Schroeter, 1988. Persistence and transport of bacteria and viruses in groundwater – a conceptual evaluation. J. Cont. Hydrol. (2) 1988, p. 171-188.

Matthess, G., E. Bedbur, K.O. Gundermann, M. Loof and D. Peters, 1991a. Vergleichende Untersuchung zum Filtrationsverhalten von Bakterien und organischen Partikeln in Porengrundwasserleitern I. Grundlagen und Methoden. Zentralblatt für Hygiene und Umweltmedizin 191, p. 53-97 (1991). Gustav Fischer Verlag Stuttgart/New York.

Matthess, G., E. Bedbur, K.O. Gundermann, M. Loof and D. Peters, 1991b. Vergleichende Untersuchung zum Filtrationsverhalten von Bakterien und organischen Partikeln in Porengrundwasserleitern II. Hydraulische, hydrochemische unde sedimentologische Systemeigenschaften, die den Filterfaktor steuern. Zentralblatt für Hygiene und Umweltmedizin 191, p. 347-395 (1991). Gustav Fischer Verlag Stuttgart/New York.

Matthess, G., S.S.D. Foster and A. Ch. Skinner, 1985. Theoretical background, hydrogeology and practice of groundwater protection zones. IAH International Contributions to Hydrogeology, Vol. 6, 1985.

McCaulou, D., R.C. Bales and R.G. Arnold, 1995. Effect of temperature-controlled motility on transport of bacteria and microspheres through saturated sediment. Wat. Res. Research Vol. 31, No. 2, p. 271-280.

McCoy, E.L. and C. Hagedorn, 1979. Quantitatively tracing bacterial transport in saturated soil systems. Water, Air, and Soil Poll. 11 (1979), p. 467-479.

McCoy, E.L. and C. Hagedorn, 1980. Transport of resistance-labeled *Escherichia coli* strains through a transition between two soils in a topographic sequence. J. Environ. Qual. (9), p. 686-691.
McDowell-Boyer, L.M., J.R. Hunt and N. Sitar, 1986. Particle transport through porous media. Water Resour. Res. 22, 1901-1921.
McFeters, G.A., and D.G. Stuart, 1972. Survival of coliform bacteria in natural waters: field and laboratory studies with membrane-filter chambers. Appl. Microbiol., Vol. 24, No. 5, p. 805-811.
McFeters, G.A., G.K. Bissonnettee, J.J. Jezeski, C.A. Thomson and D.G. Stuart, 1974. Comparative survival of indicator bacteria and enteric pathogens in well water. Appl. Microbiol., (270), p. 823-829.
Medema, G.J., P. Payment, A. Dufour, W. Robertson, M. Waite, P. Hunter, R. Kirby and Y. Andersson, 2003. Safe drinking water: an ongoing challenge. In: *Assessing microbial safety of drinking water: improving approaches and methods, Dufour et al. (eds.).* World Health Organization, 2003. ISBN 92 4 154630.
Meeussen, J.C.L., A. Scheidegger, T. Hiemstra, W.H. van Riemsdijk, and M. Borkovec, 1996. Predicting multicomponent adsorption and transport of fluoride at variable pH in a goetite-silica sand system. Environ. Sci. Technol., Vol. 30, p. 481-488.
Merkli, B., 1975. Untersuchungen ueber Mechanismen und Kinetik der Elimination von Bakterien und Viren im Grundwasser. Diss. ETH Zurich 5420.
Moreira, L., P. Agostinho, P. Vasconcellos Morais and M.S. da Costa, 1994. Survival of allochthonous bacteria in still mineral water bottled in polyvinyl chloride (PVC) and glass. J. Appl. Bacteriol. (77), p. 334-339.
Mosgiprovodkhoz, 1986. Sana'a Basin Water Resources Scheme. Vol. 3: Soils. Ministry of Agriculture and Fisheries, Yemen Arab Republic.
Murphy, E.M. and T.R. Ginn, 2000. Modelling microbial processes in porous media. Hydrogeol. J. 8, no. 1, 142-158.
Murphy, E.M., J. M. Zachara, and S. C. Smith, 1990. Influence of mineral-bound humic substances on the sorption of hydrophobic organic compounds. Environ. Sci. Technol., Vol. 24, p. 1507-1516.
Nasser, A.M., Y. Tchorch and B. Fattal, 1993. Comparative survival of *E. coli* F+bacteriophages, HAV and poliovirus in wastewater and groundwater. Wat. Sci. Tech. (27), p. 401-407.
Neihof, R., 1969. Microelectrophoresis apparatus employing palladium electrodes. J. Coll. Interface Sci., Vol. 30, No. 1, p. 128-133.
Neumann, B., 1983. Untersuchungen zur Elekrophorese als Transportmechanismus bei der Tiefinfiltration. Diss. Universität Fridericana Karlsruhe, Fakultät für Chemieingenieurwesen.
Noda, Y. and Y. Kanemasa, 1986. Determination of hydrophobicity on bacterial surfaces by nonionic surfactants. J. Bact., Vol 167, No. 3, p. 1016-1019.
Pang. L., M. Close, M. Goltz, L. Sinton, H. Davies, C. Hall and G. Stanton, 2003. Estimation of septic tank setback distances based on transport of *E. coli* and F-RNA phages. Environ. Int. (29), p. 907-921.
Parfitt, R. L., A. R. Fraser, and V.C. Farmer, 1977. Adsorption on hydrous oxides. III. Fulvic acid and humic acid on goethite, gibbsite and imogolite. J. Soil Sci., 28, 289-296.
Parkhurst, D. L. and C.A.J. Appelo, 1999. User's guide to PHREEQC (version 2)- a computer program for speciation, batch-reaction, one-dimensional transport, and inverse geochemical calculations. Water-Resources Investigations Report 99-4259. U.S. Geological Survey, U.S. Department of the Interior. Denver, Colorado, 1999.
Payment, P., M. Waite and A. Dufour, 2003. Introducing parameters for the assessment of drinking water quality. In: *Assessing microbial safety of drinking water: improving approaches and methods, Dufour et al. (eds.).* World Health Organization, 2003. ISBN 92 4 154630.
Peters, D., 1989. Der Einfluss hydraulischer und sedimentologischer Eigenschaften gleich förmiger Sande auf die Filtration ausgewählter Bakterienarten. Berichte-Reports, Geol. Pal. Inst. Univ. Kiel 29 (1989), p. 1-109.
Pieper, A.P., J.N. Ryan, R.W. Harvey, G.L. Amy, T.H. Illangasekare, and D.W. Metge, 1997. Transport and recovery of bacteriophage PRD1 in a sand and gravel aquifer: effect of sewage derived organic matter. Environ. Sci. Technol., Vol. 31, p. 1163-1170.
Powell, K.L., R.G. Taylor, A.A. Cronin, M.H. Barrett, S. Pedley, J. Sellwood, S.A. Trowsdale and D.N. Lerner, 2003. Microbial contamination of two urban sandstone aquifers in the UK. Water Res. 37 (2003), pp. 339-352.
Powelson, D.K., and A.L. Mills, 2001. Transport of *Escherichia coli* in sand columns with constant and changing water contents. J. Environ. Qual. (30), p. 238-245.
Quast, K.W., K. Lansey. R. Arnold, R. Bassett, and M. Rincon, 2006. Boron isotopes as an artificial tracer. Ground Water, Vol. 44(3), p. 453 - 466.

Rabenhorst, M.C. 1988. Determination of organic and carbonate carbon in calcareous soils using dry combustion. Soil Sci. Soc. Am. J. 52, 965-969.

Rahe, T.M., C. Hagedorn, E.L. McCoy and G.F. Kling, 1978. Transport of antibiotic-resistant *Escherichia coli* through western Oregon hillslope soils under conditions of saturated flow. J. Envrion. Qual. (7), p. 487-494.

Rajagopalan, R. and C. Tien, 1976. Trajectory analysis of deep-bed filtration with the sphere-in-a-cell porous media model. AIChE J. 22(3), p. 523-533.

Ramalho, R., A. Afonso, J. Cunha, P. Teizeira and P.A. Gibbs, 2001. Survival characteristics of pathogens inoculated into bottled mineral water. Food Control (12), p. 311-316.

Raveendran, P. and A. Amirtharajah, 1995. Role of short-range forces in particle detachment during filter backwashing. J. Environ. Eng. ,Vol. 121, No. 12, p. 860-868.

Reddy, K.R., R. Khaleel and M.R. Overcash, 1981. Behavior and transport of microbial pathogens and indicator organisms in soils treated with organic wastes. J. Environ. Qual. (10), p. 255-266.

Redman, J.A., S.B. Grant, T.M. Olson, And M.K. Estes, 2001. Pathogen filtration, heterogeneity, and the potable resue of wastewater. Environ. Sci. Technol., Vol. 35, p. 1798-1805.

Redman, J.A., S.L. Walker, and M. Elimelech, 2004. Bacterial adhesion and transport in porous media: role of the secondary energy minimum. Environ. Sci. Technol., Vol. 38, p. 1777-1785.

Reneau, R.B. and D.E. Pettry, 1975. Movement of coliform bacteria from septic tank effluent through selected coastal plain soils of Virginia. J. Environ. Qual. (4), p. 41-44.

Reneau, R.B., 1978. Influence of artificial drainage on penetration of coliform bacteria from septic tank effluents into wet tile drained soils. J. Environ, Qual. (7), p. 23-30.

Rice, E.W., C.H. Johnson, D.K. Wild and D.J. Reasoner, 1992. Survival of *Escherichia coli* O157:H7 in drinking water associated with a waterborne disease outbreak of hemorrhagic colitis. Ltt. Appl. Microbiol. (15), p. 38-40.

Rijnaarts, H.M, W. Norde, E.J. Bouwer, J. Lykklema and A.J.B. Zehnder, 1996. Bacterial deposition in porous media related to the clean bed collision efficiency and to substratum blocking by attached cells. Environ. Sci. Technol. 30 (10), p. 2869-2876.

Romero, J.C., 1970. The movement of bacteria and viruses through porous media. Ground Water (8), p. 37-48.

Runnells, D.D., 1976. Wastewater in the vadose zone of arid regions: Geochemical interactions. Ground Water, 1976, 14(6), 374-385.

Ryan, J. and M. Elimelech, 1996. Review: Colloid mobilization and transport in groundwater. Colloids and Surfaces A: Physicochemical and Engineering Aspects 107 (1996), p. 1-56.

Ryan, J.N., M. Elimelech, R.A. Ard, R.W. Harvey, and P.R. Johnson, 1999. Bacteriophage PRD1 and silica colloid transport and recovery in an iron oxide-coated sand aquifer. Environ. Sci. Technol., Vol. 33, p. 63-73.

Ryan, J.N., R.W. Harvey, D. Metge, M. Elimelech, T. Navigato, and A.P. Pieper, 2002. Field and laboratory investigations of inactivation of viruses (PRD1 and MS2) attached to iron oxide-coated quartz sand. Environ. Sci. Technol., Vol. 36, p. 2403-2413.

Sawyer, C.N., P.L. McCarthy, and G.F. Parkin, 1994. Chemistry for Environmental Engineering, 4th edition. McGraw-Hill, New York. ISBN 0-07-113908-7.

Schaub, S.A. and C.A. Sorber, 1977. Virus and bacteria removal from wastewater by rapid infiltration through the soil. Appl. Environ. Microbiol. (33), p. 609-619.

Scheffer, F. and P. Schachtschabel 2002. Lehrbuch der Bodenkunde, 15th ed., Elsevier, Amsterdam, 607 p.

Schijven, J.F. 2001. Virus removal from groundwater by soil passage. Modelling field and laboratory experiments. PhD Thesis. ISBN 90-646-4046-7. Posen and Looijen, Wageningen, The Netherlands.

Schijven, J.F. and S.M. Hassanizadeh, 2002. Virus removal by soil passage at field scale and groundwater protection of sandy aquifers. Water Sci. Technol., Vol. 46, p. 123-129.

Schijven, J.F., W. Hoogboezem, S.M. Hassanizadeh, and J.H. Peters, 1999. Modelling removal of bacteriophages MS2 and PRD1 by dune infiltration at Castricum, the Netherlands. Wat. Resour. Res., Vol. 35, p. 1101-1111.

Schijven, J.F., G. Medema, A.J. Vogelaar, and S.M. Hassanizadeh, 2000. Removal of microorganisms by deep well injection. J. Contam. Hydrol., Vol. 44, p. 301-327.

Schijven, J.F., S.M. Hassanizadeh and H.A.M. de Bruin, 2002. Two-site kinetic modeling of bacteriophages transport through columns of saturated dune sand. J. Contam. Hydrol., 2002, 57, p. 259-279.

Schijven, J.F., J.H.C. Mülschlegel, S.M. Hassanizadeh, P.F.M. Teunis, and A.M. de Roda Husman, 2006. Determination of protection zones for Dutch groundwater wells against virus contamination – Uncertainty and sensitivity analysis. J. Water Health (in press).

Scholl, M.A. and R.W. Harvey, 1992. Laboratory investigations on the role of sediment surface and groundwater chemistry in transport of bacteria through a contaminated sandy aquifer. Environ. Sci. Technol., Vol. 26, No. 7, p. 1410-1417.

Scholl, M.A., A.L. Mills, J.S. Herman and G. M. Hornberger, 1990. The influence of mineralogy and solution chemistry on the attachment of bacteria to representative aquifer minerals. J. Contam. Hydrol., 6, p. 321-336.
Schwertmann, U. and R.M. Cornell, 1991. Iron oxides in the laboratory. Preparation and Characterization. VCH Verlagsgesellschaft Weinheim, Germany. ISBN 3-527-26991-6 (VCH, Weinheim).
Shaaban, M.A., 1980. A geoelectrical study of the Sana'a groundwater basin, Yemen Arab Republic. Geol. Mijnbouw 59, p. 79-86.
Sharma, M.M., Y.I. Chang and T.F. Yen, 1985. Reversible and irreversible surface charge modification of bacteria for facilitating transport through porous media. Colloids Surf. (16), p. 193-206.
Simoni, S.F., H. Harms, T.N.P. Bosma, and A.J.B. Zehnder, 1998. Population heterogeneity affects transport of bacteria through sand columns at low flow rates. Environ. Sci. Technol., Vol. 32, p. 2100-2105.
Simunek, J., K. Huang, M. Sejna, and M.T. van Genuchten, 1998. The HYDRUS-1D software package for simulating the one-dimensional movement of water, heat, and multiple solutes in variably-saturated media – version 2.0, IGWMC-TPS-70, International Ground Water Modeling Center, Colorado School of Mines, Golden, Colorado, 202 pp.
Šimůnek, J., N.J. Jarvis, M.Th. van Genuchten and A. Gärdenäs, 2003. Review and comparison of models for describing non-equilibrium and preferential flow and transport in the vadose zone. J. Hydrol., Vol. 272, p. 14-35.
Sinnatamby, G., Mara, D.D., and McGarry, M., 1986. Sewerage: shallow systems offer hope to slums. World Water, 9, p. 39-41.
Sinton, L.W., 2001. Microbial contamination of New Zealand's aquifers. In: Groundwaters of New Zealand, edited by M.R. Rosen and P.A. White. New Zealand Hydrological Society Inc. Wellington, p. 221-251.
Sinton, L.W., M.J. Noonan, R.K. Finaly, L. Pang and M.E. Close, 2000. Transport and attenuation of bacteria and bacteriophages in an alluvial gravel aquifer. New Zealand Journal of Marine and Freshwater Research (34), p. 175-186.
Sinton, L.W., R.K. Finlay, L. Pang and D.M. Scott, 1997. Transport of bacteria and bacteriophages in irrigated effluent into and through an alluvial gravel aquifer. Water, Air and Soil Pollution 98, pp. 17-42, 1997.
Sjogren, R.E., 1994. Prolonged survival of an environmental *Escherichia coli* in laboratory soil microcosms. Water Air Soil Pollut. (75), p. 389-403.
Smith, M.S., G.W. Thomas, R.E. White and D. Ritonga, 1985. Transport of *Escherichia coli* through intact and disturbed soil columns. J. Environ. Qual. (14), p. 87-91.
Sobsey, M.D., C.H. Dean, M.E. Knuckles, and R.A. Wagner, 1980. Interaction and survival of enteric viruses in soil material. Appl. Environ. Microbiol., Vol. 40, p. 92-101.
Somasundaram, M.V., Ravindran, G. and J.H. Tellam, 1993. Groundwater pollution of the Madras urban aquifer, India. Groundwater, 31: 4-11.
Somasundaran, P. and G.E. Agar, 1967. The zero point of charge of calcite. J. Coll. Interface Sci 24, p. 433-440.
Song, L., P.R. Johnson and M. Elimelech, 1994. Kinetics of colloid deposition onto heterogeneously charged surfaces in porous media. Environ. Sci. Technol. 1994, 28, 1164-1171.
Sprent, J.I., 1987. The ecology of the nitrogen cycle. Cambridge studies in ecology. Cambridge University Press. ISBN 0 521 31052 0.
Stevik, T.K., G. Ausland, J.F. Hanssen, and P. D. Jenssen, 1999. The influence of physical and chemical factors on the transport of *E. coli* through biological filters for wastewater purification. Wat. Res. (33), p. 3701-3706.
Stevik, T.K., K. Aa, G. Ausland and J.F. Hanssen, 2004. Retention and removal of pathogenic bacteria in wastewater percolating through porous media: a review. Wat. Res. (38), p. 1355-1367.
Stumm, W. and J.J. Morgan, 1996. Aquatic Chemistry, 3rd edition. Wiley and Sons, New York.
Sun, N., M Elimelech, N-Z Sun and J.N. Ryan, 2001. A novel two-dimensional model for colloid transport in physically and geochemically heterogeneous porous media. J. Contam. Hydrol. 49, 173-199.
Tan, Y., J.T. Gannon, P. Baveye, and M. Alexander, 1994. Transport of bacteria in an aquifer sand: experiments and model simulations. Wat. Res. Research Vol. 30, No. 12, p. 3243-3252.
Tate, R.L., 1978. Cultural and environmental factors affecting the longevity of *E. coli* in histosoils. Appl. Environ. Microbiol. (35), p. 925-929.
Taylor, R.G., A.A. Cronin, S. Pedley, T.C. Atkinson, and J.A. Barker, 2004. The implications of groundwater velocity variations on microbial transport and wellhead protection: review of field evidence. FEMS Microbiology Ecology Vol. 49, p. 17-26.
Taylor, R.G., A.A. Cronin, and J. Rueedi, 2006. Groundwater flow velocities indicated by anthropogenic contaminants in urban sandstone aquifers. IAH Selected Papers on Hydrogeology, Vol. 8, p. 95-105.

Taylor, R.G., A.A. Cronin, D.N. Lerner, J.H. Tellam, S.H. Bottrell, J. Rueedi, and M.H. Barrett, 2006a. Hydrochemical evidence of the depth of penetration of anthropogenic recharge in sandstone aquifers underlying two mature cities in the UK. Applied Geochemistry 21 (2006), p. 1570-1592.
Taylor, S.W. and P.R. Jaffe, 1990. Biofilm growth and the related changes in the physical propoerties of a porous medium 2. Permeability. Water Resour. Res. 26, 2161-2169.
Thurman, E.M., 1985. Organic geochemistry of natural waters. ISBN 90-247-3143-7. Martinus Nijhoff/Dr. W. Junk Publishers, Dordrecht, The Netherlands.
Tipping, E. (1981). The adsorption of aquatic humic substances by iron oxides. Geochim. Cosmochim. Acta, 45, p. 191-199.
Tobiason, J.E. and C.R. O'Melia, 1988. Physicochemical aspects of particle removal in depth infiltration. J. AWWA ,80, p. 54-64.
Tong, M. and W.P. Johnson, 2006. Colloid population heterogeneity drives hyper-exponential deviation from classic filtration theory. Environ. Sci. Technol., in press.
Toride, N., F.J. Leij and M.Th. van Genuchten,1999. The CXTFIT code for estimating transport parameters from laboratory or field tracer experiments, version 2.1. Research Report No.137, U.S. Salinity Laboratory, U.S. Department of Agriculture, Riverside, California. Internet site: http;//www.ussl.ars.usda.gov/models/modelsmenu.htm.
Tufenkji, N. and Elimelech M., 2004. Correlation Equation for Predicting Single-Collector Efficiency in Physicochemical Filtration in Saturated Porous Media. Environ. Sci. Technol., Vol. 38, p. 529-536.
Tufenkji, N., and M. Elimelech, 2004a. Deviation from classical colloid filtration theory in the presence of repulsive DLVO interactions. Langmuir, Vol. 20, p. 10818-10828
Tufenkji, N., and M. Elimelech, 2005. Spatial distributions of Cryptosporidium oocysts in porous media: evidence for dual mode deposition. Environ. Sci. Technol. Vol. 39, p. 3620 – 3629
Tufenkji, N., and M. Elimelech, 2005a. Breakdown of colloid filtration theory: role of the secondary energy minimum and surface charge heterogeneities. Langmuir, Vol. 21, p. 841-852
Tufenkji, N. and M. Elimelech, 2005b. Reply to comment on breakdown of colloid filtration theory: role of secondary energy minimum and surface charge heterogeneities. Langmuir, Vol. 21, p. 10896-10897.
Tufenkji, N., J.A. Redman, and M. Elimelech, 2003. Interpreting deposition patterns of microbial particles in laboratory-scale column experiments. Environ. Sci. Technol., Vol. 37, p. 616-623.
Tufenkji, N., G.F. Miller, J.N. Ryan, R.W. Harvey, and M. Elimelech, 2004. Transport of Cryptosporidium oocysts in porous media: role of straining and physicochemical filtration. Environ. Sci. Technol., 2004, Vol. 38, p. 5932-5938.
van Donsel, D.J., E.E. Geldreich and N.A. Clarke, 1967. Seasonal variations in survival of indicator bacteria in soil and their contribution to storm-water pollution. Appl. Microbiol. (15), p. 1362-1370.
van Genuchten, M.Th., 1980. A closed-form equation for predicting the hydraulic conductivity of unsaturated soils. Soil Sci. Soc. Am. J. 44, 892-898.
Van Genuchten, M.Th., 1981. Analytical solutions for chemical transport with simultaneous adsorption, zero-order production and first order decay. J. Hydrol., Vol. 49, p. 213-233.
Van Genuchten, M.Th., 1985. A general approach for modeling solute transport in structured soils. International Association of Hydrogeologists, Memoirs XVII, Part 2. In: Proceedings Hydrogeology of Rocks of low permeability, 513-526.
van Genuchten, M.Th., F.J. Leij and S.R. Yates, 1991. The RETC Code for Quantifying the Hydraulic Functions of Unsaturated Soils, Version 1.0. EPA Report 600/2-91/065, U.S. Salinity Laboratory, USDA, ARS, Riverside, California. Internet site: http;//www.ussl.ars.usda.gov/models/modelsmenu.htm.
van Loosdrecht, M.C.M., J. Lyklema, W. Norde, G. Schraa and A.J.B. Zehnder, 1987a. The role of bacterial cell wall hydrophobicity in adhesion. App. Environ. Microbiol. 1987, 1893-1897.
van Loosdrecht, M.C.M., J. Lyklema, W. Norde, G. Schraa and A.J.B. Zehnder, 1987b. Electrophoretic mobility and hydrophobicity as a measure to predict the initial steps of bacterial adhesion. App. Environ. Microbiol. 1987, 1898-1901.
Veenstra, S. and J.W.A. Foppen, 1998. Pre-feasibility study for wastewater management of Sana'a city, Republic of Yemen. TNO-NITG Report no. 98-26-B, NITG-TNO, Utrecht, The Netherlands.
Viraraghavan, T., 1978. Travel of microorganisms from a septic tile. Water, Air and Soil Pollution (9), p. 355-362.
Walker, S.L., J.A. Redman and M. Elimelech, 2004. Role of cell surface lipopolysaccharides in *Escherichia coli* K12 adhesion and transport. Langmuir, Vol. 20, p. 7736-7746.
Walker, S.L., J.A. Redman, and M. Elimelech, 2005. Influence of growth phase on bacterial deposition: interaction mechanisms in packed-bed column and radial stagnation point flow

systems", Environ. Sci. Technol., Vol. 39, p. 6405 – 6411

Wang, G. and M.P. Doyle, 1998. Survival of enterohemorrhagic *Escherichia coli* O157:H7 in water. J. Food Protec. (61), p. 662-667.

Webtech360, 2003. Processing Modflow Pro, http//www.iesinet.com.

Wise, W.R., 1992. A new insight on pore structure and permeability. Water Resour. Res. 28, 189-198.

Wise, W.R., T.P. Clement and F. J. Molz, 1994. Variably saturated modelling of transient drainage: sensitivity to soil parameters. J. Hydrol. 161, 91-108.

World Health Organization, 2000-2003. Global Water Supply and Sanitation Assessment 2000 Report. http://www.who.int/docstore/water_sanitation_health/globassessment.

Yamamoto, H., H.M. Liljestrand, Y. Shimizu and M. Morita, 2002. Effects of physical-chemical characteristics on the sorption of selected endocrine disruptors by dissolved organic matter surrogates. Environ. Sci. Technol., 2003, Vol. 37, p. 2646-2657.

Yang, Y, D. Lerner, M.H. Barrett and J.H. Tellam, 1999. Quantification of groundwater recharge in the city of Nottingham, UK. Environmental Geology 38 (3), p. 183-198.

Yao, K-M, M.T. Habibian and C.R. O'Melia, 1971. Water and wastewater filtration: Concepts and Applications. Environmental Science and Technology Vol. 5, Number 11, p. 1105-1112.

Zeng, C. and P.P. Wang, 1999. MT3DMS: A modular three-dimensional multispecies transport model for simulation of advection, dispersion, and chemical reactions of contaminants in groundwater systems; documentation and user's guide. US Army Corps of Engineers Engineer Research and Development Center Contract Report SERDP-99-1.

Zhuang, J. and Y. Jin, 2003. Virus retention and transport through Al-oxide coated sand columns: effects of ionic strength and composition. J. Contam. Hydrol., Vol. 60, p. 193-209.

Ziebell, W.A., D.H. Nero, J.F. Deininger and E. McCoy, 1975. Use of bacteria in assessing waste treatment and soil disposal systems. In: Home Sewage Disposal, Proceedings of the National Home Sewage Disposal Symposium of the American Society of Agricultural Engineers, p. 58-63

Zilberbrand, M., E. Rosenthal and E. Shachnai, 2001. Impact of urbanization on hydrochemical evolution of groundwater and on unsaturated-zone gas composition in the coastal city of Tel Aviv, Israel. Journal of Contaminant Hydrology 50 (2001), p. 175-208.

SAMENVATTING

Wereldwijd en al jaren lang zorgt fecale verontreiniging van drinkwater voor grote gezondheidsproblemen. Meest voorkomende ziekten als gevolg van het innemen van fecaal verontreinigd drinkwater zijn diarrhee, cholera, typhus en schistosomiasis.

In veel landen in de wereld staat al dan niet behandeld grondwater aan de basis van de drinkwatervoorziening. De kans, dat grondwater fecaal verontreinigd is, is niet zo erg groot, omdat aquifer passage in het algemeen in staat is om pathogene microorganismen te verwijderen. Daarom wordt overal in de wereld met success gebruik gemaakt van kunstmatige infiltratie en oever infiltratie. Maar er zijn ook een groot aantal gevallen bekend, waar aquifer passage minder succesvol is. Zo tonen bijvoorbeeld studies uit Amerika aan, dat de helft van alle drinkwaterputten in dat land op z'n minst sporen vertonen van fecale verontreiniging en dat als gevolg daarvan jaarlijks zo'n 750000 tot 5.9 miljoen ziekten ontstaan, die in 1400-9400 gevallen dodelijk zijn (Macler and Merkle, 2000).

Om de aanwezigheid van pathogene microorganismen in water aan te tonen wordt meestal de concentratie van een niet-pathogene groep microorganismen bepaald. Dit zijn de zgn. fecale indicator organismen en hiervan zijn *Escherichia coli* en thermotolerante coliforme organismen de meest bekende en meest gebruikte. De detectie van deze organismen is relatief simpel, snel, betrouwbaar en goedkoop.

Alhoewel *E. coli* en thermotolerant coliforme organismen vaak wordt aangetroffen in grondwater en in drinkwaterputten, zijn er eigenlijk weinig studies, die ingaan op de wisselwerking tussen *E. coli* en sediment tijdens transport door de ondergrond en de gevolgen daarvan op de verwijdering van *E. coli* uit grondwater. Meestal wordt uitgegaan van de eenvoudige, klassieke, Colloid Filtratie Theorie (CFT; hoofdstuk 1 en hoofdstuk 2), die gebaseerd is op de bepaling van een eenvoudige eerste-orde afbraak constante. Echter, deze theorie gaat voorbij aan een aantal belangrijke processen, die mede verantwoordelijk kunnen zijn voor (bio)colloid verwijdering en die niet zonder meer beschreven kunnen worden met een eenvoudige eerste-orde constante. Meer in het bijzonder gaat het om:

- Zeefwerking van het sediment (straining);
- Geochemische heterogeniteit;
- Bacteriele populatie hetreogeniteit;
- Preferentiele stroming door de ondergrond;

In deze studie worden bovengenoemde aspekten nader onder de loep genomen. De belangrijkste resultaten daarvan worden hieronder weergegeven.

Zeefwerking (hoofdstuk 3 en 4)

De massa balans van gezeefde of gestrainde bacterien in niet afgerond ultrapuur kwartszand kan beschreven worden met

$$\frac{\partial S}{\partial t} = \frac{\theta}{\rho_{bulk}} \frac{a}{x} (C - C^{1-b} x^b a_c^{bd} S) \qquad (1)$$

waarbij $a = 0.05$ $b = 0.717$ en $d = 1.20$ .Deze nieuw afgeleide massabalans, die de zeefwerking van bacterien door sediment beschrijft als een gemodificeerde Langmuir retentie, was afhankelijk van de *E. coli* concentratie in het porie water, C , de afgelegde afstand x , de korrelgrootte a_c en 3 constanten. De waarden van deze constanten bleken van toepassing voor een groot aantal korrelgrootten, transport afstanden en injectie concentraties en werden verkregen op basis van kolom experimenten. 'column extrusion' experimenten, 'flow reversal' experimenten en model fitting.

Zeefwerking is ook bestudeerd m.b.v. kolommen met daarin een mengsel van kwartszand met verschillende korrelgroottes. Voor beide studies vonden we een maximaal porievolume , dat beschikbaar was voor straining en dat zich snel vulde bij zeer hoge bacterie concentraties. Het maximale porievolume in de studie met het ultrapure niet-afgeronde kwartszand was 0.01% van het totale kolom volume en 0.21-0.35% in de studie met de kwartszand mengsels. Deze laatste waarden kwamen goed overeen met berekende waarden, bepaald op basis van puur geometrische overwegingen en ook met waarden bepaald aan de hand van porie grootte verdelingen. Het belang van straining op veldschaal kan grofweg worden bepaald op basis van de relatie tussen porie grootte van een sediment en straining. Porie grootte verdelingen kunnen relatief eenvoudig worden bepaald met de alom gebruikte van Genuchten bodem parameters.

Geochemische heterogeniteit van de ondergrond (hoofdstuk 5, 6 en 7)

Uit een set kolom experimenten met verschillende sedimenten (kwarts, goethiet, calciet en actief kool; hoofdstuk 5) blijkt, dat geochemische heterogeniteit de retentie van *E. coli* kan beinvloeden. Actief kool bleek veel beter in staat *E. coli* vast te houden dan calciet en goethiet. Kwartszand bleek het minst in staat tot retentie van *E. coli*. Echter, (bio)colloiden in de ondergrond reizen veelal in een pluim van afvalwater, waarin opgelost organisch materiaal (OOM) aanwezig is, zoals humus materiaal, polysacchariden, polyphenolen, proteinen, lipiden en heterogene organische moleculen (Fujita et al., 1996; Ma et al., 2001; Imai et al., 2002; Ilani et al, 2005) met concentraties varierend van 1-100 mg/L, afhankelijk van het type afvalwater zuivering. Uit ons onderzoek blijkt, dat door de aanwezigheid van OOM de hechting van *E. coli* 2 tot 80 keer minder was dan in gevallen zonder OOM, afhankelijk van de chemische samenstelling van het water, waarin *E. coli* gesuspendeerd was (hoofdstuk 6). De oorzaak van deze verwijderingsreductie door de hechting van OOM bleek een combinatie te zijn van:

- Verandering van (positieve) lading op het mineraaloppervlak door de aanwezigheid van (negatief) geladen OOM;
- Competitie tussen OOM en tweewaardige kationen voor de beschikbare hechtings-plaatsen op het mineraaloppervlak;
- Door de grootte van de OOM moleculen werd het *E. coli* fysiek onmogelijk gemaakt om maximaal gebruik te maken van het attractieve van der Waals krachtveld, dat aanwezig was tussen collector en *E. coli*. Hierdoor nam hechting waarschijnlijk af;

Alle hierboven genoemde mechanismen traden op in de experimenten en hieruit bleek, dat geochemische heterogeniteit, als oorzaak van hechting van biocolloiden minder belangrijk was voor mineraaloppervlakken, die al langduring in aanraking waren geweest met organisch materiaal.

Tot een zelfde conclusie werd gekomen na het analyseren van experimenten met PRD1 en kolommen met kwartszand mengsels en kwartszand gecoat met goethiet (hoofdstuk 7). Hier werd PRD1 niet alleen gebruikt om de hechtingskarakteristieken nader te bepalen, maar ook om de juistheid van het maken van conservatieve aannamen met betrekking tot de hechting van virussen voor het berekenen van beschermingszones voor grondwaterwinningen te evalueren (Schijven et al., 2006). Door de aanwezigheid van OOM, namen PRD1 effluent concentraties toe met 5 log eenheden en hechting aan kwartszand en aan goethiet werd vrijwel compleet geelimineerd. Deze resultaten impliceren, dat als OOM aanwezig is, virussen ver getransporteerd kunnen worden vanwege geringe hechting, omdat OOM kennelijk hechtingsplaatsen bezet houdt. Daarom is ook in het geval vna dit specifieke virus geochemische heterogeniteit minder belangrijk bij mineraaloppervlakken, die al langduring in aanraking zijn geweest met organisch materiaal.

Populatie heterogeniteit (hoofdstuk 8)

De variatie in hechtingsefficiency van een *E. coli* populatie kan tenminste een factor 100 bedragen. Deze variatie kan aanleiding zijn voor verschillende afstand afhankelijke gehechte concentratie patronen. In een serie kolomexperimenten met ronde glasbolletjes met verschillende diameter, name de hechting van *E. coli* hyper-exponentieel af: in de nabijheid van het injectiepunt bleken zeer veel bacterien te hechten en verder weg van het injectiepunt heel weinig. Op basis van curve-fitting, bleek dat 60% van de bacterie populatie snel hechtte, terwijl 40% van diezelfde populatie 100 keer minder snel hechtte. Ter bevestiging van dit resultaat zijn een aantal kolomexperimenten uitgevoerd met *E. coli* subpopulaties bestaande uit langzame hechters met lage hechtingsefficiencies. Hieruit bleek, dat doorbraak concentraties in deze gevallen aanmerkelijk hoger waren dan voor *E. coli* populaties bestaande uit zowel snelle en langzame hechters. Dus, heterogeniteit van *E. coli* bacterien onderling kon inderdaad verantwoordelijk worden gehouden voor hyper-exponentiele patronen van gehechte bacterie concentraties als functie van de getransporteerde afstand.

Preferentiele stroming en de casus van Sana'a, Jemen (hoofdstuk 9, 10 en 11)

In Sana'a, de hoofdstad van de Arabische Republiek Jemen, werd de kwaliteit van het grondwater (hoofdstuk 10) gekarakteriseerd door hoge concentraties van bijna alle belangrijke kat- en anionen. Dit werd vooral veroorzaakt door de ongecontroleerde diffuse afvalwater wegzijging naar de aquifers via zgn. 'pit latrines'. Op basis van conservatief mengen van afval water met niet verontreinigd grondwater uit de alluvial aquifer, kon worden vastgesteld, dat calcium in grondwater relatief verrijkt was, terwijl met name natrium, kalium en ammonium relatief afgenomen waren in grondwater. Op basis hiervan kon worden geconcludeerd, dat nitrificatie van ammonium, aanwezig in afvalwater, en kation-uitwisseling hadden plaatsgevonden. In een periode van 5 jaar, tussen 1995 en 2000, namen concentraties van bijna alle kat- en anionen verder toe, terwijl de pH afnam, beiden door de voortdurende infiltratie van afvalwater. Een indicatief 1D hydrochemisch transport model van een 200 m dikke aquifer kolom onder Sana'a, gaf aan, dat ongeveer 30% van ammonium aanwezig in onbehandeld afvalwater was geoxideerd tot nitraat, inclusief de produktie van zuur. Ongeveer 70% van het ammonium was uitgewisseld voor calcium en, in mindere mate, magnesium, aanwezig op het uitwisselcomplex van de bodem.

Infiltratie van afvalwater in het slecht gesorteerde alluviale aquifer materiaal onder Sana'a is een complex proces, zowel fysisch als hydrochemisch. Door de componenten van de diepteafhankelijke *E. coli* massabalans te analyseren, konden vertikale transport mechanismen worden geconcentualiseerd (hoofdstuk 11). Ondiep grondwater (0-100 m diep) was verontreinigd met fecale *E. coli* in concentraties varierend van 0-10^5 KVE/100 mL (KVE: Kolonie Vormende Eenheden). Op grotere diepte (200-300 m) werden geen noemenswaardige *E. coli* concentraties gevonden. Echter, in kolom experimenten was de verwijdering van bacterien 2-5 log eenheden per 0.5 cm kolom sediment dikte. Op basis hiervan kon geconcludeerd worden, dat juist omdat er relatief veel *E. coli* werd aangetroffen in de ondiepere delen van de alluviale aquifer, het transport in dit deel van de aquifer hoogstwaarschijnlijk had plaatsgevonden middels een met elkaar in verbinding staand netwerk van porien met een diameter, die groot genoeg was voor het transport van bacterien. Transport in zo'n netwerk moet snel geweest zijn, aangezien er sprake was van aanzienlijke afsterving van *E. coli* met de tijd, die middels afstervingsexperimenten in het laboratorium bepaald op 0.15 d^{-1}. Gecombineerd snel microbieel transport en infiltratie van afvalwater, inclusief kation-uitwisseling en nitrificatie, kon goed gesimuleerd worden met een 1D transport model van een 200 m dikke aquifer kolom, die gekenmerkt werd door dual porosity.

In de periode 2001-2005 is in Sana'a het management van afvalwater drastisch veranderd doordat hele delen van de stad zijn aangesloten op een riolering (hoofdstuk 9). De effecten van deze maatregel op de grondwater kwaliteit op lange termijn in de aquifers onder Sana'a zijn geanalyseerd. Hierbij is specifiek aandacht besteed aan het bepalen van de waterkwaliteit van de puttenvelden van het drinkwaterbedrijf van Sana'a. De resultaten, verkegen met een tijdsafhankelijk grondwaterstromingsmodel, gaven aan, dat in het jaar 2020 ten gevolge van de aanleg van de riolering de grootte van het gebied met vervuild grondwater drastisch verminderd zal zijn, maar ook, dat het opruimen van de vervuilde vlek zeer langzaam gaat. *E. coli* concentraties zijn door de aanleg van de riolering al enorm afgenomen in 2004. Het effect op de chemische grondwater kwaliteit in de nabijheid van de puttenvelden is niet groot, omdat stroming vanuit de stad niet richting de onttrekkingsputten is. Een onaangenaam effect van de aanleg van de riolering is de verlaging van de grondwaterspiegel in delen van de stad vanwege de verminderde wegzijging van afvalwater. Hierdoor vallen veel putten in de stad droog. Een van de grootste gebruikers van het grondwater in de stad is de stedelijke landbouw, en de daarvan voor hun voedselvoorziening afhankelijke, meestal vergrijsde populatie. Er bestaat een kans, dat een deel van deze stedelijke landbouwgebieden zullen verdwijnen als gevolg van de aanleg van de riolering.

CURRICULUM VITAE

Jan Willem (Anton) Foppen werd op 5 augustus 1965 geboren te Amstelveen in Noord-Holland. Hij behaalde zijn VWO diploma in 1984 in Amstelveen en in de periode 1985-1990 studeerde hij Aardwetenschappen aan de Vrije Universiteit in Amsterdam. Vanaf 1990 was hij werkzaam voor Natuurmonumenten en de VU en vanaf 1991 tot 1998 voor NITG-TNO. Via TNO vertrok hij in 1995 naar Jemen, waar hij betrokken was bij het SAWAS projekt. Het doel van dit project was om watervoorraden voor de drinkwater-voorziening van Sana'a te identificeren en binnen dit project hield hij zich bezig met de watervoorraden in de nabijheid van de stad en in het Sana'a Basin. Daar werd zijn interesse gewekt voor microbiologische verontreiniging van grondwater en de onttrekking daarvan door de vele duizenden pompputten in Sana'a en omgeving.
Vanaf 1998 was hij werkzaam bij IHE-Delft, dat later, vanaf 2003, UNESCO-IHE is gaan heten. Via het IHE was hij tot begin 2000 gestationeerd bij de Faculty of Engineering van de Sana'a University voor het Sana'a University Support (SUS) project, dat tot doel had om de kenniscapaciteit van de Water Engineering Department te doen toenemen. Menigmaal werden toen diverse lokaties in en om de stad bezocht om monsters te verzamelen, die dan later in het universiteits-laboratorium geanalyseerd konden worden. Bij terugkomst naar Delft in mei 2000 hield hij zich bezig met de nazorg van het SUS-project, diverse projecten in de Arabische regio (Palestina, Libanon, Jordanie, Syrie, Egypte, Saoedie-Arabie, Oman, Bahrein, Kuweit, Verenigde Arabische Emiraten), gericht op kennis-ontwikkeling in de water sector, en met onderwijs in Delft.
Halverwege 2001 ontstond het idee om per jaar een of twee onderzoeksprojecten van de M.Sc. studenten op het IHE in het kader van de microbiologische verontreiniging van grondwater te plaatsen. Verschillende studenten hebben hierin geparticipeerd en op basis van de resultaten van dit werk kon hij zich vanaf juni 2005 full-time bezig gaan houden met het onderzoek naar het transport van *E. coli* in verzadigde sedimenten.